W9-BXU-686

National Audubon Society®
Field Guide to
North American Reptiles and
Amphibians

A Chanticleer Press Edition

National Audubon Society®
Field Guide to
North American Reptiles and
Amphibians

John L. Behler, Curator of Herpetology,
New York Zoological Society, and
F. Wayne King, Director,
Florida State Museum

Alfred A. Knopf, New York

This is a Borzoi Book.
Published by Alfred A. Knopf, Inc.

Copyright © 1979 by Chanticleer
Press, Inc. All rights reserved under
International and Pan-American
Copyright Conventions. Published in
the United States by Alfred A. Knopf,
Inc., New York, and simultaneously in
Canada by Random House of Canada,
Limited, Toronto. Distributed by
Random House, Inc., New York.

www.randomhouse.com

Knopf, Borzoi Books, and the
colophon are registered trademarks
of Random House, Inc.

Prepared and produced by
Chanticleer Press, Inc., New York.

Color reproductions by Nievergelt Repro
AG, Zurich, Switzerland. Typeset in
Garamond by Dix Type Inc., Syracuse,
New York. Printed and bound by Toppan
Printing Co. Ltd., Tokyo, Japan.

Published December 1979
Twenty-third printing, March 2006

Library of Congress Cataloging-in-
Publication Number: 79-2217
ISBN: 0-394-50824-6

CONTENTS

10 How to Use This Guide
11 Introduction
14 Parts of Reptiles and Amphibians

Part I Color Plates
27 Key to the Color Plates
28 Thumb Tab Guide
Plates
1–144 Salamanders
145–255 Frogs and Toads
256–261 Crocodilians
262–330 Turtles
331–456 Lizards
457–657 Snakes

Part II Text Descriptions
265 *Amphibians*
267 Salamanders
357 Frogs and Toads
425 *Reptiles*
427 Crocodilians
433 Turtles
487 Lizards
581 Snakes

Part III Appendices
701 Glossary
705 Picture Credits
709 Index

NATIONAL AUDUBON SOCIETY

The mission of NATIONAL AUDUBON SOCIETY, *founded in 1905, is to conserve and restore natural ecosystems, focusing on birds, other wildlife, and their habitats for the benefit of humanity and the earth's biological diversity.*

One of the largest, most effective environmental organizations, AUDUBON has nearly 550,000 members, numerous state offices and nature centers, and 500 + chapters in the United States and Latin America, plus a professional staff of scientists, educators, and policy analysts. Through its nationwide sanctuary system AUDUBON manages 160,000 acres of critical wildlife habitat and unique natural areas for birds, wild animals, and rare plant life.

The award-winning *Audubon* magazine, which is sent to all members, carries outstanding articles and color photography on wildlife, nature, environmental issues, and conservation news. AUDUBON also publishes *Audubon Adventures,* a children's newspaper reaching 450,000 students. Through its ecology camps and workshops in Maine, Connecticut, and Wyoming, AUDUBON offers nature education for teachers, families, and children; through *Audubon Expedition Institute* in Belfast, Maine, AUDUBON offers unique, traveling undergraduate and graduate degree programs in Environmental Education.

AUDUBON sponsors books and on-line nature activities, plus travel programs to exotic places like Antarctica, Africa, Baja California, the Galápagos Islands, and Patagonia. For information about how to become an AUDUBON member, subscribe to *Audubon Adventures,* or to learn more about any of its programs, please contact:

NATIONAL AUDUBON SOCIETY
Membership Dept.
700 Broadway
New York, NY 10003
(800) 274-4201
(212) 979-3000
http://www.audubon.org/

ACKNOWLEDGMENTS

The authors wish to express their sincere appreciation to the following for reviewing parts of the text, and for providing information, guidance, and encouragement: K. Adler, W. B.Allen, R. G. Arndt, R. E. Ashton, W. Auffenberg, R. W. Axtell, J. P. Bacon, B. Banta, L. E. Bayless, J. K. Bowler, R. Brandner, R. A. Brandon, P. Brazaitis, E. D. Brodie, B. C. Brown, P. M. Burchfield, J. Burger, R. B. Bury, H. W. Campbell, C. C. Carpenter, A. Carr, J. L. Christiansen, C. J. Cole, D. Collins, J. T. Collins, D. Dietz, J. R. Dixon, J. L. Dobie, J. F. Douglass, P. C. Dumas, H. A. Dundee, D. Ehrenfeld, T. Ellis, C. H. Ernst, G. W. Folkerts, B. Foster, M. J. Fouquette, W. Frair, C. Gans, F. R. Gehlbach, J. W. Gibbons, J. Gorman, T. E. Graham, J. D. Groves, D. A. Hammer, D. Haynes, R. Highton, J. A. Holman, W. Holmstrom, A. C. Hulse, V. H. Hutchinson, J. B. Iverson, D. R. Jackson, T. Joanen, F. W. Judd, R. L. Lardie, A. Larson, R. L. Lattis, H. E. Lawler, J. F. Lynch, H. McCrystal, D. Meany, J. S. Mecham, A. Meylan, M. B. Mittleman, D. Moll, E. O. Moll, R. R. Montanucci, R. H. Mount, M. A. Nickerson, L. D. Ober,

Acknowledgments

J. C. Ogden, D. Pettus, M. V.
Plummer, F. H. Pough, P. C. H.
Pritchard, F. L. Rose, S. N. Salthe,
A. Schwartz, M. E. Seidel, C. R.
Shoop, E. N. Smith, P. P. Smith,
J. Soto, S. S. Sweet, W. W. Tanner,
S. G. Tilley, D. W. Tinkle, B. D.
Valentine, R. C. Vogt, D. B. Wake,
A. O. Wasserman, R. G. Webb,
H. M. Wilbur, L. D. Wilson, C. L.
Yntema, G. Zug, and R. G. Zweifel.
Our respective wives, Debbie and
Sherry, were active in every phase of the
work, checking references, typing,
reviewing rough drafts, and helping
with the correspondence. Only
contractual restrictions prevent us from
listing them as co-authors.
The completion of this field guide
would have been impossible without
the help of the staff at Chanticleer
Press. We would especially like to
thank Paul Steiner, Gudrun Buettner,
and Milton Rugoff for their creative
contributions; Barbara Williams,
Richard Christopher, and Susan
Rayfield for their support, patience,
and guidance; and Carol Nehring,
Susan Woolhiser, Ray Patient, Dean
Gibson, and Helga Lose for their
special efforts in the production of this
book.

THE AUTHORS

John L. Behler is Curator of
Herpetology at the New York
Zoological Society's Bronx Zoo. Since
he joined the staff of the Society in
1970, his popular articles on reptiles
and amphibians have appeared regularly
in *Animal Kingdom* magazine. In
addition to his ongoing work with
native northeastern reptiles and
amphibians, Behler has participated in
numerous national and international
studies, including the United Nations
Development Programme's 1975
expedition to Papua, New Guinea, to
examine that country's crocodilian
conservation program.
F. Wayne King is Director of the
Florida State Museum and an
internationally recognized expert on the
conservation of endangered animals,
especially reptiles and amphibians. He
is Deputy Chairman for the Species
Survival Commission of the
International Union for the
Conservation of Nature and Natural
Resources, former Director of
Conservation for the New York
Zoological Society, and a past Curator
of Herpetology at the New York
Zoological Society's Bronx Zoo. He has
written numerous scientific and popular
articles on reptiles and amphibians as
well as other wildlife.

HOW TO USE THIS GUIDE

Example 1 At dusk, after a spring shower, you find
Stout-bodied a chunky, shiny black, 7″ salamander
Salamander near with yellow spots near a woodland pond
a Woodland Pond in the eastern part of our range.

1. In the Thumb Tab Guide preceding the
   color plates you find that your specimen
   most closely matches the silhouette for
   Mole Salamanders, plates 37–62.
2. Turning to the color plates, you find
   several photographs resembling your
   specimen, but only one black species
   with yellow spots—the Spotted
   Salamander. The caption gives you the
   species length and the page number of
   the species description.
3. The species description identifies the
   Spotted Salamander as living around
   woodland ponds in your area.

Example 2 Under a flat rock on a wooded hillside
Unpatterned in Kansas you discover a small, slender,
Snake on a shiny grayish-black snake with no
Wooded Hillside surface pattern. The belly is yellow
with black spots, the underside of the
tail red, and a band encircles the neck.

1. In the Thumb Tab Guide your
   specimen matches the silhouette for
   Plain Snakes, plates 457–489.
2. Turning to the plates you find 4
   illustrations resembling your specimen
   —all Ringneck Snakes.
3. The species description indicates that of
   several Ringneck Snake subspecies, only
   one is in your range—the Prairie.

INTRODUCTION

Observing reptiles and amphibians in the wild, like bird watching, requires skill, practice, and patience. As experience is gained, the observer becomes more proficient at spotting animals and locating their retreats. Like a birder who can identify a small bird darting about, the experienced watcher of amphibians and reptiles is able to identify chorusing frogs in the dead of night, a snake slipping through a damp meadow, or a turtle basking on a log in a canal.

Early in their youth, many youngsters become fascinated with reptiles and amphibians. It is not unusual for inquisitive children to find the "lower vertebrates" intriguing and to bring home a frog, a snake, or a turtle to keep as a pet. More and more parents have shown themselves willing to encourage such an interest, recognizing that it can become an absorbing hobby and valuable way of learning to know and care for living creatures. Directly or indirectly, this book is written for all readers, both young and old, who want to learn more about our herpetofauna.

Geographical Scope Most guides to reptiles and amphibians have been regional in scope or have limited their coverage to a particular group of animals. This is the first guide to cover, in one volume, all native and introduced reptiles and amphibians found in North American north of Mexico.

Photographs as an Identification Guide

There are two ways of illustrating field guides, each with its advantages and disadvantages. We have chosen to use color photographs instead of drawings because we feel they best capture the form, natural stance and beauty of reptiles and amphibians. In recent years, many amateur and professional herpetologists have trained their lenses on these creatures. As a result, we have been able to obtain outstanding portrait photographs of nearly every reptile and amphibian in our range. We believe that such photographs add an exciting new dimension to these guides and provide a fascinating gallery of animal pictures.

In general, we have chosen photographs that show typical animals. Where a species is highly variable, we have often selected several photographs to show this range. Line drawings augment the species descriptions in the rare instances where no photograph was available.

Organization of the Color Plates

The visual section of this guide is organized according to the following basic groups of amphibians and reptiles:

Salamanders
Frogs and Toads
Crocodilians
Turtles
Lizards
Snakes

Within each group the photographs are further subdivided by body shape, then by color. The different body shapes are shown in a silhouette on a thumb tab on the left side of each double page of color plates. Thus, the Salamander group is organized in 4 parts, each with its own thumb tab: Aquatic Salamanders, Newts, Mole Salamanders, and Lungless Salamanders; the Frog and Toad group is divided into 4 parts: Treefrogs, True Frogs, Narrow-mouthed Frogs, and Toads; and so on. Since only 19 species of snakes in our range are venomous, and most of these occur within the pit viper subfamily, we have divided the snakes into 2 major categories: Snakes, and Pit Vipers. Because snakes have similar body shapes but varied colors and patterns, the thumb tabs emphasize the latter. Thus, the larger Snake category is divided into 4 subgroups: Plain, Two-toned, Striped, and Patterned, each represented by the same shape on the thumb tab but with different key markings.

Silhouette and Thumb Tab Guide

Preceding the photographs is a section showing silhouettes of animals typical of their groups. For example, a silhouette of a treefrog represents the group called Treefrogs, Chorus Frogs and Cricket Frogs; a silhouette of a skink represents Skinks and Alligator Lizards. To make it easy to locate each group the silhouette typical of the group has been inset on a thumb tab at the left of each double page of color plates. A table of all these silhouettes precedes the color plates and serves as a guide to the color section. Once you have found one or more photographs of the species you are seeking, you can check the identification by turning to the text.

Captions

The caption under each photograph gives the plate number and common name of the pictured species (or subspecies) followed by the size range of adults of the species shown. When known, the gender of the species is also indicated. Occasionally, an immature form, such as an amphibian larva or a juvenile or hatchling reptile, is shown.

Parts of Amphibians

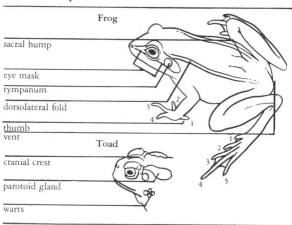

Frog

sacral hump

eye mask
tympanum
dorsolateral fold
thumb
vent

Toad

cranial crest

parotoid gland

warts

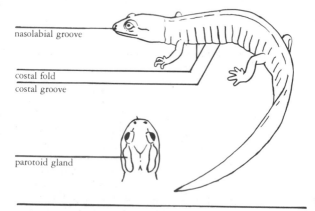

Salamander

nasolabial groove

costal fold
costal groove

parotoid gland

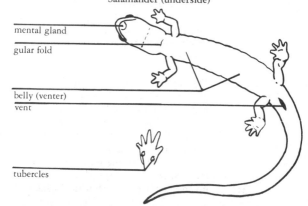

Salamander (underside)

mental gland
gular fold

belly (venter)
vent

tubercles

The caption closes with the number of the text page on which the species is described.

Types of Amphibians

Two of the three major amphibian groups occur in our range and are represented by 194 species; 112 salamanders, and 82 frogs and toads. Salamanders superficially resemble lizards in that they have slender bodies, long tails, distinct body regions, and, usually, front and hind legs of nearly equal size. Like related amphibians, they have a moist scaleless skin. As the salamander and lizard drawings in these pages show, salamanders lack the scales, claws, and external ear openings of the reptile group.

Frogs and toads are easily identified by the structure of their hind feet: true frogs have webbed toes (a); treefrogs have toe pads and webbing (b); toads have tubercles and no webbing (c); spadefoot toads have a horny projection —the spade—and reduced webbing (d). In the accompanying drawings note the anatomical differences displayed by this group—the presence or absence of warts, parotoid glands, dorsolateral folds and other distinguishing characteristics.

Types of Reptiles

Some 283 species of native and introduced reptiles occur in our area: three crocodilians, 49 turtles, 115 lizards, 1 amphisbaenid, and 115 snakes.

Size alone may identify the giants of the reptile world, the crocodilians. Their massive heads with protruding eyes and nostrils, and heavily muscled, compressed tails distinguish them from their scaly-skinned relations, the lizards. Turtles, with their distinctive shells, are recognized by everyone. Shell characteristics are generally used to distinguish species.

Snakes, the most specialized group of reptiles, are distinctive in having an

Parts of Reptiles

Turtle

Scutes

carapace (upper shell)

plastron (lower shell)

a-abdominal
al-anal
c-costal
f-femoral
g-gular
h-humeral
m-marginal
n-nuchal
p-pectoral
v-vertebral

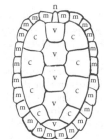
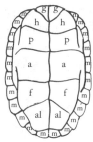

American Crocodile

hump

protruding 4th tooth

narrow snout

American Alligator

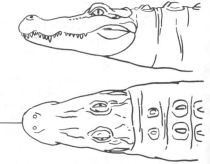

broad snout

a

b

c

elongated, scaly body. Unlike lizards, they lack limbs, external ear openings, and eyelids. Snakes and lizards have 3 basic types of scales: smooth (a), keeled (b), and granular (c). A drawing appears in the Introduction showing how to count snake scale rows. Our single amphisbaenid occurs in Florida and has an elongated limbless body resembling an earthworm. Body scales are fused into rings encircling the body.

Classification of Amphibians and Reptiles

Biologists divide plants and animals into major groupings called phyla. The phylum Chordata includes all fishes, amphibians, reptiles, birds, and mammals. Phyla are divided into classes, amphibians belong to the class Amphibia, reptiles to Reptilia. Classes are subdivided into orders, orders into families, families into genera (singular: genus), and genera into species (singular: species). The species is the basic unit of our classification system and is generally what people have in mind when they talk about a "kind" of animal. A species is a population of animals that possess common characteristics and freely interbreed in nature and produce fertile offspring. Although species occasionally crossbreed, the hybrid offspring are usually sterile. The scientific name of a species may consist of 2 or 3 Latin words. The first word, the genus, is capitalized while the second word, the species, is not capitalized. Thus, for example, the formal name of the Milk Snake is *Lampropelitis triangulum*. As it happens, the Milk Snake ranges over a broad geographical area, and populations in one part of the range often look different from those in another area. These populations are called geographic races or subspecies. A third name is added to designate such subspecies; hence, the Milk Snake found in the Northeast is known as the Eastern Milk Snake, *Lampropeltis*

Parts of Reptiles

Lizard head (side view)

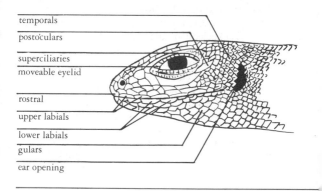

temporals
postoculars
superciliaries
moveable eyelid
rostral
upper labials
lower labials
gulars
ear opening

Snake head (side view)

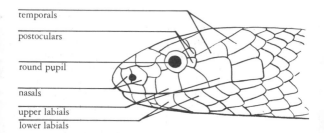

temporals
postoculars
round pupil
nasals
upper labials
lower labials

Pit Viper (side view)

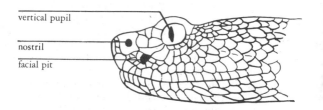

vertical pupil
nostril
facial pit

Lizard head (top view)

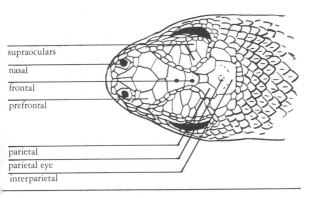

supraoculars

nasal

frontal

prefrontal

parietal

parietal eye

interparietal

Snake head (top view)

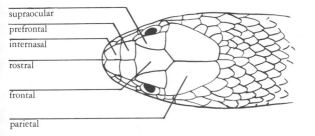

supraocular

prefrontal

internasal

rostral

frontal

parietal

Snake scales

ventral scales

method of counting back scales (midbody)

anal plate divided

anal plate single

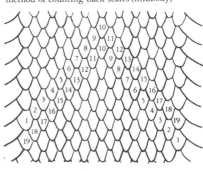

triangulum triangulum, while the population in midland America is called the Red Milk Snake, *Lampropeltis triangulum syspila.* Where the ranges of such populations overlap, their characteristics intergrade, or blend, showing a mixture of colors or patterns from each of the subspecies. The formal name for these intergrading subspecies is expressed as *Lampropeltis triangulum triangulum* x *syspila,* that is, *triangulum* "crossed" with *syspila.*

Organization of
the Text

The text contains descriptions of each amphibian and reptile found within the geographical range of this guide. Amphibians are treated first, followed by reptiles. Within each of these major groups—called classes—the species descriptions are arranged scientifically by order or suborder:

Salamanders
Frogs and Toads
Crocodilians
Turtles
Lizards
Snakes

Within each order, the text is arranged by family and within that, by genera and species, both of which are arranged in alphabetical order. This organization places descriptions of closely related species near one another, thus allowing for convenient comparison.
The following details are given in each species account:

Common Name

Common names for North American reptiles and amphibians were standardized in a 1956 publication appearing in *Copeia,* the Journal of the American Society of Ichthyologists and Herpetologists, and were subsequently revised in Roger Conant's, *A Field Guide to Reptiles and Amphibians of Eastern and Central North America* and Robert C. Stebbins' *A Field Guide to Western Reptiles and Amphibians.* With few exceptions, we follow these common names.

Scientific names We have generally used the forms given in *Standard Common and Current Scientific Names for North American Amphibians and Reptiles,* published in 1978 by the Society for the Study of Amphibians and Reptiles. For introduced species the scientific names are those currently used by the herpetological community.

Size Measurements (in both the English and metric systems) noted at the beginning of each description are the approximate size at which a species reaches sexual maturity and the maximum adult length. Sizes given for salamanders, lizards, and snakes are total lengths—a straight-line measurement from tip of snout to tip of tail; those for frogs and toads are straight-line measurements from snout to vent, and for turtles straight-line measurements of the length of the upper shell, or carapace.

Description This section summarizes the key visible features of the species such as shape, color, pattern, type of scale, and any other characteristics that may facilitate identification. The description will also enable the reader to distinguish the form from closely related or similar appearing species. Each description is based on a composite of specimens throughout the range of the species, and therefore highly variable species may not exactly conform to the description. The technical terms used in the text are defined in the glossary or illustrated by labeled drawings in the Introduction.

Voice Mating calls of each frog and toad are distinctive and can be extremely helpful in making identifications. A brief description of the voice precedes all frog and toad breeding descriptions.

Breeding We have given up-to-date information on reproductive cycles, courtship, size of the clutch and of the hatchling, and the age of sexual maturity.

Habitat The living as well as the physical

environment in which a species is found constitutes its habitat. Although wide-ranging forms may occupy many different situations, many species are quite restricted in their habitat requirements. Knowing the typical habitat of a species can aid in identification. It also directs the naturalist to localities where he may see a given species. Finally, it provides information for anyone who wishes to maintain a reptile or amphibian in captivity.

Range Each species entry includes a general description of the geographical range. A range map is also provided in the margin of each species account, with the exception of the far-ranging sea turtles. Each map shows the approximate range of the species within the area covered by the book. A species will not be evenly distributed in the areas marked on the map since the various sections of its habitat may be separated by many miles. Occasionally, species may be encountered beyond the limits of their known range. Some of these are legitimate extensions to the range; others represent a pet—perhaps picked up during a summer vacation or purchased on a whim—and "liberated" in an alien environment.

Subspecies The common and scientific name of each recognized subspecies within our range is provided, along with a brief description and information concerning its distribution. When species have races that occur beyond the limits of this guide, the total number of subspecies is given, followed by the number occurring within the scope of this book. For example, 23 subspecies of the Milk Snake are recognized but only 9 are found north of Mexico; these are described here in detail. Sometimes subspecies, such as those of the Milk Snake, are quite distinctive, but others, such as those of the Smooth Earth Snake, may be impossible to distinguish without knowing what area they come from. Where the ranges of subspecies merge, the characteristics of one subspecies blend into those of another; such specimens are often referred to as "intergrades." In regions marked by abrupt topographical variations, the intergrade zones are narrow; in relatively unbroken terrain they may be quite broad.

Comments Each species account is closed by a paragraph of notes, which may include information on the microhabitat of the animal, activity patterns, diet, longevity, status in the wild, and items of general interest.

⊗ This symbol indicates poisonous reptile species or families.

Observing and Collecting Amphibians and Reptiles Those who wish to develop their skills as an observer of our herpetofauna will find the notes on the habitat, breeding, activity periods and retreats of a species especially useful. In spring, visit a pond or marsh or patrol back roads near them on rainy nights. With experience even the smallest creature can be spotted in the headlights of an auto or in the beam of a flashlight. In this fashion one may track down the focal points of the breeding activities of many salamanders and frogs. The same techniques may be used to find nocturnally active snakes and lizards on highways in wetlands, along streambeds, or on arid flatlands. By day, turtles may be seen basking atop a tussock in a marsh or stacked on logs in the shallows of a lake or river. Flat rocks, logs or refuse along a pond may be turned to reveal hiding places. While salamanders, frogs, and most lizards and small snakes may be handled safely, larger snakes and lizards, along with mud, musk, snapping, softshells, and larger basking turtles can deliver a painful bite. Be sure you can recognize *all* the venomous reptiles in your area before you attempt to capture any snake. If you do encounter a venomous species do not attempt to collect it. Such equipment as snake hooks and tongs are available commercially, and the herpetology department of a local zoological park or museum can direct you to a supplier.

Some states have rigid regulations protecting wildlife and a license may be required for collecting certain amphibians and reptiles. Your state fish and game or conservation department will provide you with information concerning protected rare or endangered species and those protected by closed seasons.

Conservation Status

Amphibians and reptiles, like other wildlife, have suffered from habitat destruction, as well as from pesticide poisoning and other pollutants introduced into their environments. In some areas the reduction of their numbers has been dramatic.

Commercial collectors have taken their toll of specimens to satisfy a pet market that demands tens of thousands of native and foreign amphibians and reptiles each year. Many of these are purchased as novelty items and die within a few weeks. Or they are "liberated" into an alien habitat where few survive. Some survive only at the expense of another form. In southern Florida, for example, many exotic species have become established and in some areas have replaced their native competitor. If you decide to release a specimen after you have collected it, return it as close as possible to the original capture site. If it is not native to your area, do not release it, but find someone to adopt it or donate it to a wild-life center or zoological park.

Be kind to the habitat you visit. Do not leave a trail of split-open logs or ravaged den sites. A good conservationist leaves an animal's habitat the way he found it.

Part I
Color Plates

Key to the Color Plates

The color plates on the following pages are divided into 6 groups:

Salamanders
Frogs and Toads
Crocodilians
Turtles
Lizards
Snakes

Within each group, the species are arranged into subgroups based on shape and—in the case of snakes—color and pattern.

Thumb Tabs Each subgroup is represented by a thumb tab silhouette of a typical member, which appears at the left edge of each double page of plates. A key to the thumb tab organization appears on the pages preceding the color plates. Together the thumb tabs provide a quick and convenient index to the color plates.

Thumb Tab	Salamanders	Plate Nos.
	Aquatic Salamanders	1-24
	Newts	25-36
	Mole Salamanders	37-62
	Lungless Salamanders	63-144

Thumb Tab	Frogs and Toads	Plate Nos.
	Treefrogs, Chorus Frogs, and Cricket Frogs	145-186
	True Frogs	187-216
	Narrow-mouthed Frogs	217-222
	Toads	223-255

Thumb Tab	Crocodilians/Turtles	Plate Nos.
	Crocodilians	256-261
	Marine Turtles	262-267
	Softshell Turtles	268-273
	Pond, Marsh, and Box Turtles	274-309
	Musk, Mud, and Snapping Turtles	310-327
	Tortoises	328-330

Thumb Tab	Reptiles	Plate Nos.
	Large-bodied and Spiny Lizards	331-390
	Geckos and Night Lizards	391-408
	Whiptails	409-420
	Skinks and Alligator Lizards	421-450
	Worm and Legless Lizards	451-456

Snakes

Pit Vipers

Thumb Tab	Subgroup	Plate Nos.
	Plain	457-489
	Two-toned	490-504
	Striped	505-546
	Patterned	547-615
	Coral Snakes (venomous)	616-618
	Copperhead, Cottonmouth, and Rattlesnakes	619-657

The color plates on the following pages
are numbered to correspond with the
numbers preceding the text
descriptions. The reptiles and
amphibians shown are adults
unless otherwise indicated by the
caption.

Salamanders

Salamanders are a diverse group of amphibians, ranging from aquatic eel-like species that reach more than 3′ in length to small chunky-bodied land-dwellers that average 6″ from head to tip of tail. Moisture is absolutely essential to salamander survival. Some species are totally aquatic, while others are terrestrial and live near water or other moist situations, under rocks, woodland debris, or in damp rotting logs. Salamanders are voiceless, secretive, and typically nocturnal, only rarely active during daylight hours. Under cover of darkness, especially during or after a shower, they may be seen prowling their haunts in search of prey or en route to breeding sites.

1 Valdina Farms Salamander, 2–3″, *p. 328*

2 Comal Blind Salamander, 2–3⅜″, *p. 327*

4   Texas Salamander, 1⅞–3⅞″, *p. 326*

5   San Marcos Salamander, 1⅝–2″, *p. 325*

7 Georgia Blind Salamander, 2–3″, *p. 331*

8 Tennessee Cave Salamander, 4–8⅞″, *p. 329*

10 Oklahoma Salamander, 2–3⅛″, *p. 328*

11 Comal Blind Salamander, 2–3⅜″, *p. 327*

13   **Lesser Siren,** 7–27″, *p. 272*

14   **Greater Siren,** 19¾–38½″, *p. 273*

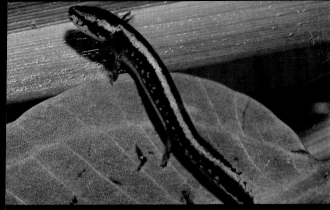

16 Mudpuppy, 8–17″, *young, ⅞–8″, p. 283*

17 Dwarf Siren, 4–9⅞″, *p. 271*

18 Two-toed Amphiuma, 18–45¾″, *p. 285*

19 Dwarf Waterdog, 4½–7½″, *p. 284*

20 Mudpuppy, 8–17″, *p. 283*

21 Neuse River Waterdog, 6½–10⅞″, *p. 282*

22   Tiger Salamander, *larva, p. 298*

23   Pacific Giant Salamander, *larva, p. 300*

24   Hellbender, 12–29⅛″, *p. 269*

25 Striped Newt, 2⅛–3⅛″, *p. 276*

26 Red-spotted Newt, 2⅜–5½″, *p. 276*

27 Broken-striped Newt, 2⅝–5½″, *p. 276*

28   Black-spotted Newt, 2⅞–4¼", p. 275

29   Red-spotted Newt, *red eft*, 1⅜–3⅜", p. 276

30   Broken-striped Newt, *eft*, 1⅜–3⅜", p. 276

31 Rough-skinned Newt, 5–8½″, *p. 278*

32 California Newt, 5–7¾″, *p. 280*

33 California Newt, 5–7¾″, *p. 280*

34 Red-bellied Newt, 5⅝–7⅝″, *p. 279*

35 Rough-skinned Newt, 5–8½″, *p. 278*

37 **Eastern Tiger Salamander,** 6–13⅜″, *p. 298*

38 **Tiger Salamander,** 6–13⅜″, *p. 298*

40 Tiger Salamander, 6–13⅜″, *p. 298*

41 Pacific Giant Salamander, 7–11¾″, *p. 300*

43 California Tiger Salamander, 6–8½″, p. 289

44 Marbled Salamander, 3½–5″, p. 295

46 Barred Tiger Salamander, 6–13⅜″, *p. 298*

47 Ringed Salamander, 5½–9¼″, *p. 288*

49   Western Long-toed Salamander, *4–6⅝", p. 293*

50   Santa Cruz Long-toed Salamander, *4–6⅝", p. 293*

52 Eastern Long-toed Salamander, 4–6⅝", *p. 293*

53 Southern Long-toed Salamander, 4–6⅝", *p. 293*

54 Spotted Salamander, 6–9¾", *p. 294*

55 Tremblay's Salamander, 3¾–6⅜″, *p. 299*

56 Jefferson Salamander, 4¾–8¼″, *p. 291*

57 Brown Salamander, 5⅝–8⅝″, *p. 290*

58   Blue-spotted Salamander, 3–5″, p. 292

59   Silvery Salamander, 5⅛–7¾″, p. 296

61 Mole Salamander, 3¼–4¾", *p. 296*

62 Small-mouthed Salamander, 4½–7", *p. 297*

64 Black-bellied Salamander, 3½–8¼″, *p. 316*

65 Seal Salamander, 3–5⅞″, *p. 315*

66 Pygmy Salamander, 1½–2″, *p. 318*

67 Del Norte Salamander, 4½–6″, *p. 338*

68 Red Hills Salamander, 4–10″, *p. 335*

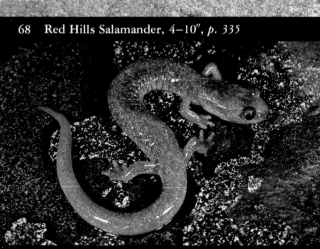

69 Jemez Mountains Salamander, 3¾–5⅝″, *p. 344*

70 Valley and Ridge Salamander, 3¼–5⅜″, *p. 341*

71 Red-backed Salamander, 2½–5″, *p. 336*

72 Ravine Salamander, 3⅛–5⅝″, *p. 346*

73   Channel Islands Slender Salamander, 5¼″, *p. 308*

74   Tehachapi Slender Salamander, 3⅝–4¾″, *p. 310*

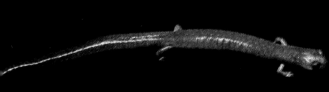

75   Garden Slender Salamander, 3¼–6⅜″, *p. 307*

76 California Slender Salamander, 3–5½″, *p. 307*

77 Relictual Slender Salamander, 2¾–4¾″, *p. 309*

78 Oregon Slender Salamander, 3⅜–4¼″, *p. 310*

79 Siskiyou Mountain Salamander, 3⅞–5½", *p. 348*

80 Wehrle's Salamander, 4¾–6¼", *p. 350*

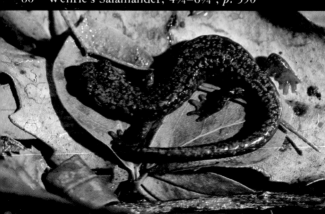

81 Sacramento Mountain Salamander, 3–4½", *p. 305*

82 Weller's Salamander, 2½–3⅝", *p. 351*

83 Clouded Salamander, 3–5¼", *p. 303*

84 Larch Mountain Salamander, 3–4", *p. 342*

85   Green Salamander, 3⅛–5½″, *p. 302*

86   Mount Lyell Salamander, 2¾–4½″, *p. 333*

87   Shasta Salamander, 3–4¼″, *p. 333*

88 Northern Two-lined Salamander, 2½–4¾″, *p. 320*

89 Northern Dusky Salamander, 2½–5½″, *p. 313*

91 Ouachita Dusky Salamander, 3⅛–7″, *p. 312*

92 Black Mountain Dusky Salamander, 3–6⅝″, *p. 31*

94   Seepage Salamander, 1½–2¼″, *p. 311*

95   Mountain Dusky Salamander, 2¾–4⅜″, *p. 316*

97　Limestone Salamander, 2¾–4⅜″, *p. 332*

98　Gulf Coast Mud Salamander, 2⅞–7⅝″, *p. 352*

100 Olympic Salamander, 3–4⅝", *p. 301*

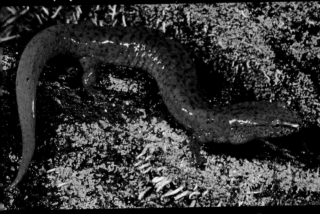

101 Blue Ridge Spring Salamander, 4¼–8⅝", *p. 329*

103　Four-toed Salamander, 2–4", *p. 331*

104　Oregon Ensatina, 3–5⅞", *p. 319*

106 **Arboreal Salamander,** 4¼–7¼", *p. 305*

107 **Sierra Nevada Ensatina,** 3–5⅞", *p. 319*

108 **Large-blotched Ensatina,** 3–5⅞", *p. 319*

109   Dwarf Salamander, 2⅛–3½", *p. 326*

110   Dunn's Salamander, 3⅞–5⅞", *p. 338*

111   Coeur d'Alene Salamander, 3¾–4⅞", *p. 349*

12 Mountain Dusky Salamander, 2¾–4⅜", *p. 316*

113 Dark-sided Salamander, 3⅞–7⅞", *p. 323*

115 Rich Mountain Salamander, 4–6¼", *p. 345*

116 Yonahlossee Salamander, 4¼–7½", *p. 352*

118   Southern Red-backed Salamander, 4⅞″, *p. 347*

119   Western Red-backed Salamander, 4½″, *p. 349*

120   Mountain Dusky Salamander, 2¾–4⅜″, *p. 316*

121 Blue Ridge Two-lined Salamander, 4¾″, *p. 320*

122 Cave Salamander, 4⅞–7⅛″, *p. 324*

123 Long-tailed Salamander, 3⅞–7⅞″, *p. 323*

124   Mountain Dusky Salamander, 2¾–4⅜″, *p. 316*

125   Junaluska Salamander, 2½–3½″, *p. 322*

127 Blue Ridge Spring Salamander, 4¼–8⅝″, *p. 329*

128 Gulf Coast Mud Salamander, 2⅞–7⅝″, *p. 352*

130 Kentucky Spring Salamander, 4¼–8⅝", *p. 329*

131 Midland Mud Salamander, 2⅞–7⅝", *p. 352*

133　Appalachian Woodland Salamander, 7¼″, *p. 341*

134　Imitator Salamander, 2¾–4″, *p. 314*

136   Appalachian Woodland Salamander, 7¼″, *p. 341*

137   Mountain Dusky Salamander, 2¾–4⅜″, *p. 316*

139 White-spotted Salamander, 4–6³⁄₁₆″, *p. 346*

140 Slimy Salamander, 4½–8⅛″, *p. 340*

142 Speckled Black Salamander, 3⅞–6½″, *p. 304*

143 Crevice Salamander, 5–8⅝″, *p. 343*

Frogs and Toads

Frogs and toads are 4-legged tailless amphibians. The vast majority of frogs have smooth skin moist to the touch, while toads are rough, warty, and dry. Most species get about by hopping or leaping, but treefrogs are walkers and climbers. Some toads both walk and hop. Frogs and toads usually are found in moist or wet habitats. Most species are vocal, and during the breeding season choruses of males can be heard in the evening calling females to the mating ponds.

145 Barking Treefrog, 2–2¾″, *p. 408*

146 Green Treefrog, 1¼–2½″, *p. 405*

148 Pacific Treefrog, ¾–2″, *p. 408*

149 Pine Barrens Treefrog, 1⅛–2″, *p. 402*

150 Mountain Treefrog, ¾–2¼″, *p. 406*

151   Barking Treefrog, 2–2¾″, *p. 408*

152   Common Gray Treefrog, 1¼–2⅜″, *p. 404*

153   Northern Cricket Frog, ⅝–1½″, *p. 400*

54 Barking Frog, 2½–3¾", *p. 420*

55 Cuban Treefrog, 1½–5½", *p. 410*

157 Common Gray Treefrog, 1¼–2⅜″, *p. 404*

158 California Treefrog, 1–2″, *p. 404*

160   Cope's Gray Treefrog, 1¼–2⅜", *p. 404*

161   Southern Cricket Frog, ⅝–1¼", *p. 401*

162   Florida Cricket Frog, ⅝–1¼", *p. 401*

163   Mountain Chorus Frog, 1–1½″, *p. 411*

164   Pine Woods Treefrog, 1–1¾″, *p. 407*

165   Tailed Frog, 1–2″, *p. 359*

166 White-lipped Frog, 1⅜–2″, *p. 420*

167 Rio Grande Chirping Frog, ⅝–1″, *p. 421*

169 Puerto Rican Coqui, 1 5/16″–2¼″, *p. 418*

170 Pacific Treefrog, ¾–2″, *p. 408*

172 Little Grass Frog, ½–⅝″, *p. 410*

173 Spring Peeper, ¾–1⅜″, *p. 406*

174 Squirrel Treefrog, ⅞–1⅝″, *p. 409*

175   Brimley's Chorus Frog, 1–1¼″, *p. 412*

176   Ornate Chorus Frog, 1–1⁷⁄₁₆″, *p. 413*

177   Ornate Chorus Frog, 1–1⁷⁄₁₆″, *p. 413*

178   Cuban Treefrog, 1½–5½″, *p. 410*

179   Western Chorus Frog, ¾–1½″, *p. 415*

180   Spotted Chorus Frog, ¾–1¼″, *p. 412*

181   Mexican Treefrog, 2–3⅝″, *p. 416*

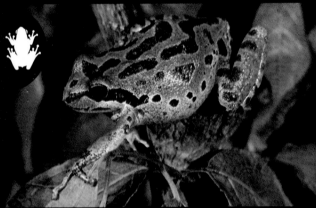

182   Pacific Treefrog, ¾–2″, *p. 408*

183   Southern Chorus Frog, ¾–1¼″, *p. 413*

184   Mexican Treefrog, *2–3⅝″, p. 416*

85   Illinois Chorus Frog, *1–1⅞″, p. 414*

187   **Bullfrog,** 3½–8″, *p. 372*

188   **Pig Frog,** 3¼–6⅜″, *p. 374*

190 Bullfrog, *female,* 3½–8″, *p. 372*

191 Southern Leopard Frog, 2–5″, *p. 379*

192 Northern Leopard Frog, 2–5″, *p. 377*

193   Tarahumara Frog, 2⅜–4½″, *p. 381*

194   Florida Crawfish Frog, 2¼–4½″, *p. 367*

195   Dusky Crawfish Frog, 2¼–4½″, *p. 367*

196   Rio Grande Leopard Frog, 2¼–4½″, *p. 370*

197   Plains Leopard Frog, 2–4⅜″, *p. 370*

199 Cascades Frog, 1¾–2⁵⁄₁₆″, *p. 371*

200 Mink Frog, 1⅞–3″, *p. 379*

202   **Plains Leopard Frog,** 2–4⅜″, *p. 370*

203   **Northern Leopard Frog,** 2–5″, *p. 377*

205   Carpenter Frog, 1⅝–2⅝″, p. 381

206   Spotted Frog, 2–4″, p. 378

208   Northern Red-legged Frog, 2–5⅜″, *p. 369*

209   Foothill Yellow-legged Frog, 1⅝–3″, *p. 371*

210   River Frog, 3¼–5⁵⁄₁₆″, *p. 375*

211 Wood Frog, 1⅜–3¼", *p. 380*

212 Wood Frog, 1⅜–3¼", *p. 380*

213 Bronze Frog, 2⅛–4", *p. 373*

214 Wood Frogs, *breeding, p. 380*

215 Red-legged Frog, 2–5⅜″, *p. 369*

216 Wood Frog, *female,* 1⅜–3¼″, *p. 380*

217   Mexican Burrowing Toad, 2–3½″, *p. 361*

218   Sheep Frog, 1–1¾″, *p. 385*

219   Eastern Narrow-mouthed Frog, ⅞–1½″, *p. 383*

220 Mexican Burrowing Toad, 2–3½", *p. 361*

221 Great Plains Narrow-mouthed Frog, 1⅝", *p. 384*

222 African Clawed Frog, 2–3¾", *p. 423*

223 Southwestern Toad, 2–3″, *p. 394*

224 Woodhouse's Toad, 2½–5″, *p. 398*

225 Colorado River Toad, 3–7″, *p. 386*

226 Boreal Toad, 2½–5″, *p. 388*

227 Yosemite Toad, 1¾–3″, *p. 389*

228 Houston Toad, 2–3¼″, *p. 392*

229 Western Spadefoot, 1½–2½″, *p. 364*

230 New Mexico Spadefoot, 1½–2½″, *p. 364*

231 Plains Spadefoot, 1½–2½″, *p. 362*

232   Great Basin Spadefoot, 1½–2″, *p. 366*

233   Eastern Spadefoot, 1¾–3¼″, *p. 365*

235 Southwestern Toad, 2–3", *p. 394*

236 Southern Toad, 1⅝–4½", *p. 397*

238   Texas Toad, 2–3⅝″, *p. 396*

239   Canadian Toad, 2–3¼″, *p. 391*

240   Western Toad, 2½–5″, *p. 388*

241 Southern Toad, 1⅝–4½″, *p. 397*

242 Giant Toad, 4–9½″, *p. 393*

243 California Toad, 2½–5″, *p. 388*

244 Yosemite Toad, *female,* 1¾–3″, *p. 389*

245 Black Toad, 1¾–2⅜″, *p. 391*

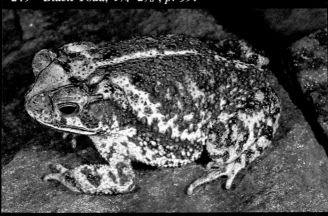

246 Gulf Coast Toad, 2–5⅛″, *p. 397*

247 **Great Plains Toad,** 2–4½″, *p. 389*

248 **Fowler's Toad,** 2½–5″, *p. 398*

249 **Southwestern Woodhouse's Toad,** 5″, *p. 398*

250 Oak Toad, ¾–1¼″, *p. 395*

251 Sonoran Green Toad, 1½–2¼″, *p. 396*

252 Couch's Spadefoot, 2¼–3½″, *p. 363*

253   Western Spadefoot, 1½–2½″, *p. 364*

254   Green Toad, 1¼–2⅛″, *p. 390*

255   Green Toad, 1¼–2⅛″, *p. 390*

Crocodilians

Alligators and crocodiles are the giants of the reptile world, the sole surviving descendants of the awesome archosaurs. Their well-armored bodies, massive heads with protruding eyes and nostrils, and heavily muscled, compressed tails set crocodilians apart from other reptiles. Two species, the American Alligator and the American Crocodile, are native to North America. The crocodile has a very limited range and is rarely observed. The American Alligator is frequently sighted in the southeastern United States—sunning itself on the banks of a canal, pond, or stream, or swimming slowly, legs held close to the body, through a flooded marsh or swamp.

256   American Alligator, *juvenile, p. 429*

257   Spectacled Caiman, *juvenile, p. 430*

258   American Crocodile, *juvenile, p. 431*

59 American Alligator, 6′–19′2″, *p. 429*

60 Spectacled Caiman, 4′–8′8″, *p. 430*

Turtles

Structurally, turtles are unmistakable.
With their protective shell and horny
beak they cannot be confused with any
other animal. Turtles come in many
sizes, shapes and colors and occupy a
diversity of habitats. Marine turtles,
with paddle-shaped flippers, are the
largest North American forms. They
are sometimes seen while boating or
caught in fishing nets. Occasionally
females may be observed coming ashore
to rest. Tortoises, our high-domed
turtles with elephantine hind feet, are
terrestrial and adapted for existence in
hot dry habitats. But the majority of
North American turtles are associated
with fresh water. Some prefer shallow,
quiet weed-choked waters, others swift-
moving rivers. Solitary individuals may
be sighted resting in warm shallows
with part of the shell exposed to the
sun or perched on a tussock in the
midst of a marsh. Gregarious baskers
may provide a real spectacle—a dozen
or more turtles on a single log, stacked
one upon another like fallen dominoes.

262 Atlantic Ridley, 23½–29½″, *p. 479*

263 Leatherback, 50–84″, *p. 481*

265   Loggerhead, 31–48", *p. 475*

266   Hawksbill, 30–36", *p. 478*

268   Midland Smooth Softshell, *female*, 14", p. 484

269   Gulf Coast Smooth Softshell, *juvenile*, p. 484

271 Spiny Softshell, *female*, 6½–18″, *p. 485*

272 Florida Softshell, 5⅞–12⅞″, *p. 483*

274   Ouachita Map Turtle, 3½–10¾″, *p. 464*

275   Ringed Sawback, 3–8½″, *p. 463*

277 Sabine Map Turtle, 3½–10¾″, *p. 464*

278 Mississippi Map Turtle, 3½–10″, *p. 462*

280 Map Turtle, 4–10¾″, *p. 461*

281 Black-knobbed Sawback, 3–7½″, *p. 462*

283 Barbour's Map Turtle, 3½–12¾", *p. 458*

284 Alabama Map Turtle, 3½–11½", *p. 465*

285 Chicken Turtle, 4–10", *p. 457*

286   Red-eared Turtle, 5–11⅜″, *p. 452*

287   River Cooter, 5¾–16⅜″, *p. 447*

288   Florida Cooter, 7½–15⅞″, *p. 448*

289 Yellow-bellied Turtle, 5–11⅜″, *p. 452*

290 Spotted Turtle, 3½–5″, *p. 453*

292   Alabama Red-bellied Turtle, 8–13¼″, *p. 446*

293   Western Painted Turtle, 4–9⅞″, *p. 450*

295 Red-bellied Turtle, 10–15¾", *p. 451*

296 Florida Red-bellied Turtle, 8–13⅜", *p. 449*

297 Southern Painted Turtle, 4–9⅞", *p. 450*

298   Ornate Diamondback Terrapin, 4–5½", *p. 466*

299   Northern Diamondback Terrapin, 4–5½", *p. 466*

300   Florida East Coast Terrapin, 4–5½", *p. 466*

301 Bog Turtle, 3–4½″, *p. 456*

302 Wood Turtle, 5–9″, *p. 454*

304 Three-toed Box Turtle, 4–8½", *p. 468*

305 Desert Box Turtle, 4–5¾", *p. 469*

307   Ornate Box Turtle, 4–5¾″, *p. 469*

308   Florida Box Turtle, 4–8½″, *p. 468*

310 Razor-backed Musk Turtle, 4–5⅞", *p. 443*

311 Loggerhead Musk Turtle, 3⅛–5¼", *p. 444*

313 Yellow Mud Turtle, 3½–6⅜″, *p. 439*

314 Sonora Mud Turtle, 3⅛–6½″, *p. 441*

315 Mexican Mud Turtle, 3¾–6⅝″, *p. 440*

316 Flattened Musk Turtle, 3–4½", *p. 443*

317 Striped Mud Turtle, 3–4¾", *p. 438*

318 Eastern Mud Turtle, 3–4⅞", *p. 441*

319 Stinkpot, 3–5⅜", *p. 445*

320 Mississippi Mud Turtle, 3–4⅞", *p. 441*

322   Snapping Turtle, 8–18½", *p. 435*

323   Snapping Turtle, 8–18½", *p. 435*

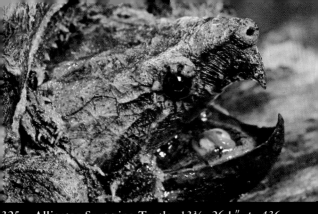

325   Alligator Snapping Turtle, 13⅜–26+″, *p. 436*

326   Alligator Snapping Turtle, 13⅜–26+″, *p. 436*

328 Desert Tortoise, 9¼–14½", p. 471

329 Berlandier's Tortoise, 4½–8¾", p. 472

Lizards

Lizards exhibit a greater diversity in size, shape, color, and behavior than any other group of reptiles. Typically they have a dry scaly skin, 4 legs, and a long tail. However, the legs of burrowing species may be reduced in size or altogether missing. These legless forms usually can be distinguished from snakes by the presence of eyelids or external ear openings.

Lizards are found in a wide variety of habitats—from arid deserts to moist forests and the wet fringes of marshes and streams—and from subterranean burrows to rocky outcrops and the leafy tops of trees. In our range, lizards are most abundant in dry habitats with open scattered vegetation. Typically they are active during the warm morning and late afternoon hours, but retire during the midday heat. In the southern United States some species of geckos are entirely nocturnal. Most North American species lay eggs in moist earth, leaf litter, or rotten logs, but some give birth to living young. ⊗ The symbol before plate number 332 indicates that this species is poisonous.

331 Chuckwalla, 11–16½", p. 518

⊗ 332 Gila Monster, 18–24", p. 546

334 Flat-tailed Horned Lizard; 3–4¾″, *p. 516*

335 Regal Horned Lizard, 3½–6½″, *p. 518*

336 Round-tailed Horned Lizard, 3–4¼″, *p. 516*

337 Mountain Short-horned Lizard, 2½–5⅞″, *p. 515*

338 Desert Short-horned Lizard, 2½–5⅞″, *p. 515*

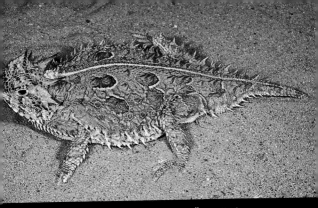

340   Texas Horned Lizard, 2½–7⅛″, *p. 513*

341   Southern Desert Horned Lizard, 3–5⅜″, *p. 517*

343 Desert Fringe-toed Lizard, 5–7", *p. 532*

344 Mojave Fringe-toed Lizard, 5–7", *p. 533*

346 Speckled Earless Lizard, 4–5⅛″, p. 510

347 Blunt-nosed Leopard Lizard, 8–9¼″, p. 507

348 Northern Spot-tailed Earless Lizard, 6″, p. 509

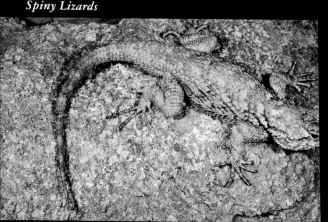

349 Clark's Spiny Lizard, 7½–12¹³⁄₁₆″, *p. 519*

350 Desert Spiny Lizard, 7–12″, *p. 523*

351 Granite Spiny Lizard, 7½–10⅝″, *p. 527*

352 Blue Spiny Lizard, 5–14¼", *p. 520*

353 Yarrow's Spiny Lizard, 5–8¾", *p. 522*

354 Crevice Spiny Lizard, 8½–11¼", *p. 527*

355   Collared Lizard, 8–14″, *p. 503*

356   Collared Lizard, *female*, 8–14″, *p. 503*

357   Leopard Lizard, 8½–15⅛″, *p. 508*

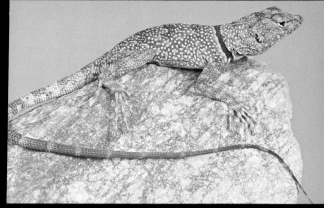

358   Banded Rock Lizard, 8½–11½″, *p. 513*

359   Desert Collared Lizard, 6–13″, *p. 504*

360   Reticulate Collared Lizard, 8–16¾″, *p. 505*

361 Greater Earless Lizard, 3¼–7¼″, *p. 503*

362 Zebra-tailed Lizard, 6–9⅛″, *p. 502*

363 Side-blotched Lizard, 4–6⅜″, *p. 537*

364 Curly-tailed Lizard, 7–10½″, p. 512

365 Keeled Earless Lizard, 4½–5⁹⁄₁₆″, p. 511

367 Bunch Grass Lizard, 2–3¾", *p. 528*

368 Canyon Lizard, *female,* 4½–6¼", *p. 524*

370   Bleached Earless Lizard, 4–5⅛″, *p. 510*

371   Mesquite Lizard, *female,* 4–6⅞″, *p. 522*

372   Clark's Spiny Lizard, 7½–12¹³⁄₁₆″, *p. 519*

373   Texas Rose-bellied Lizard, 3¾–5½″, *p. 530*

374   Striped Plateau Lizard, 4–6⅞″, *p. 531*

375   Northern Prairie Lizard, 3½–7½″, *p. 529*

376 Florida Scrub Lizard, 3½–5⅜″, *p. 531*

377 Northern Sagebrush Lizard, 5–6⅞₁₆″, *p. 521*

379 Western Fence Lizard, 6–9¼", *p. 525*

380 Long-tailed Brush Lizard, 5¾–7¼", *p. 534*

382 Small-scaled Tree Lizard, 3¾–4¾", p. 535

383 Green Anole, 5–8", p. 497

385 Green Anole, 5–8″, *p. 497*

386 Green Lizard, 9–15¼″, *p. 566*

388 Common Iguana, 40–79", *p. 511*

389 Knight Anole, 13–19⅜", *p. 500*

391 Leaf-toed Gecko, 4–5″, *p. 494*

392 Banded Gecko, 4½–6″, *p. 491*

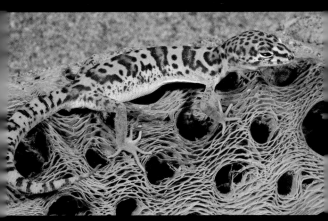

394 Texas Banded Gecko, 4–4¾", *p. 490*

395 Tucson Banded Gecko, 4½–6", *p. 491*

396 Ashy Gecko, *juvenile, p. 495*

397 Mediterranean Gecko, 4–5″, *p. 493*

398 Ashy Gecko, 2½–2⅞″, *p. 495*

399 Reef Gecko, *juvenile, p. 496*

400 **Reef Gecko,** 1¾–2½″, *p. 496*

401 **Indo-Pacific Gecko,** 4–5½″, *p. 492*

402 **Yellow-headed Gecko,** 2¾–3½″, *p. 492*

403   Arizona Night Lizard, 3¾–5¹⁄₁₆″, *p. 551*

404   Granite Night Lizard, 4–5⅝″, *p. 550*

405   Granite Night Lizard, 4–5⅝″, *p. 550*

406 Desert Night Lizard, 3¾–5¹⁄₁₆″, *p. 551*

407 Granite Night Lizard, 4–5⅝″, *p. 550*

409 Orange-throated Whiptail, 6⁷⁄₁₆–8⁵⁄₁₆″, *p. 557*

410 Sonoran Spotted Whiptail, 8–11⅓″, *p. 561*

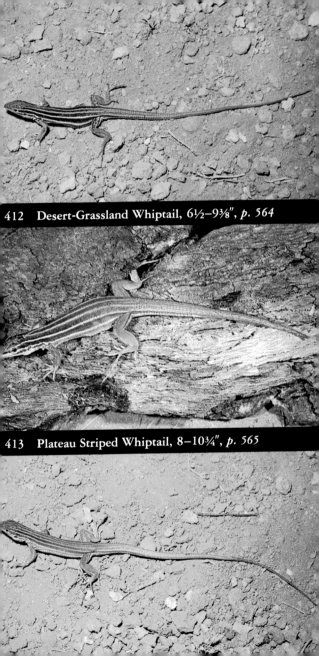

412   Desert-Grassland Whiptail, 6½–9⅜", p. 564

413   Plateau Striped Whiptail, 8–10¾", p. 565

415   New Mexican Whiptail, 8–11⅞″, *p. 559*

416   Giant Spotted Whiptail, 11–17¾″, *p. 554*

418 Chihuahuan Spotted Whiptail, *fem.*, 12⅜″, p. 555

419 Checkered Whiptail, 11–15½″, p. 562

421   Western Skink, 6½–9⁵⁄₁₆″, *p. 576*

422   Many-lined Skink, 5–7⅝″, *p. 574*

424　Broad-headed Skink, *female*, 6½–12¾", *p. 573*

425　Coal Skink, 5–7", *p. 568*

426　Southeastern Five-lined Skink, 5½–8½", *p. 572*

427   Five-lined Skink, 5–8¹⁄₁₆″, *p. 570*

428   Southern Prairie Skink, 5–8⅛″, *p. 575*

429   Southern Coal Skink, 5–7″, *p. 568*

430 Gilbert's Skink, 7⅛–12⅞", *p. 571*

431 Broad-headed Skink, 6½–12¾", *p. 573*

433 Ground Skink, 3–5⅛", p. 578

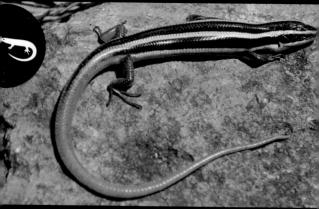

434 Arizona Skink, 7⅛–12⅞", p. 571

436　Mole Skink, 3½–6½″, *p. 569*

437　Five-lined Skink, 5–8⅟₁₆″, *p. 570*

439 Mountain Skink, 5–7⅞″, *p.* 577

440 Blue-tailed Mole Skink, 3½–6½″, *p.* 569

442   Western Skink, *juvenile, p. 576*

443   Five-lined Skink, *juvenile, p. 570*

445 Southern Alligator Lizard, 10–16⅞", *p. 542*

446 Arizona Alligator Lizard, 7½–12½", *p. 541*

448 Northern Alligator Lizard, 8¾–13″, *p. 539*

449 Southern Alligator Lizard, 10–16⅞″, *p. 542*

450 Sand Skink, 4–5⅛″, *p. 578*

451   Worm Lizard, 7–16″, *p. 580*

452   California Legless Lizard, 6–9¼″, *p. 548*

453   Eastern Glass Lizard, 18–42⅝″, *p. 544*

454 Island Glass Lizard, 15–24″, *p. 544*

455 Western Slender Glass Lizard, 22–42″, *p. 543*

456 Eastern Glass Lizard, 18–42⅝″, *p. 544*

Snakes

Snakes have elongated scaly bodies with no legs, ear openings, or eyelids. They range from the 6″ wormlike blind snakes to the 8½″-long Eastern Indigo Snake. 115 species occur north of Mexico. Some snakes are active during the day, others only at night. As daytime temperatures rise during summer months, nocturnal activity increases, and in some areas they are commonly seen crossing roads between sunset and midnight. Snakes may be found in trees, underground, on land, or in water from sea level to more than 10,000′ (3,050 m) in the western mountains. In the spring and fall they can be seen basking on branches overhanging water, among rocks, or on top of grassy tufts. Over most of our range, snakes emerge from hibernation in April and remain active until falling autumn temperatures force them to return to their winter retreat.

⊗ The symbol before plate numbers 616–651 indicates poisonous species.

457   Western Blind Snake, 7–16″, *p. 584*

458   Peninsula Crowned Snake, 5–7⅔″, *p. 660*

460 Plains Black-headed Snake, 7–14¾", *p. 658*

461 Central Florida Crowned Snake, 5–7⅔", *p. 660*

463 **Flat-headed Snake**, 7–9⅝″, *p.* 658

464 **New Mexico Blind Snake**, 5–10¾″, *p.* 583

466   Southeastern Crowned Snake, 5¼–13″, *p.* 657

467   Western Earth Snake, 7–13¼″, *p.* 679

468   Southern Black Racer, 34–77″, *p.* 596

469   Eastern Coachwhip, 36–102″, *p. 628*

470   Rough Earth Snake, 7–12¾″, *p. 678*

471   Sharp-tailed Snake, 10–19″, *p. 599*

72 Rubber Boa, 14–33″, *p. 586*

73 Rough Earth Snake, 7–12¾″, *p. 678*

475   Smooth Green Snake, 14–26″, *p. 640*

476   Western Smooth Green Snake, 14–26″, *p. 640*

478   Western Yellow-bellied Racer, 34–77″, *p. 596*

479   Green Rat Snake, 24–50″, *p. 608*

480   Eastern Yellow-bellied Racer, 34–77″, *p. 596*

481   Yellow-bellied Water Snake, 30–62″, *p. 633*

482   Florida Green Water Snake, 30–74″, *p. 632*

483   Black Kingsnake, 36–82″, *p. 618*

484   Rat Snake, 34–101″, *p. 605*

485   Eastern Hognose Snake, 20–45½″, *p. 614*

487 Swamp Snake, 10–18½″, *p. 652*

488 Black Pine Snake, 48–100″, *p. 644*

490   Red-bellied Water Snake, 30–62″, *p. 633*

491   Red Coachwhip, 36–102″, *p. 628*

492   Eastern Mud Snake, 38–81″, *p. 609*

493   Western Worm Snake, 8–14¾″, *p. 591*

494   North Florida Swamp Snake, 10–18½″, *p. 652*

495   Southern Ringneck Snake, 10–30″, *p. 600*

496 Pacific Ringneck Snake, 10–30″, *p. 600*

497 Prairie Ringneck Snake, 10–30″, *p. 600*

499 Ground Snake, 8–19″, *p. 653*

500 California Black-headed Snake, 7–15″, *p. 659*

502   Ground Snake, 8–19″, *p. 653*

503   Queen Snake, 16–36¾″, *p. 648*

504   Two-striped Garter Snake, 18–57″, *p. 664*

505   Northern Red-bellied Snake, 8–16″, *p. 655*

506   Northern Red-bellied Snake, 8–16″, *p. 655*

507   Lined Snake, 7½–21″, *p. 677*

508 Desert Rosy Boa, 24–42″, *p. 587*

509 Baird's Rat Snake, 34–101″, *p. 605*

511   Wandering Garter Snake, 18–42″, *p. 667*

512   Northwestern Garter Snake, 15–26″, *p. 669*

514 Big Bend Patch-nosed Snake, 22–40″, *p. 650*

515 Checkered Garter Snake, 14–42½″, *p. 669*

516 Texas Patch-nosed Snake, 22–47″, *p. 650*

517 Sonora Whipsnake, 30–67″, *p. 627*

518 California Striped Racer, 30–60″, *p. 630*

519 Graham's Crayfish Snake, 18–47″, *p. 646*

20 Blue-striped Ribbon Snake, 18–40", *p. 672*

521 Striped Whipsnake, 40–72", *p. 631*

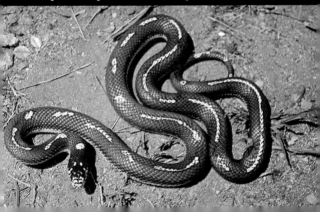

523   **Trans-Pecos Rat Snake,** 34–66″, *p. 607*

524   **Gray Rat Snake,** 34–101″, *p. 605*

526 Yellow Rat Snake, 34–101″, *p. 605*

527 Desert Patch-nosed Snake, 22–45″, *p. 651*

529   Butler's Garter Snake, 15–27¼", *p. 664*

530   Common Garter Snake, 18–51⅝", *p. 674*

532   Eastern Ribbon Snake, 18–40″, *p. 672*

533   Texas Garter Snake, 18–51⅝″, *p. 674*

534   Western Plains Garter Snake, 20–40″, *p. 671*

535   Red-spotted Garter Snake, 18–51⅝″, *p. 674*

536   Black-necked Garter Snake, 16–43″, *p. 666*

537   Pacific Gopher Snake, 48–100″, *p. 644*

538   Blue-striped Garter Snake, 18–51⅝″, p. 674

539   Eastern Garter Snake, 18–51⅝″, p. 674

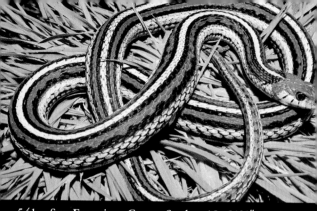

541   San Francisco Garter Snake, 18–51⅝", *p. 674*

542   Red-sided Garter Snake, 18–51⅝", *p. 674*

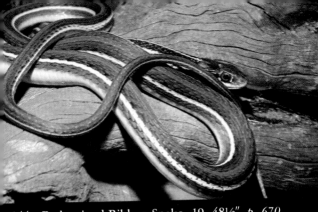

544 Red-striped Ribbon Snake, 19–48½″, p. 670

545 California Red-sided Garter Snake, 51⅝″, p. 674

547 Mexican Hook-nosed Snake, 9–19″, *p. 611*

548 Narrow-headed Garter Snake, 18–34″, *p. 672*

550 Brown Snake, 10–20¾", *p. 654*

551 Kirtland's Snake, 14–24½", *p. 596*

552 Banded Water Snake, 16–62½", *p. 634*

553   Red Coachwhip, 36–102″, *p. 628*

554   Western Coachwhip, 36–102″, *p. 628*

555   Ground Snake, 8–19″, *p. 653*

556 Western Coachwhip, 36–102″, *p. 628*

557 Prairie Kingsnake, 30–52⅛″, *p. 617*

558 Lined Coachwhip, 36–102″, *p. 628*

559   Speckled Racer, 30–50″, *p. 603*

560   Speckled Kingsnake, 36–82″, *p. 618*

561   Desert Kingsnake, 36–82″, *p. 618*

562 Broad-banded Water Snake, 16–62½″, *p. 634*

563 Eastern Hognose Snake, 20–45½″, *p. 614*

564 Eastern Fox Snake, 34–70½″, *p. 608*

565   Eastern Hognose Snake, 20–45½", *p. 614*

566   Glossy Snake, 26–70", *p. 590*

567   Brown Water Snake, 28–69", *p. 639*

568   Texas Lyre Snake, 24–47¾", *p. 676*

569   Prairie Kingsnake, 30–52⅛", *p. 617*

571 Western Leaf-nosed Snake, 12¾–20″, *p. 643*

572 Plains Hognose Snake, 16–35¼″, *p. 613*

573 Bullsnake, 48–100″, *p. 644*

574 Diamondback Water Snake, 30–63″, *p. 636*

575 Great Basin Gopher Snake, 48–100″, *p. 644*

576 Eastern Garter Snake, 18–51⅝″, *p. 674*

577 Kansas Glossy Snake, 26–70″, *p. 590*

578 Atlantic Salt Marsh Snake, 16–62½″, *p. 634*

579 Mangrove Water Snake, 16–62½″, *p. 634*

580   Northern Water Snake, 22–53″, *p. 637*

581   Gray Rat Snake, 34–101″, *p. 605*

583   Western Leaf-nosed Snake, 12¾–20″, *p. 643*

584   Short-tailed Snake, 14–25¾″, *p. 654*

585   Southern Hognose Snake, 14–24″, *p. 615*

586 Night Snake, 12–26″, _p. 616_

587 Glossy Snake, 26–70″, _p. 590_

589 Pima Leaf-nosed Snake, 12–20″, *p. 642*

590 Florida Kingsnake, 36–82″, *p. 618*

591 Northern Pine Snake, 48–100″, *p. 644*

592   California Kingsnake, 36–82″, *p. 618*

593   Western Long-nosed Snake, 22–41″, *p. 649*

594   Eastern Kingsnake, 36–82″, *p. 618*

595 Scarlet Snake, 14–32¼", *p. 592*

596 Northern Scarlet Snake, 14–32¼", *p. 592*

598 Arizona Mountain Kingsnake, 20–41″, *p. 621*

599 California Mountain Kingsnake, 20–40″, *p. 625*

600 Central Plains Milk Snake, 18–36″, *p. 622*

601   Gray-banded Kingsnake, 24–47½″, *p. 620*

602   Gray-banded Kingsnake, 24–47½″, *p. 620*

603   Gray-banded Kingsnake, 24–47½″, *p. 620*

604 Western Shovel-nosed Snake, 10–17″, *p. 594*

605 Banded Sand Snake, 7–10″, *p. 593*

607   Scarlet Snake, 14–32¼″, *p. 592*

608   Corn Snake, 24–72″, *p. 604*

609   Long-nosed Snake, 22–41″, *p. 649*

610 Organ Pipe Shovel-nosed Snake, 15½″, p. 595

611 Ground Snake, 8–19″, p. 653

612 Tucson Shovel-nosed Snake, 10–17″, p. 594

613   New Mexico Milk Snake, 14–24¾″, *p. 622*

614   Louisiana Milk Snake, 16–31″, *p. 622*

615   Red Milk Snake, 21–42″, *p. 622*

616 Arizona Coral Snake, 13–21", *p. 680*

617 Texas Coral Snake, 22–47½", *p. 681*

⊗ 619   Timber Rattlesnake, 35–74½″, *p. 688*

⊗ 620   Timber Rattlesnake, 35–74½″, *p. 688*

⊗ 621   Arizona Black Rattlesnake, 16–64″, *p. 694*

622 Mojave Rattlesnake, 24–51″, *p. 692*

623 Great Basin Rattlesnake 16–64″, *p. 694*

624 Eastern Diamondback Rattlesnake, 96″, *p. 685*

⊗ 625   **Western Pigmy Rattlesnake**, 15–30⅞", *p. 697*

◈ 626   **Black-tailed Rattlesnake**, 28–49½", *p. 690*

◊ 628 **Tiger Rattlesnake**, 20–36″, *p.* 693

⊗ 629 **Prairie Rattlesnake**, 16–64″, *p.* 694

⊗ 630 **Northern Pacific Rattlesnake**, 16–64″, *p.* 694

⊗ 631 Hopi Rattlesnake, 16–64″, *p. 694*

⊗ 632 Desert Massasauga, 18–39½″, *p. 696*

634 Colorado Desert Sidewinder, 17–32⅜", *p.* 687

635 Southwestern Speckled Rattlesnake, 52", *p.* 690

636 Mottled Rock Rattlesnake, 16–32⅝", *p.* 689

⊗ 637 Twin-spotted Rattlesnake, 12–26″, *p. 691*

⊗ 638 Western Massasauga, 18–39½″, *p. 696*

640 Banded Rock Rattlesnake, 16–32⅝", *p. 689*

641 Carolina Pigmy Rattlesnake, 15–30⅞", *p. 697*

⊗ 643 Arizona Ridge-nosed Rattlesnake, 24″, *p. 695*

⊗ 644 Red Diamond Rattlesnake, 29–64″, *p. 692*

⊗ 645 Pigmy Rattlesnake, 15–30⅞″, *p. 697*

646   Panamint Speckled Rattlesnake, 23–52″, *p. 690*

647   Sonora Sidewinder, 17–32⅜″, *p. 687*

⊗ 649 Osage Copperhead, 22–53″, *p. 683*

⊗ 650 Southern Copperhead, 22–53″, *p. 683*

652 Trans-Pecos Copperhead, 22–53″, *p. 683*

653 Timber Rattlesnake, 35–74½″, *p. 688*

⊗ 655   Osage Copperhead, 22–53″, *p. 683*

⊗ 656   Western Cottonmouth, 20–74½″, *p. 684*

⊗ 657   Florida Cottonmouth, 20–74½″, *p. 684*

Part II
Text Descriptions

The number preceding each species
description in the following pages
corresponds to the number of the
illustration in the color plates section.
If the description has no plate number,
it is illllustrated by a drawing that
accompanies the text.

AMPHIBIANS
(class Amphibia)

Living amphibians comprise 3 major groups, 2 of which occur in our range: the salamanders (order Caudata) and the frogs and toads (order Salientia). While the oldest fossil salamanders and frogs can only be traced back about 150 million years, their ancestors evolved from the fishes of the early Devonian Period, appearing about 350 million years ago. The amphibians were the first vertebrates to face the rigors of life on land. They solved locomotion and air breathing problems but remained vulnerable to the dehydrating effects of terrestrial existence and were never able to divorce themselves completely from the aqautic environment. However, they gave rise to reptiles whose scaly skin and shelled eggs offered protection against excessive water loss, thus enabling them to advance into an unoccupied arid habitat.

Salamanders
(order Caudata)

Approximately 350 salamander species
are known from North and South
America and the north temperate zones
of Europe, Asia, and North Africa.
They belong to 8 families, 7 of which
are represented in North America. The
sirens, mole salamanders, and
amphiumas are found nowhere else.
112 species occur north of
Mexico.

Secretive, typically nocturnal, and
voiceless, salamanders are not nearly as
familiar to us as their moist-skinned
relatives, the frogs and toads. Since
they have slender bodies, long tails,
distinct body regions, and usually
possess front and hind legs of nearly
equal size, salamanders are often
mistaken for lizards. However, they
lack the scaly skin, claws, and external
ear openings which characterize the
reptiles they superficially resemble.
Adults, like their larvae, are
carnivorous.

Salamanders display a wide range of
courtship patterns. With the exception
of the Hellbender and Asian species,
fertilization is internal. It is
accomplished without copulation.
Spermatophores, gelatinous pyramidal
structures which are capped with a
packet of sperm, are deposited by the
male. The female retrieves a sperm cap
with her cloacal lips and the jellylike
eggs are fertilized as they pass through
the cloaca. Most salamanders lay their
eggs in water. Those that do not do so
select a sheltered, moist cavity for egg
deposition. Larvae hatching into water
bear external gills that are lost during
transformation. The larvae of species
that lay eggs on land bypass the aquatic
stage and hatch as miniature replicas of
the adults.

GIANT SALAMANDER FAMILY
(Cryptobranchidae)

Two genera: 1 species of *Cryptobranchus* in North America and 2 endangered species of *Andrias* in Asia, including the largest living salamander, which may surpass 5′ (1.5 m) in length. All giant salamanders are strictly aquatic. Fertilization is external. Gender is difficult to determine except during the breeding season when males develop a swollen ring around the cloaca and females a swollen abdomen.

24 Hellbender
(*Cryptobranchus alleganiensis*)
Subspecies: Allegheny, Ozark

Description: 12–29⅛″ (30.5–74 cm). A giant among salamanders; *totally aquatic*. Body and head flattened; loose flap of skin along lower sides of body. Single pair of circular gill openings on neck. Gray or olive-brown above, with or without dark mottling or spotting. Belly lighter, with few markings. Male smaller than female. Male has swollen ridge around vent during breeding season.

Breeding: Late August to early September in North; September to early November in South. At night males prepare saucer-shaped nest cavity beneath large, flat rocks or submerged logs. Female lays 200–500 yellowish eggs in long strings, forming tangled mass; male positions himself beside or above her and sprays milt. Male guards nest. Larvae hatch in 2–3 months at 1⅟16″ (27 mm).

Habitat: Clear fast-flowing streams and rivers with rocky bottoms.

Range: Sw. New York to n. Alabama and Georgia. Separate populations in Missouri and in Susquehanna River (New York and Pennsylvania).

Subspecies: Allegheny (*C. a. alleganiensis*), spotting variable on body and tail; range as noted except absent in se. Missouri and ne. Arkansas. Ozark (*C. a. bishopi*), heavily blotched on back and tail, gill slits small and occasionally closed in adults; Black River system and north fork of White River in se. Missouri and ne. Arkansas.

Commonly called Allegheny Alligator or Devil Dog. Fishermen often encounter Hellbenders while searching for insect bait under flat river rocks. Folklore has it that Hellbenders smear fishing lines with slime, drive game fish away, and inflict poisonous bites. In fact, they are harmless; they feed on crayfish, snails, and worms. Captive longevity, 29 years. Long-term survival threatened by dam construction and pollution.

SIREN FAMILY
(Sirenidae)

Two genera limited to North America: 2 species of *Siren* and 1 of *Pseudobranchus.*

Sirens are aquatic permanent larvae and are easily mistaken for eels as they have long bodies with external gills and gill slits, no hind limbs, and tiny forelegs. Sirens are primarily carnivorous; they forage at night for prey. When their watery habitat dries up, sirens aestivate in mud burrows; glands in the skin secrete a mucus cocoon to protect their bodies from desiccation.

Gender cannot be determined visually.

17 Dwarf Siren
(*Pseudobranchus striatus*)
Subspecies: Broad-striped, Slender, Narrow-striped, Gulf Hammock, Everglades

Description: 4–9⅞" (10.2–25.1 cm). Smallest siren. Aquatic, gilled throughout life; slender, eel-like. *Brown or light gray above, with light stripes on sides.* External gills, single gill slit, 3 toes on forefeet; tail finned, tip compressed. Costal grooves, 29–37.

Breeding: In spring. Eggs laid singly on roots of water plants. Larvae hatch about a month later; ⅝" (14 mm).

Habitat: Shallow ditches, cypress swamps, weed-choked ponds, especially water hyacinth.

Range: Coastal plain of South Carolina and Georgia; Florida, except w. panhandle.

Subspecies: Broad-striped (*P. s. striatus*), short and stocky, broad dark-brown stripe down back with thin light vertebral stripe, flanked by broad yellow or buff stripe; s. South Carolina to se. Georgia.
Slender (*P. s. spheniscus*), head narrow, snout wedge-shaped, two distinct tan or yellow stripes on sides; sw. Georgia

and adjacent Florida panhandle. Narrow-striped (*P. s. axanthus*), slender, with shortened head, side stripes narrow and pale gray; Okefenokee Swamp, Georgia south to Lake Okeechobee, Florida. Gulf Hammock (*P. s. lustricolus*), stout and large, head flattened, snout blunt with 3 narrow light stripes within wide dark stripe down back, and 2 side stripes, top one orange-brown, bottom one silvery-white; Gulf Hammock region, Florida. Everglades (*P. s. belli*), small and slender, head long and narrow, stripes like Gulf Hammock, except side stripes buff; s. Florida.

Secretive. They dwell among water-hyacinth roots and amid debris at pond bottom, feeding on tiny invertebrates. During droughts they encase themselves in mud beneath the pond bottom. When caught, they make faint yelping noises.

13 Lesser Siren
(*Siren intermedia*)
Subspecies: Eastern, Western, Rio Grande

Description: 7–27″ (17.8–68.6 cm). Aquatic; slender eel-like body; brown, gray, or blackish above, sometimes with dark spotting. External gills, 3 gill slits; 4 toes on front limbs. *Tail compressed, with fin;* tail tip pointed. Costal grooves, 31–38. Male larger than female.

Breeding: Nests in winter and early spring. Female lays about 200 eggs in a sheltered cavity. Hatching larvae are ½″ (11 mm) long, mature in 2 years.

Habitat: Warm, shallow, quiet waters; swamps, sloughs, weedy ponds.

Range: Coastal plain from se. North Carolina to c. Florida, west to parts of Texas and Oklahoma, north to sw. Michigan.

273

Subspecies: Eastern (*S. i. intermedia*), black or brown with minute black dots, costal grooves 31–34; coastal plain, from se. N. Carolina to w. Alabama. Western (*S. i. nettingi*), brown, olive, or gray with tiny black dots, costal grooves 34–36; w. Alabama to e. Texas, north in Mississippi Valley to Michigan. Rio Grande (*S. i. texana*), largest ssp., dark gray and unpatterned to light or brownish gray with many dark flecks, costal grooves 36–38; lower Rio Grande Valley.

The Lesser Siren emits clicking sound when excited, yelps when captured, and squirms vigorously when handled. It eats aquatic invertebrates and plants.

14 Greater Siren
(*Siren lacertina*)

Description: 19¾–38½" (50–97.8 cm). Aquatic; stout, eel-like body; gray or olive above, sometimes with dark spots on head, back, and sides; *sides lighter, with many faint greenish-yellow dashes and blotches.*
External gills, 3 gill slits; 4 front toes on limbs. Tail compressed, with fin; tail tip rounded. Costal grooves, 36–39.

Breeding: Eggs laid February to March. Larvae hatch April to May, are ⅝" (16 mm) long.

Habitat: Shallow, muddy-bottomed, weed-choked water.

Range: Coastal plain from District of Columbia south through Florida, s. Alabama.

Nocturnal. Sirens spend the day under debris or rocks, burrowed in mud or thick vegetation. Young are often seen amid water-hyacinth roots. Adults are sometimes caught at night by bait fishermen. When drought dries up their habitat, sirens aestivate in mud burrows; their skin glands secrete a moisture-sealing cocoon over the body. They eat snails, insect larvae, small fish, and aquatic plants. Captive longevity about 25 years.

NEWT FAMILY
(Salamandridae)

Fifteen genera worldwide; 2 genera of 6 species in the New World: *Notophthalmus,* the eastern newts, inhabit eastern North America, and *Taricha,* the western newts, are found along the Pacific coast.

The skin of newts is not slimy but rough-textured unlike that of other salamanders; costal grooves are indistinct.

As the breeding season approaches, male newts develop swollen vents, broadly keeled tails, and enlarged hind legs with black, horny structures on inner surfaces of the thighs and toe tips. In male western newts, the skin becomes smooth.

Eastern newts are primarily aquatic, western newts terrestrial. Transforming eastern newts leave the water as brightly·colored forms called efts that live on the forest floor for 1–3 years, and then return to the water and assume adult characteristics.

Occasionally the larvae change directly into adults, skipping the eft stage. Western newts, by contrast, do not undergo an eft stage. They hatch as aquatic larvae, then transform into land-dwelling adults that return to the water at breeding time.

28 Black-spotted Newt
(*Notophthalmus meridionalis*)

Description: 2⅞–4¼" (7.1–11 cm). Aquatic. *Covered with black spots; no red spotting.* Olive-green above; back edged by narrow, often broken, yellow line. Sides light blue-green, belly yellowish-orange to orange.

Breeding: Can occur any time, depending on availability of water.

Habitat: Quiet stretches of streams with

submerged vegetation; permanent and
temporary ponds, and ditches.

Range: Coastal plain of s. Texas south into
Mexico.

Subspecies: One in our range, *N. m. meridionalis.*

No eft stage. When ponds dry up, both
young and adults are forced to find
shelter on land; they may be found
under rocks and rocky ledges.

25 Striped Newt
(*Notophthalmus perstriatus*)

Description: 2⅛–3⅛" (5.2–7.9 cm). Small,
slender, aquatic salamander, with 2 red
stripes running down body, often
broken on head and tail. Olive-green to
dark brown above; *belly yellow, with
sparse black specks.* A few red spots on
sides. Land-dwelling eft is orange-red
but striped like adult.

Breeding: Late winter to spring.

Habitat: Flatwoods, hammock ponds, and
drainage ditches.

Range: S. Georgia to c. Florida.

Efts are commonly found in woodlands
near breeding ponds after a downpour.
Sexually mature larvae with gills are
occasionally seen in ponds, along with
similarly sized aquatic adults, fully
transformed and lacking gills.

26, 27, 29, 30 Eastern Newt
(*Notophthalmus viridescens*)
Subspecies: Red-spotted, Broken-
striped, Central, Peninsula

Description: 2⅝–5½" (6.5–14 cm). Aquatic and
terrestrial forms. *Aquatic adult
yellowish-brown or olive-green to dark
brown above, yellow below;* back and belly
both peppered with small black spots.
Land-dwelling eft is orange-red to reddish-

brown; varies in size from 1⅜–3⅜"
(3.5–8.6 cm). Costal grooves, indistinct.

Breeding: Late winter to early spring. As season
approaches, male develops enlarged
hind legs, with black, horny structures
on inner surfaces of thighs and toe tips,
swollen vents, and broadly keeled tails.
Female lays 200–400 eggs singly, on
submerged vegetation. Incubation
period 3–8 weeks. Hatching larvae
average ⅜" (8 mm). In late summer or
early fall they transform to aquatic
subadults or efts.

Habitat: Ponds and lakes with dense submerged
vegetation, quiet stretches or
backwaters of streams, swamps,
ditches, and neighboring damp
woodlands.

Range: Nova Scotia to Florida and west in a
swath to sw. Ontario and Texas.

Subspecies: Red-spotted (*N. v. viridescens*), back
with series of black-bordered orange-red
spots in adult and eft stages; Nova
Scotia west to Great Lakes, south to
nw. South Carolina, c. Georgia, and
Alabama. Broken-striped (*N. v.
dorsalis*), broken black-bordered red
stripe from head to base of tail on either
side of midline, eft reddish-brown with
back stripes incompletely bordered by
black; coastal plain, ne. North Carolina
and se. South Carolina. Central (*N. v.
louisianensis*), back and belly color
sharply contrast, red spots usually
absent or small and incompletely ringed
with black, efts with reduced spotting;
sw. Ontario and w. Michigan south to
the Gulf, east through s. Alabama, n.
Florida, s. Georgia, and s. South
Carolina. Peninsula (*N. v. piaropicola*),
dark olive to dark brown above, red
spots absent or small, faint and
unringed, belly heavily peppered with
fine black spots; peninsular Florida.

Adult newts are often seen foraging in
shallow water. They prey voraciously on
worms, insects, small crustaceans and
molluscs, amphibian eggs, and larvae.

Searching for eggs, they visit the spawning beds of fish. Newts secrete toxic substances through the skin and so are avoided by fish and other predators. Efts can be found on the forest floor after a shower. A hungry one may consume 2,000 springtails.

31, 35 Rough-skinned Newt
(*Taricha granulosa*)
Subspecies: Northern, Crater Lake

Description: 5–8½" (12.7–21.6 cm). Warty skin, light brown to black above, with sharply contrasting yellow to orange belly. Breeding male temporarily develops smooth skin, swollen vent, compressed tail, and toes tipped with black horny layer. Small eyes with *dark lower lids.*

Breeding: December to July in quiet waters; October to November at higher elevations. Eggs laid singly on aquatic plants or submerged twigs, hatching in 5–10 weeks. Larvae transform late summer at about 2" (5.1 cm) or overwinter and transform following June or July at 3" (7.6 cm).

Habitat: Ponds, lakes, and slow-moving streams with submerged vegetation and adjacent humid forests or grasslands.

Range: Pacific coast, from Santa Cruz County, California, to se. Alaska. Sea level to 9,000' (2,743 m). Isolated population near Moscow, Idaho; believed to be introduced.

Subspecies: Northern (*T. g. granulosa*), color typically uniform above, although populations in higher altitudes may have random dark blotches, belly with little or no blotching; range as indicated. Crater Lake (*T. g. mazamae*), belly with heavy dark blotching; Crater Lake National Park, Oregon.

Most aquatic Pacific newt. On land it may be seen wandering abroad on cool

humid days, searching for
invertebrates. When threatened, it
strikes a warning posture.

34, 36 Red-bellied Newt
(*Taricha rivularis*)

Description: 5⅜–7⅝" (14.3–19.4 cm). A western
species, with *tomato-red belly* and *large
dark-brown eyes*. Black or very dark
brown above. Male's vent crossed by
dark stripe; stripe weak or absent in
female.

Breeding: Males arrive at breeding site before
females, in February. From March to
April, females lay 12 or so flat clusters
of 6–16 eggs each on undersides of
rocks in fast-moving mountain brooks.
Hatching larvae are about ½" (11 mm);
transform late summer to early fall at 2"
(5.1 cm).

Habitat: Cool mountain streams and
surrounding coastal redwood forest.

Range: Coastal California north of San
Francisco Bay to s. Humboldt County.

Rarely seen. After transformation,
juveniles disappear underground until
reaching sexual maturity, in about 5
years. Adults may be encountered
during autumn rains, when they
emerge from subterranean retreats and
move toward the section of the stream
where they hatched—perhaps a mile or
more away. Males locate females in
water by chemical attractants released
by females. When threatened, the
Redbelly strikes a warning posture,
arching its back and raising head and
tail to display its vividly colored belly.
This behavior may alert predators to its
poisonous skin secretion. May live 15
or more years in the wild.

32, 33 California Newt
(*Taricha torosa*)
Subspecies: Coast Range, Sierra

Description: 5–7¾" (12.7–19.7 cm). Resembles
Rough-skinned Newt but has larger
eyes and *light-colored lower eyelids.*
Tan to reddish-brown above, yellow to
orange below. Back and belly colors
blend. Breeding male has smooth skin,
swollen vents, compressed tail, and
toes tipped with black, horny layer.

Breeding: December to May. Female lays one or
two dozen eggs in spherical masses, on
aquatic plants or submerged forest
litter. Larvae hatch at about ½" (11
mm), transform in fall at 2¼" (5.7 cm)
or following spring when larger.

Habitat: Quiet streams, ponds, and lakes and
surrounding evergreen and oak forests
along coast. Fast-moving streams
through digger pine and blue oak
communities in Sierra Nevada foothills.

Range: Coastal California from San Diego to
Mendocino County; also w. slope of
Sierra Nevada. Separate population at
Squaw Creek, Shasta County. Near sea
level to 7,000' (2,134 m).

Subspecies: Coast Range (*T. t. torosa*), brown
above, light yellow to orange below;
San Diego to Mendocino County,
California. Sierra (*T. t. sierrae*), reddish-
brown to dark brown above, deep
orange below; w. slope of Sierra
Nevada.

During rainy season it may be seen
abroad during the day. Dry periods are
passed under moist forest litter and in
rodent burrows. Like other Pacific
newts, this species strikes a warning
posture when threatened, revealing its
brightly colored underbelly.

MUDPUPPY AND WATERDOG FAMILY
(Proteidae)

Two genera: *Necturus*, the mudpuppies and waterdogs, with 5 species in eastern North America; and 1 species of *Proteus*, a blind cave dweller in Europe. Mudpuppies and water dogs are aquatic permanent larvae, characterized by deep red plumelike gills, 4 toes on both front and hind feet, and strongly compressed tails. The male's vent is lined with tiny projections, bears 2 fleshy lobes, and is followed by a transverse groove.
Fertilization is internal; eggs are laid on the undersides of stones or logs on stream bottoms. The female guards the eggs until hatching.

Alabama Waterdog
(*Necturus alabamensis*)

Description: 6–8½" (15.2–21.6 cm). Aquatic; bushy red gills. *Hind feet have 4 toes,* tail compressed. Mottled or spotted red-brown, dark brown, or black above. *White belly.*

Breeding: Courtship fall to winter, nesting in spring. Lays eggs on undersurface of shelter in stream bed. Larvae hatch in 4–6 weeks.

Habitat: Medium-sized to large streams with beds lined by forest litter.

Range: Poorly known; ne. Mississippi to nc. Georgia south to Gulf.

Their days are spent near shelter or under debris, but at night these waterdogs are opportunistic feeders, searching out crayfish, worms, snails, and small fish. They are eaten by large predaceous fish.

Gulf Coast Waterdog
(*Necturus beyeri*)

Description: 6¼–8¾" (16–22.2 cm). Aquatic, heavily spotted, with *deep-red, feathery gills.* Hind feet have 4 toes; tail is compressed. Dark brown above, with light-tan network; back and belly marked with many circular or oval dark-brown spots.

Breeding: April to June. Clutches of 37–67 eggs.

Habitat: Sandy-bottomed, spring-fed streams.

Range: Poorly known; mostly c. Louisiana to e. Texas.

During inactive periods, these waterdogs hide under rocks on stream bottoms or in holes in the stream bank. The popular belief that they can bark like dogs is erroneous.

21 Neuse River Waterdog
(*Necturus lewisi*)

Description: 6½–10⅞" (16.5–27.6 cm). Aquatic, with *4-toed feet, deep-red bushy gills,* and markedly compressed tail. Rusty-brown above, dull brown or gray below. Many well-defined brown to blue-black *spots above and below.*

Breeding: Larvae hatch June to July.

Habitat: Large, relatively deep rivers; from shallows to depths.

Range: Neuse and Tar river systems, North Carolina.

This creature is occasionally caught by fishermen, some of whom mistakenly regard it as dangerous or poisonous. It can bite but usually doesn't cause much pain. It eats crayfish, snails, and insects.

16, 20 Mudpuppy
(*Necturus maculosus*)
Subspecies: Mudpuppy, Lake
Winnebago, Louisiana Waterdog

Description: 8–17″ (20.3–43.2 cm). Large aquatic
salamander, with *feathery maroon gills,
4-toed feet,* compressed tail. Gray to
rusty-brown above, with fuzzy-edged
dark-blue spots. Belly gray, with dark
spots. Costal grooves, 15–16, not well
developed.

Breeding: April to June. Female lays 30–190
eggs, singly attached to underside of
stone or log; larvae hatch in 5–9 weeks,
at ⅛″ (22 mm). Maturity is reached in
4–6 years.

Habitat: Lakes, rivers, and streams of all
descriptions, from muddy, weed-
choked shallows to a record depth of
90′ (27.4 m) in cold Lake Michigan.

Range: Se. Manitoba to s. Quebec, south to n.
Georgia and Louisiana. Introduced into
large New England rivers.

Subspecies:  Mudpuppy (*N. m. maculosus*), rust-
brown above with large blue-black
spots, belly gray with spotting; se.
Manitoba to s. Quebec, south to s.
Missouri and n. Georgia. Lake
Winnebago (*N. m. stictus*), dark gray
above with many small dots; ne.
Wisconsin and adjacent upper
Michigan peninsula. Louisiana
Waterdog (*N. m. louisianensis*),
yellowish-brown to reddish above with
many large dark spots, center line of
belly grayish white, unspotted;
Arkansas River drainage, se. Kansas
and s. Missouri south to nc. Louisiana.

Animals from cold, clear, highly
oxygenated water have short gills; those
from warm, muddy water, long bushy
gills. Nocturnal; feeds on worms,
crayfish, insects, and small fish.

19 Dwarf Waterdog
(*Necturus punctatus*)

Description: 4½–7½" (11.4–18.9 cm). Smallest of
the waterdogs. Bushy narrow gills;
compressed tail. *All feet have 4 toes.*
Dark brown or slate-gray to black
above; no black spots. Belly gray,
bluish-white along midline.

Breeding: Habits uncertain.

Habitat: Slow-moving muddy or sand-bottomed
streams and associated deep irrigation
ditches.

Range: Coastal plain, se. Virginia into Georgia
(may extend westward along Gulf
coastal plain).

The natural history of this species is
even less known than that of other
waterdogs. The extent of its range
especially needs attention.

AMPHIUMA FAMILY
(Amphiumidae)

The smallest family of salamanders:
only 1 genus, *Amphiuma,* with 3
aquatic species restricted to the
southeastern United States.
Amphiumas are eel-like and bear 4
useless tiny limbs, each with 1–3 toes.
Larvae hatch with external gills and do
not transform completely. Adults lose
gills but retain one pair of gill slits.
Amphiumas are primarily nocturnal
and carnivorous. Courtship takes place
in water; fertilization is internal. The
female lays a long beadlike string of
eggs in a muddy depression in shallow
water, guarding them until they hatch.

15, 18 Two-toed Amphiuma
(*Amphiuma means*)

Description: 18–45¾″ (45.7–116.2 cm). Aquatic
eel-like salamander; *4 tiny legs each with
2 toes.* Uniformly dark gray to grayish
brown above; belly lighter. Costal
grooves average 58.

Breeding: Lays eggs June to July in N. Carolina
and n. Florida, January to February in
s. Florida. Female lays about 200 eggs
in a damp cavity beneath debris;
remains coiled about them during
incubation—about 5 months.
Hatchlings are about 2⅛″ (54 mm)
long, with light-colored gills lost soon
after hatching.

Habitat: Acid waters of swamps, bayous,
drainage ditches.

Range: Coastal plain from se. Virginia to
Florida and e. Louisiana.

Nocturnal. Amphiumas are ill-
tempered and can inflict a nasty bite.
Their slippery skins make them
difficult to handle. This species prowls
shallows for crayfish, frogs, small
snakes, and fish. It may leave water

temporarily if weather is wet enough.
For shelter it digs burrows in muddy
bottoms or invades the burrows of other
marine creatures. Long-lived; one is
known to have survived 27 years in
captivity.

One-toed Amphiuma
(*Amphiuma pholeter*)

Description: 8½–13" (21.6–33 cm). Aquatic eel-
like salamander, with *4 tiny legs each
with one toe.* Grayish-brown above; belly
lighter.

Breeding: Habits unknown.

Habitat: Muck-bottomed ponds, intermittent
streams.

Range: Gulf hammock region of panhandle and
peninsular Florida; adjacent Georgia.

First collected in 1950 by herpetologist
W. T. Neill, this dwarf amphiuma
seems more secretive than other
amphiumas. Adapted for digging and
tunneling, it tends to stick to its
muddy burrows. Hence it is rarely
observed in the wild, and its ways of
life remain uncertain.

Three-toed Amphiuma
(*Amphiuma tridactylum*)

Description: 18–41¾" (45.7–106 cm). Aquatic eel-
like salamander; *4 tiny legs, each with 3
toes.* Dark brown above, belly light
gray. Costal grooves average 62.

Breeding: Internal fertilization. Mates December
to June; nests April to October. About
200 eggs are laid in single strand that
becomes tangled in cavity.

Habitat: Bottomland marshes and lakes, bayous,
cypress sloughs, and streams in hilly
regions. Frequently occupies crayfish
burrows.

Range: W. Alabama into Texas, north through

Mississippi valley to extreme se.
Missouri and adjacent Kentucky.

Nocturnal. This amphiuma can be
found after dark in shallows, poking its
head out of debris or bottom mud, or
foraging for crayfish, frogs, small
snakes and fish. Like other amphiumas,
Three-toes are commonly caught by
fishermen, who detest them as a
nuisance preying on virtually
everything that swims, including other
amphiumas. In turn they are preyed
upon by mudsnakes and cottonmouths.
Amphiumas rarely leave the water but
may travel short distances overland
during rainstorms.

MOLE SALAMANDER FAMILY
(Ambystomidae)

Three genera: *Ambystoma, Dicamptodon,* and *Rhyacotriton.* All occur in North America, from southeastern Alaska and Labrador to the southern edge of the Mexican Plateau. 18 species in our range.

Adult mole salamanders are characterized by robust bodies and limbs and short blunt heads. Lack of a nasolabial groove between lip and nostrils distinguishes moles from lungless salamanders. During the breeding season males develop a swelling around the vent. Larvae have wide heads with long plumelike gills and well-developed tail fins. Larvae of some species do not transform but breed in the larval form.

Courtship and breeding usually take place in ponds in late winter or early spring; fertilization is internal. Adult mole salamanders are typically terrestrial and confirmed burrowers. Both larvae and adults are carnivorous.

47 Ringed Salamander
(*Ambystoma annulatum*)

Description: 5½–9¼″ (14–23.5 cm). Small head. Slender body; deep brown to black above. *Narrow cream to yellow bands encircle body and tail* and join light-gray streak along lower side. Belly slate, with scattered light spots. Costal grooves, 15.

Breeding: Mates and nests in shallow ponds or temporary pools after first heavy autumn rains. Masses of 10–20 eggs are attached to submerged vegetation. Larvae hatch in October, transform in late May.

Habitat: Damp forested areas or clearings; Ozark Plateau and Ouachita Mountains.

Range: C. Missouri southwest to w. Arkansas and e. Oklahoma.

Many may be seen at pools during the short breeding season; otherwise, it is secretive and seldom encountered aboveground.

43 California Tiger Salamander
(*Ambystoma californiense*)

Description: 6–8½" (15.2–21.6 cm). Black above, with cream to yellow oval spots on head, body, and tail. Belly grayish, occasionally with a few small, dull-yellow spots. *Tubercles on feet; toe tips pinkish.* Costal grooves, usually 12.

Breeding: During the rainy season, January to February, in temporary ponds. Eggs laid singly, on plants. Hatching larvae ½" (11 mm); transform in about 4 months.

Habitat: Subterranean retreats near ponds in grasslands and open woodlands.

Range: West of the Sierra Nevada between Sonoma and Santa Barbara counties, California.

Considered a subspecies of the Tiger Salamander by some authorities. Larvae eat snails, Pacific Tree Frogs, and Red-legged Frog tadpoles.

42 Flatwoods Salamander
(*Ambystoma cingulatum*)

Description: 3½–5" (8.9–12.9 cm). Head and limbs relatively small; blackish above; often patterned with gray netlike marks or many small flecks. *Belly black with scattered gray spots* or many tiny gray flecks. Costal grooves, 13–16, usually 15.

Breeding: October to November. Autumn showers prompt migration to breeding sites—dry pond basins and roadside ditches. Female lays up to 160 eggs, singly, in small groups, under debris or on bare soil at bottom of basin. Larvae hatch 3–5 weeks later, at about ½″ (13 mm); become blackish with conspicuous light stripes; transform March to April at 2¾″ (70 mm).

Habitat: Flatwoods dominated by longleaf pine or slash pine and wire grass.

Range: S. Carolina south to Okefenokee Swamp in Georgia and Florida, west to se. Mississippi.

Adults are most often seen during breeding migration or under debris near breeding areas. They often choose nesting sites near crayfish burrows. Drainage and land clearing have greatly reduced their habitat.

57 Northwestern Salamander
(Ambystoma gracile)
Subspecies: Brown, British Columbia

Description: 5⅝–8⅝″ (14.3–21.9 cm). Robust; gray-brown to chocolate-brown above. *Parotoid glands form large oval swelling behind each eye.* Rounded rough ridge along upper edge of tail. Belly light brown. No tubercles on feet. Costal grooves, 10–12.

Breeding: January to July, late in northern areas and at high elevations. Female lays small, compact masses of 15–35 eggs or large, elongated masses of 100–200 that adhere to debris in lakes, ponds, and slow-moving streams. Larvae hatch in 2–4 weeks at ⅝″ (16 mm) and transform 1–2 years later at 3–3½″ (76–89 mm).

Habitat: Humid sites, open grassland to dense forest; often beneath debris along stream banks; sea level to 10,000′ (3,048 m).

Range: Pacific coast from Gualala River, California, north to extreme se. Alaska.

Subspecies: Brown (*A. g. gracile*), uniform color, 2 joints on fourth toe of hind feet; Vancouver Island and British Columbia to n. California. British Columbia (*A. g. decorticatum*), with many light flecks, 3 joints on fourth toe; se. Alaska to British Columbia.

Secretive and rarely seen except during breeding season. When molested, adults secrete a sticky, white, mildly irritating substance from glands on head, body, and tail. Montane populations are often neotenic.

56 Jefferson Salamander
(*Ambystoma jeffersonianum*)

Description: 4¾–8¼″ (12.1–21 cm). Long and slender; snout wide, digits long. *Dark brown or brownish-gray*, often with bluish flecks on limbs and lower sides of body. Belly lighter, area around vent gray. Costal grooves, 12.

Breeding: Migrates to ponds March to April. Female lays 10–20 cylindrical masses of 15 eggs each, attaching them to slender twigs underwater. Larvae hatch in 30–45 days, are ½″ (13 mm) long; transform July to September at 2–3″ (51–76 mm).

Habitat: Deciduous forests; under debris near swamps and ponds.

Range: W. New England and s. New York to Virginia and Indiana.

In the past, Jefferson and Blue-spotted salamanders interbred and created 2 hybrid all-female species—the Silvery (*A. platineum*) and Tremblay's (*A. tremblayi*). Jefferson males breed with Silvery females. The male's sperm only stimulates egg development; its genetic material is not contributed.

58 Blue-spotted Salamander
(*Ambystoma laterale*)

Description: 3–5″ (7.6–12.9 cm). Slender, like
Jefferson Salamander, but smaller,
shorter-legged, narrower-snouted, and
darker. *Grayish-black to bluish-black
above, with large bluish-white flecks,*
especially on lower sides. Belly lighter,
with flecking; area around vent black.
Costal grooves, 12.

Breeding: March to April, in ponds. Eggs laid
singly or in small masses of 6–10 eggs
on debris on pond bottom.

Habitat: Deciduous forests.

Range: Atlantic coast, Quebec to New Jersey
and throughout the Great Lakes region.

In the past, Blue-spotted and Jefferson
salamanders interbred and created 2
hybrid all-female species—Tremblay's
(*A. tremblayi*) and the Silvery (*A.
platineum*). Blue-spotted males breed
with Tremblay's females. The male's
sperm only stimulates egg
development; its genetic material is not
contributed.

60 Mabee's Salamander
(*Ambystoma mabeei*)

Description: 3–3⅞″ (7.6–10 cm). Small and stout,
with narrow head and short tail. *Deep
brown to black above,* with indistinct
light flecks; sides lighter, with many
whitish flecks. Belly brown, with few
flecks. Costal grooves, 13.

Breeding: January to March, in acidic, fishless
ponds in or near pine stands. Egg
masses are attached to water plants or
submerged twigs. Larvae transform in
May, are 2⅜″ (6 cm) long.

Habitat: Tupelo and cypress bottoms in pine
woods, open fields, and lowland
deciduous forest.

Range: Coastal plain, North and South
Carolina.

First collected in 1923 by W. B. Mabee, it can be found beneath logs or other debris near or on dry bottoms of breeding ponds. Mabee's larvae are prey to Tiger Salamander larvae.

49, 50, 52, 53 Long-toed Salamander
(*Ambystoma macrodactylum*)
Subspecies: Western, Eastern, Santa Cruz, Northern, Southern

Description: 4–6⅝" (10–17 cm). Slender, with long toes. *Dark brown to black above,* with back stripe made of many light blotches. Belly sooty to dark brown. Tubercles on feet. Costal grooves, 12–13.

Breeding: January to June, depending on latitude and elevation; sometimes before ice is out; in temporary or permanent ponds. Eggs laid singly on spike rushes near surface of water; or in small clusters adhering to vegetation or undersides of logs in deepest part of pond. Hatching larvae are ½" (11 mm) long; transform June to August or following summer, at 2–4" (48–98 mm).

Habitat: Arid sagebrush communities to moist evergreen forests and alpine meadows; sea level to 9,000' (2,743 m).

Range: Tuolumne County, California, to se. Alaska and northeast to w. Montana. Separate populations in Santa Cruz and Monterey counties, California.

Subspecies:  Western (*A. m. macrodactylum*), dull greenish to yellowish back stripe, reduced to scattered flecks on head; wc. Oregon and w. Washington north to Vancouver Island. Eastern (*A. m. columbianum*), fused bright yellow to tan blotches form back stripe, head spotted; w. Idaho, c. and e. Oregon and Washington north to se. Alaska. Santa Cruz (*A. m. croceum*), black with series of yellow to orange markings on back, endangered; Santa Cruz and Monterey counties, California. Northern (*A. m.*

krausei), unbroken yellow back stripe, yellow spot on eyelids; e. Idaho and w. Montana north to se. British Columbia and sw. Alberta. Southern (*A. m. sigillatum*), fused yellow blotches form broken back stripe, small spots on head; n. California and sw. Oregon. Where the ranges of these ssp. overlap, intergradations of characteristics occur.

Usually seen under logs or debris near pools, it sometimes shares breeding sites with the Northwestern Salamander. Long-toed's egg masses do not support the growth of algae; Northwestern's do. The Santa Cruz population is endangered. Its remaining habitat, a few small ponds, is protected by law.

51, 54 Spotted Salamander
(*Ambystoma maculatum*)

Description: 6–9¾" (15.2–24.8 cm). Stoutly built; black, blue-black, dark gray, or dark brown above, with 2 irregular rows of round, yellow or orange spots beginning on head and extending to tail tip. *Belly slate-gray.* Costal grooves, usually 12.

Breeding: March to April in North, January to February in the Great Smokies, December to February in South. Heavy rains and warming temperatures prompt migration to breeding ponds. Female lays 1 or more compact, clear or milky egg masses, 2½–4" (6.4–10.2 cm) in diameter, each containing about 100 eggs, that adhere to submerged branches. Larvae hatch in 1–2 months, are ½" (13 mm) long; transform in 2–4 months at 2½" (64 mm).

Habitat: Hardwood forests and hillsides around pools and flooded depressions.

Range: South-central Ontario to Nova Scotia, south to Georgia and e. Texas.

This species spends most of the time underground, so adults are rarely encountered. Spotted Salamanders often share a breeding pond with Marbled Salamanders; their larvae are commonly seen together. Acid rains have so polluted the water in some Northeast ponds that eggs cannot develop and populations have died out. Developing egg masses turn green from a beneficial algae. May live 20 years.

44, 45 Marbled Salamander
(*Ambystoma opacum*)

Description: 3½–5″ (8.9–12.7 cm). Chunky; dark gray to black above, with *bold white or silvery crossbands*. Belly black. Costal grooves, 11–12. Recently transformed juveniles are dark gray to brown, with light flecks. Male brighter than female.

Breeding: September to October in the North, October to December in South. Mates and nests on land; female lays 50–200 eggs, one at a time, in sheltered depression that later fills with rainwater. Larvae hatch at ¾″ (19 mm); transform in 4–6 months at about 2¾″ (70 mm).

Habitat: Woodlands, from low swampy areas to relatively dry hillsides.

Range: S. New Hampshire to n. Florida, west to e. Texas, north to lakes Michigan and Erie.

The nesting female typically curls herself around the eggs while waiting for rain to fill nest cavity. The larvae usually hatch a few days after inundation. If autumn rains are scant, eggs may not hatch until spring.

59 Silvery Salamander
(*Ambystoma platineum*)

Description: 5⅛–7¾" (12.0–19.9 cm). Long and slender; brownish gray with many small silvery-blue spots on back and sides; *area around vent gray*.

Breeding: March to April. Female lays cylindrical egg masses (like those of Jefferson Salamander), and attaches them to twigs underwater.

Habitat: Deciduous forest.

Range: Sc. Michigan and adjacent Indiana and Ohio; w. Massachusetts south to n. New Jersey.

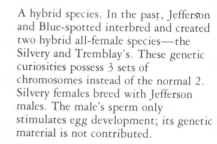

A hybrid species. In the past, Jefferson and Blue-spotted interbred and created two hybrid all-female species—the Silvery and Tremblay's. These genetic curiosities possess 3 sets of chromosomes instead of the normal 2. Silvery females breed with Jefferson males. The male's sperm only stimulates egg development; its genetic material is not contributed.

61 Mole Salamander
(*Ambystoma talpoideum*)

Description: 3¼–4¾" (8.1–12.2 cm). Chunky; body and tail short; head and limbs seem abnormally large. Brown or gray to black above, with *many bluish-white flecks*. Belly gray, with light blotches. Tail crest often bears light area. Costal grooves, 10–11.

Breeding: December to February after heavy rains and sharp temperature drop, in shallow ponds and flooded depressions. Female lays 200–400 eggs in small, fragile masses of 10–40 eggs each. Larvae transform late spring to early summer or following spring; occasionally neotenic.

Habitat: Beneath debris and leaf litter or in burrows in flatwoods and bottomland

forests, near floodplains and low-lying
areas.

Range: South Carolina to n. Florida west to e.
Texas, north in Mississippi Basin to se.
Oklahoma and s. Illinois. Separate
populations in sw. North Carolina and
e. Tennessee.

Talpoideum, meaning "molelike," is a
good name for this confirmed burrower.
Breeding sites are sometimes shared
with Marbled, Spotted, or Tiger
salamanders.

62 Small-mouthed Salamander
(*Ambystoma texanum*)

Description: 4½–7" (11.4–17.8 cm). Head and
mouth tiny. Dark brown to black
above, with many gray to grayish-
yellow lichenlike patches, some
unmarked. *Belly black, often with tiny
flecks.* Costal grooves, 14–15.

Breeding: Late January to April, in streams,
pools, and flooded ditches. Female lays
up to 700 eggs, singly or in small
clusters, attached to sticks, grass, or
undersides of flat stones propped
against bank. Larvae hatch at ½" (13
mm); transform May to June at about
1⅝" (40 mm).

Habitat: Moist pine woodlands and deciduous
forest bottomlands to tallgrass prairie
and farming areas; subterranean, near
temporary ponds, along streams.

Range: Ohio south to the Gulf, west to
Kansas, Oklahoma, and Texas.

When disturbed, the Small-mouthed
raises its tail and waves it back and
forth. A shy and sensitive animal, it
shares breeding pools with Spotted and
Marbled salamanders.

22, 37, 38, 39, **Tiger Salamander**
40, 46, 48 (*Ambystoma tigrinum*)
Subspecies: Eastern, Barred, Arizona,
Blotched, Gray, Sonora

Description: 6–13⅜" (15.2–40 cm). World's larges[t]
land-dwelling salamander. Stoutly
built, with *broad head* and small eyes.
Color and pattern extremely variable—
large light spots, bars, or blotches on
dark background or network of spots or
lighter background. Tubercles on soles
of feet. Costal grooves, 11–14 (usually
12–13).

Breeding: Prompted by rain; in North and higher
elevations, eggs laid March to June; in
South, December to February; in
Southwest, July to August. Mates in
temporary pools, fishless ponds, stream
backwaters, and lakes soon after ice is
out. Egg masses adhere to submerged
debris. Hatching larvae are 9⁄16" (14
mm) long; transform June to August at
about 4" (90–123 mm).

Habitat: Varied: arid sagebrush plains, pine
barrens, mountain forests, and damp
meadows where ground is easily
burrowed; also in mammal and
invertebrate burrows; sea level to
11,000' (3,353 m).

Range: Widespread from c. Alberta and
Saskatchewan, south to Florida and
Mexico, but absent from New England,
Appalachian Mountains, Far West.

Subspecies: Eastern (*A. t. tigrinum*), dark with
olive spots; east coast; also c. Ohio
to nw. Minnesota and south to Gulf.
Barred (*A. t. mavortium*), dark with
yellow crossbars or blotches; ne.
Nebraska to extreme se. Wyoming,
south to sc. Texas and New Mexico,
and Mexico. Arizona (*A. t. nebulosum*),
gray with small dark marks; w.
Colorado and Utah to sc. New Mexico
and c. Arizona. Blotched (*A. t.
melanostictum*), dark with yellow to olive
blotches or netlike lines; extreme s.
British Columbia, e. Washington and
c. Alberta southeast to s. Wyoming

and nw. Nebraska. Gray (*A. t. diaboli*), light olive to dark brown with small dark spots; s. Saskatchewan and s. Manitoba to Minnesota. Sonora (*A. t. stebbinsi*), yellowish spots, belly brown with a few yellow spots; Santa Cruz County, Arizona.

Often seen at night after heavy rains, especially during breeding season; they live beneath debris near water or in crayfish or mammal burrows. They are voracious consumers of earthworms, large insects, small mice, and amphibians. In the West, Tigers are often neotenic; some reach more than 13″ (33 cm) in length.

55 Tremblay's Salamander
(*Ambystoma tremblayi*)

Description: 3¾–6⅜″ (9.3–16 cm). Long and slender; dark gray to gray-black with many bluish-white markings; *area around vent black.*

Breeding: March to April. Eggs are laid singly or in small masses of 6–10 eggs on debris on pond bottom.

Habitat: Deciduous forest.

Range: N. Wisconsin; n. Indiana, n. Ohio and s. Michigan east through s. Quebec to New England coastal plain.

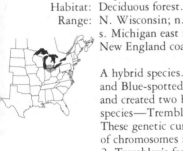

A hybrid species. In the past Jefferson and Blue-spotted salamanders interbred and created two hybrid all-female species—Tremblay's and the Silvery. These genetic curiosities possess 3 sets of chromosomes instead of the normal 2. Tremblay's females mate with Blue-spotted males. The male's sperm only stimulates egg development; its genetic material is not contributed.

Cope's Giant Salamander
(*Dicamptodon copei*)

Description: 4⅞–7½" (12.4–19.1 cm). Similar to Pacific Giant Salamander larvae, but never transforms to terrestrial stage; smaller overall, with narrower head, shorter limbs. Brown above, with patches of yellowish-tan covering clusters of white skin glands. *Belly dark bluish-gray.* Costal grooves, 12–13, inconspicuous.

Breeding: Habits largely unknown.

Habitat: Streams and rivers in humid coastal forests.

Range: Olympic Peninsula, Washington.

No transformed adult has ever been encountered in nature. In the lab, mature larvae can be induced to transform by thyroid treatments.

23, 41 **Pacific Giant Salamander**
(*Dicamptodon ensatus*)

Description: 7–11¾" (17.8–30 cm). Robust and smooth-skinned. Brown or purplish, with black mottling. Belly light brown to yellowish-white. No foot tubercles; 3 segments on 4th toe of hind foot. *Costal grooves, 12–13, indistinct.*

Breeding: Terrestrial adults breed in spring, in river headwaters. Eggs laid singly, on submerged timber. Hatching larvae, about ⅝" (16 mm), may transform during or following 2nd year at 3¼–6" (8.9–15.2 cm). Neotenic larvae mature at about 8" (20.3 cm).

Habitat: Rivers, their tributaries, and surrounding cool, humid forests.

Range: Extreme sw. British Columbia south along coast to Santa Cruz County, California; Rocky Mountains in Idaho and extreme wc. Montana.

Most salamanders are voiceless, but the Pacific Giant has been known to emit a

low-pitched yelp when captured. Land-dwelling adults live under logs, rocks, and forest litter but are sometimes seen crawling on the surface or even · climbing in bushes or trees to 8 feet (2.4 m). They eat large insects, mice, salamanders, and garter snakes. Voracious larvae cannabalize smaller larvae and eat Tailed Frog tadpoles and insects.

100 Olympic Salamander
(*Rhyacotriton olympicus*)
Subspecies: Northern, Southern

Description: 3–4⅝" (7.6–11.7 cm). Smallest mole salamander. *Prominent eyes on small head;* slim body, short tail. Plain chocolate-brown or mottled olive above. Belly yellowish-green or yellow-orange, with variable black flecking. Costal grooves, 14–15. Males have conspicuous, squarish vent lobes.

Breeding: Spring to summer. Clutch size less than 15 eggs. Larvae transform at about 2½" (6.4 cm).

Habitat: Coastal forest in cold, well-shaded permanent streams and spring seepages.

Range: Olympic Peninsula, Washington, to Mendocino County, California.

Subspecies: Northern (*R. o. olympicus*), Brown above with white speckles on sides, belly yellow-orange, sometimes with sparse black flecking; w. Washington to nw. Oregon. Southern (*R. o. variegatus*), olive with brown mottling above, belly yellowish green with heavy black flecking; nw. California, sw. Oregon. These ssp. intergrade in a diagonal zone from wc. Oregon to Lewis County, Washington.

Found at the edge or within the splash zone of fast-moving streams amid moss-covered rock rubble. The Tailed Frog often shares its habitat. It eats small insects and spiders.

LUNGLESS SALAMANDER FAMILY
(Plethodontidae)

Largest family of salamanders, it is believed to have originated in eastern North America. There are 23 genera and about 215 species, 80 of which occur in our range.

Species of the genera *Plethodon* and *Batrachoseps* are long and slender; *Pseudotriton* and *Desmognathus* are robust. All are lungless and breathe through thin moist skin. All have a nasolabial groove—a small narrow gland-lined slit between nostril and upper lip. Costal grooves are well defined. Most are terrestrial species that lay eggs on land. Others are fully aquatic, partly aquatic, or completely terrestrial.

Lungless salamanders conduct an elaborate courtship. Males rub and prod females, and females may straddle the male's tail ("tail walk") while he moves forward and drops a spermatophore (sperm case) for her to retrieve. Males have a mental gland under the chin, and, during breeding season, the area surrounding their vent is greatly enlarged. Fertilization is internal. Females often coil about and guard egg clutch during its development, with hatchlings appearing as miniature replicas of adults.

---

85 **Green Salamander**
(*Aneides aeneus*)

Description: 3⅛–5½" (7.9–14 cm). *Black above, with green or greenish-yellow patches.* Head looks swollen behind eyes. Toe tips expanded, squarish. Costal grooves, 14–15.

Breeding: May to August. Cluster of 10–20 sticky eggs is suspended in crevice by short mucus strands. Female broods

eggs until they hatch; 12–13 weeks.
Hatchlings are ⅞″ (22 mm).

Habitat: Damp, but not wet, sheltered, narrow crevices along face of sandstone outcrop; also underneath loose bark of rotting hardwood trees and stumps. To 4,400′ (1,341 m).

Range: Extreme sw. Pennsylvania through Allegheny and Cumberland mountains to c. Alabama; Blue Ridge Mountains of North and South Carolina.

Nocturnal; it hides in rock crevices or rotting trees by day. By night it climbs about on vertical surfaces, seeking beetles, mosquitoes, and ants. Hatchlings look like miniature adults.

83 Clouded Salamander
(*Aneides ferreus*)

Description: 3–5¼″ (7.6–13.3 cm). Slim, long-legged climbing salamander, with expanded, squarish-tipped toes. Two color phases: uniformly dark brown above; or brown "clouded" with greenish-gray, ash, or coppery blotches. *Belly whitish to brownish, with white flecking.* Costal grooves, 16–17.

Breeding: Late spring to early summer; lays 8–17 eggs singly on roof or side of cavity in rotting log or under bark. Eggs are suspended separately on mucus stalks, sometimes get tangled. Female guards eggs, which hatch fall to early winter.

Habitat: Humid coastal redwood, Douglas fir, and Port Orford cedar forests; frequents margins of clearings. To 5,400′ (1,646 m) in w. slopes of Cascades.

Range: Coastal California and Oregon, from Mendocino County to Columbia River; Vancouver Island.

These agile climbers are seen 20 feet (7 m) and higher in trees. They spend the dry season deep within logs. May be abundant in clearings, where they hide under bark of standing or fallen trees, in talus piles, or in moist rock crevices.

---

142 Black Salamander
(*Aneides flavipunctatus*)
Subspecies: Speckled, Santa Cruz

Description: 3⅞–6½" (9.8–16.7 cm). Slim-bodied and black, with *triangular-shaped head*. Spotted or frosted above; belly gray-black. Teeth of upper jaw project beyond lip and can be felt by lightly stroking snout tip. Toe tips rounded. Base of juvenile's limbs pale yellow. Costal grooves, 14–15.

Breeding: Late spring to summer. Lays 1–2 dozen eggs, each attached by short stalk to common base, in soil cavity or among rocks; hatch in fall. Female guards nest.

Habitat: Evergreen to deciduous forest and coastal prairies. Rarely above 2,000′ (610 m). Southern populations prefer moist woodlands along streams and seepages; northern populations, grassy areas; in far north, moss-covered rockslides.

Range: Extreme sw. Oregon and n. California.

Subspecies: Speckled (*A. f. flavipunctatus*), usually with white or pale yellow spots; Sonoma and Napa counties, California to Jackson County, Oregon. Santa Cruz (*A. f. niger*), uniformly black above, or with small white spots; Santa Cruz, Santa Clara, and San Mateo counties, California.

In southerly areas this salamander is active most of the year; it may be encountered under woodland debris or rocks. Farther north, where it lives away from water, aestication occurs during the dry season, April to October; it returns to the surface during fall rains.

81 Sacramento Mountain Salamander
(*Aneides hardii*)

Description: 3–4½" (7.6–11.4 cm). Smallest species of *Aneides;* slender and short-legged. *Upper teeth project beyond lip.* Light to dark brown above, most with some greenish-gray to bronze mottling. Toe tips rounded. Juveniles have brown-bronze dorsal stripe. Costal grooves, 14–15.

Breeding: July to August. Lays 4–6 eggs, suspending them from top of cavity in rotting fir log. Female guards eggs until hatched.

Habitat: Douglas fir, Engelmann spruce, and white fir forests between 8,500–11,000' (2,591–3,353 m); prefers slopes facing north or east.

Range: Sacramento, Capitan, and White mountains, New Mexico.

They dwell in rotting logs and under forest litter or in talus rubble. A hostile environment allows limited surface activity between late June and August; during summer rains these salamanders emerge from retreats. Ants and beetle larvae are favorite foods.

106 Arboreal Salamander
(*Aneides lugubris*)

Description: 4¼–7¼" (10.8–18.4 cm). Climbing; grayish to chocolate brown with oversized head. Cream or yellowish spots usually seen above; *belly creamy-white.* Tail somewhat prehensile; toe tips expanded, squarish. Costal grooves, 15–16.

Breeding: Late spring to early summer. 1–2 dozen eggs laid in tree hollow, rotten log, or earthen cavity. Female broods eggs, which hatch in 3–4 months.

Habitat: Live oak woodlands along coast to yellow pine and black oak forests in foothills.

Range: Coast Ranges of California from Humboldt County to Baja California. Also, c. Sierra Nevada foothills; South Farallon, Santa Catalina, and Los Coronados islands.

Champion climber among salamanders. One was found in a red tree mouse's nest 60 feet (18.3 m) up. Rarely seen during summer dry season, it hides in moist tree hollows, rodent burrows, caves, and damp basements. Several dozen may share a retreat. It surfaces during fall rains. At night it forages for insects in trees and on the ground amid leaf litter. It may squeak like a mouse when handled and bites.

Desert Slender Salamander
(*Batrachoseps aridus*)

Description: 2¼–3¾" (5.4–9.6 cm). Body and tail elongated, slim. *Feet have only 4 toes.* Blackish-maroon above, with indistinct band of silvery to brassy flecks from snout to near tail tip. Belly blackish-maroon; underside of tail flesh-colored. Costal grooves, 16–19.

Breeding: Habits uncertain. No aquatic larval stage.

Habitat: Water seepage area and talus slope of shaded desert canyon, 2,500′ (762 m).

Range: Hidden Palm Canyon, Riverside County, California.

First found in 1969. Water seepage below sheetlike limestone deposits and talus provides sanctuary for this endangered species. Because of its limited habitat and fragile nature, the Desert Slender is protected by federal and state laws.

76 California Slender Salamander
(*Batrachoseps attenuatus*)

Description: 3–5½″ (7.6–14 cm). Elongated body and tail. Small narrow feet; only 4 toes. Sooty or black above, with broad yellow, brownish, or reddish band on back; colors vary with locality. *Belly black or dusky, finely speckled with white.* Costal grooves, 18–22.

Breeding: Late fall and winter. Clutches of 4–21 eggs are seen in pockets beneath rocks or logs; hatch in spring. No aquatic larval stage.

Habitat: Coastal mountains and interior foothills; redwood forests, grasslands with scattered trees.

Range: Extreme sw. Oregon south along Coast Ranges to sw. California; also, along base of Sierra Nevada adjacent to Sacramento–San Joaquin Valley.

Common, especially during rainy periods. Spends its days in moist forest litter or in channels of rotted-out tree roots; forages at night for worms and small arthropods. It occurs with the Ensatina Salamander over much of its range. Specimens from southern California are now recognized as Black-bellied Slender Salamander, *B. nigriventris.*

75 Garden Slender Salamander
(*Batrachoseps major*)

Description: 3¼–6⅜″ (8.3–16.2 cm). Slender with long tail. Feet have only 4 toes. Pale brown to reddish-brown above, rarely with band on back. Belly whitish or light gray, peppered with minute black flecks. *Head very narrow;* tail length about 1⅓ times snout-to-vent length. Costal grooves, 18–20.

Breeding: December to January; female lays string of 10–20 eggs under stones or woodland debris. No aquatic larval stage; tiny replicas of adults hatch when ¾" (19 mm) long.

Habitat: Coastal live-oak woodlands and open chaparral; sandy or gravelly soil along canyon bottoms, washes, bases of open grass-covered hills.

Range: Santa Catalina Island and adjacent s. California along the coast into Baja California.

Best observed under rocks and woodland litter, November to April, after rain showers and during cool, humid weather. Like other slender salamanders, the Garden Slender may coil snakewise. When touched, it usually goes into violent contortions and flips into the air; its tail may break off in the process.

73 Channel Islands Slender Salamander
(*Batrachoseps pacificus*)

Description: 3–5¼" (7.6–13.5 cm). Slender body and tail; feet have only 4 toes; pale brown to reddish-brown above, occasionally with brick-red dashes forming indistinct central bands. Belly whitish or light gray, peppered with minute black flecks. *Head broad;* tail length about same as snout-to-vent length. Costal grooves, 18–20.

Breeding: Eggs laid late fall and winter; hatching occurs in spring. No aquatic larval stage.

Habitat: Damp situations; along streams, in canyon mouths and pine forests.

Range: Anacapa, Santa Cruz, Santa Rosa, and San Miguel islands, California.

These creatures are found under stones, rotting logs, and bark. Inactive during the summer dry season, they emerge from retreats during late fall rains. On Santa Cruz they share their habitat with the Relictual Slender Salamander.

77 Relictual Slender Salamander
(*Batrachoseps relictus*)

Description: 2¾–4¾" (6.8–12 cm). Elongated body
and tail. Feet have only 4 toes. *Black
above,* with yellow, red, brown, or
brownish-black dorsal band. Underside
gray-black, with many tiny white dots.
Costal grooves, 16–20.

Breeding: Habits uncertain. No aquatic larval
stage.

Habitat: Under logs and rocks near spring
seepages or streamlets, but sometimes
at considerable distance from water. Sea
level to 8,000' (2,438 m).

Range: In the Sierra Nevada, Kern to Mariposa
counties, California; also w. Monterey
and north San Luis Obispo counties;
Santa Cruz Island; and Baja California.

The Relictual Slender is thought to be
close to the ancestral form that gave rise
to the other slender salamander species.

Kern Canyon Slender Salamander
(*Batrachoseps simatus*)

Description: 3⅝–4⅞" (9.2–12.5 cm). Slim,
elongated body and tail. Feet have only
4 toes. Dark brown above, with bronze
and light reddish-brown dashes and
patches that may form an imperfect
dorsal band. *Sides and underside black,*
with many tiny white dots. Costal
grooves, 20–21.

Breeding: Habits uncertain. No aquatic larval
stage.

Habitat: North-facing slopes; pine-oak-chaparral
community, 1,500–2,000' (457–
610 m).

Range: South side of Kern River Canyon, Kern
County, California.

The Kern was first collected in 1960 by
herpetologist Arden Brame. It lives
under rocks near moist stream beds and
in or beneath rotting pine or oak logs.

74 Tehachapi Slender Salamander
(Batrachoseps stebbinsi)

Description: 3⅝–4¾" (9–12.2 cm). Slim, elongated body and tail. Large feet with only 4 toes; noticeably webbed. Black above, with large, scattered red, dark-brown, or beige patches forming an indistinct band on back. *Underside dark gray to black,* often with large white patches. Costal grooves, 18–19.

Breeding: Habits uncertain. No aquatic larval stage.

Habitat: Rock talus along streams; sycamore-oak-buckeye-Digger pine community, 2,500–3,000' (762–914 m); and montane pine forest meadows, 8,300' (2,530 m).

Range: Paiute and Tehachapi mountains, Kern County, and Sequoia National Forest, Tulare County, California.

First collected in 1957, this species is named in honor of Robert Stebbins, a major contributor to our knowledge of western herpetofauna. He found it in a rockslide covered with fallen leaves and a thick growth of miner's lettuce. Its life history remains largely unknown.

78 Oregon Slender Salamander
(Batrachoseps wrighti)

Description: 3⅜–4¼" (8.6–10.8 cm). Slender. *Feet have only 4 toes.* Dark brown above, with yellowish-brown or reddish flecks forming band from snout to tail tip. Belly black, with *large white spots and blotches.* Costal grooves, 16–17.

Breeding: String of about 10 eggs is laid in June, often under bark of rotting log, hatching in October.

Habitat: Moist Douglas fir, maple, and red cedar woodlands; to 3,000' (914 m).

Range: Nc. Oregon, along Columbia River in Hood River and Multnomah counties,

south on w. slope of Cascades into Lane County.

Colonies are widely scattered, rarely abundant in any one area. Adults become active in April or May. They dwell in moss-covered logs, rotting stumps, and under rocks or pieces of bark near spring seeps.

94 Seepage Salamander
(*Desmognathus aeneus*)

Description:
: 1½–2¼" (3.8–5.7 cm). Very small. *Pale line runs from eye to angle of jaw.* Wide tan, yellow, or reddish-brown stripe above has dark center line continuous with Y-shaped mark or series of dark smudges on back. Belly plain to mottled brown and white. Tail rounded, half of total length. Costal grooves, 13–14.

Breeding:
: Usually April to June; late winter and spring in Alabama. Eggs, 5–17, are laid under moss clump or in small depression and are attended by female; hatch in 5–7 weeks; hatchlings ½" (11 mm). No aquatic larval stage. Sexually mature in 2 years.

Habitat:
: Moist hardwood or mixed forests near small creeks, springs, and seepage areas; damp, shaded ravines. Near sea level to 4,500' (1,372 m).

Range:
: Sw. North Carolina to central Alabama; also wc. Alabama.

This species includes what were formerly called the Cherokee Salamander and the Alabama Salamander. The Cherokee, from North Carolina and Georgia, has straight-edged reddish-brown stripes. The Alabama has wavy-edged tan or yellow stripes. All Seepage Salamanders frequent leaf litter; they relish springtails, beetle larvae, mites.

63 Southern Dusky Salamander
(*Desmognathus auriculatus*)

Description: 3–6⅜" (7.6–16.2 cm). Medium-sized, dark-brown or black dusky; *belly speckled with white.* Sides have a conspicuous row or two of round whitish or reddish spots. Tail keeled on top and compressed. Costal grooves, 14.

Breeding: September to October. In Alabama, females have been seen guarding their nests under logs.

Habitat: Mucky margins of swampy lakes, floodplain sloughs, and streams; in ravine streams where pockets of organic debris collect.

Range: Coastal plain from se. Virginia to c. Florida, west to e. Texas.

During the day these duskies hide under debris or burrow in muck. When their habitat dries up, they gather under the crust in low spots. Like the Northern Dusky, the Southern can lighten or darken its color to camouflage itself. It shares its habitat with the Rusty Mud Salamander.

91 Ouachita Dusky Salamander
(*Desmognathus brimleyorum*)

Description: 3⅛–7" (7.8–17.8 cm). Large and sturdy; one of the most aquatic duskies. Pale line runs from eye to angle of jaw. Juveniles have blotched backs, show faint row of pale spots along each side. Adults become uniformly brown above; sides have bicolored appearance, dark top color meeting light belly color midway along sides. *Tail tip keeled, compressed.* Costal grooves, 14.

Breeding: Habits uncertain.

Habitat: Rocky, gravelly streams; 400–2,600' (122–792 m).

Range: Ouachita Mountains in se. Oklahoma and sw. Arkansas.

Rarely found totally out of water, adult Ouachita Duskies dwell under large rocks in streams; juveniles inhabit wet gravel or rock rubble of seepages or streamlets.

89 Dusky Salamander
(*Desmognathus fuscus*)
Subspecies: Northern, Spotted

Description: 2½–5½" (6.4–14.1 cm). The most common dusky. Pale line from eye to angle of jaw. Juveniles have 5–8 pairs of round yellowish or reddish spots on back. Adults tan or dark brown above, plain or mottled; some show alternating pairs of oval to rhombic blotches, often fused to form stripe. Pattern becomes obscured by dark pigment with age. *Tail triangular,* sharply keeled, compressed. Costal grooves, 14.

Breeding: June to September; lays compact, grapelike cluster of 1–3 dozen eggs near water, beneath rocks, in rotting logs, stream-bank cavities. Larvae hatch in 6–13 weeks at ⅝" (16 mm); transform in 6–13 months at 1½" (38 mm). Mature in 3–4 years.

Habitat: Rock-strewn woodland creeks, seepages, and springs in northern areas; floodplains, sloughs, and mucky sites along upland streams in southern areas. Near sea level to about 5,300' (1,615 m).

Range: S. New Brunswick and se. Quebec southwest to Louisiana.

Subspecies: Northern (*D. f. fuscus*), round yellow spots on young that quickly fade as animal ages; s. New Brunswick to e. Kentucky and Carolinas. Spotted (*D. f. conanti*), adults retain remnants of juvenile red or golden spotting; s. Illinois, w. Tennessee, w. South Carolina to the Gulf.

Where ranges coincide, Dusky, Seal, and Black-bellied salamanders often share the same stream, but Dusky likes higher stream bank elevations. It is often found with the Red Salamander. It eats insect larvae, sow bugs, and earthworms.

134 Imitator Salamander
(*Desmognathus imitator*)

Description: 2¾–4″ (7–10.2 cm). Pale line from eye to angle of jaw. Highly variable in coloration; usually has dark, rippling back stripes, rarely a single wide back stripe. Older individuals often become dark and patternless. *Yellow to red cheek patches common.*

Breeding: Unknown.

Habitat: Cool, moist red spruce–Fraser fir and hardwood forests; chiefly between 4,000–5,500′ (1,219–1,676 m).

Range: Restricted to Great Smoky Mountains, North Carolina–Tennessee border; nearby Balsam Mountain, North Carolina.

Seen under rocks and rotting logs on forest floor, or amid wet gravel and leaf litter along edge of seepages, springs, or intermittent brooks. The Imitator often mimics the Appalachian Woodland Salamander and is also easily confused with the Mountain Dusky Salamander. In fact, Imitator and Mountain Dusky are so similar in color pattern and body structure that laboratory analysis is often the only way to distinguish them. The Imitator prefers higher elevations so attention to collecting site is critical in making a field identification.

65 Seal Salamander
(*Desmognathus monticola*)
Subspecies: Appalachian, Virginia

Description: 3–5⅞" (7.6–14.9 cm). A stout,
variably patterned species. Pale stripe
from eye to angle of jaw. Light brown
or grayish above, with numerous dark-
brown or black streaks, blotches, or
reticulations surrounded by paler areas.
Belly light-colored, with or without
blotches. Sides dark above, light-
speckled below, the transition abrupt.
Tail sharply keeled, compressed; tip pointed.
Costal grooves, 14.

Breeding: Females guarding nests are seen June to
October. About 1–3 dozen eggs are
singly attached to undersides of rocks in
seepages. Larvae begin hatching in
September at ¾" (19 mm) and
transform at about 1¾" (44 mm).

Habitat: Banks of mountain streams, small rocky
spring-fed brooks in hardwood-shaded
ravines. Near sea level to 5,450' (1,661
m).

Range: Sw. Pennsylvania southwest in uplands
to c. Alabama; also sw. Alabama.

Subspecies:  Appalachian (*D. m. monticola*), dark
markings highly variable and
numerous; w. Pennsylvania
southwestward to extreme w. Florida.
Virginia (*D. m. jeffersoni*), dark
markings reduced to scattered small
round spots or patternless; Blue Ridge
Mountains of Virginia.

This stream-bank dweller is never far
from water and quickly takes to it when
disturbed. By day it hides under rocks
or in burrows, dining on passing ants,
beetles, and occasionally other
salamanders. Dusky or Black-bellied
salamanders may share its habitat.

95, 112, 120,   Mountain Dusky Salamander
124, 137 (*Desmognathus ochrophaeus*)

Description: 2¾–4⅜" (7–11.1 cm). Varies greatly
in color and pattern. Pale line from eye
to angle of jaw. *Tail rounded;* half of
total length. Northern forms: gray,
brown, olive, yellow, or orange; have
wide, straight-edged, dark-bordered
stripe marked with V shapes down back
and tail. Southern forms: dark margins
of stripe usually wavy, but vary from
nearly straight to zigzagged; dark bars
cross stripe. Costal grooves, 14.

Breeding: Grapelike cluster of 1–2 dozen eggs
laid spring to fall, attached by female
to moss or rotting logs near water.
Female attends nest. Larvae hatch
summer to fall at about ⅝" (14 mm);
transform in 2–8 months at about 1"
(25 mm). Sexually mature in 3–4
years.

Habitat: Uplands from 600–6,500' (183–
1,981 m). At low elevations, stays close
to streams, springs, and seepage areas.
At higher elevations, favors cool, moist
floors of spruce-fir forests.

Range: West of Hudson River in New York to
ne. Georgia; also ne. Alabama.

Entire populations may congregate
around springheads and seepages during
winter. These sites also provide
brooding areas for females and an
aquatic habitat for larvae. It is also seen
on wet cliff faces. It eats small flies,
beetles, mites. Salamanders from areas
near the Great Smokies may have
reddish cheek patches, mimicking the
Appalachian Woodland Salamander.

64   Black-bellied Salamander
(*Desmognathus quadramaculatus*)

Description: 3½–8¼" (8.9–20.8 cm). Largest, most
robust, and most aquatic dusky
salamander. Pale line from eye to angle

of jaw. Black above, with greenish blotches; *belly uniformly dark brown or black;* with yellow flecking in young. *Two rows of light dots on each side.* Tail sharply keeled on top. Costal grooves, 14.

Breeding: June to July. Female lays 15–40 eggs, singly, on undersides of stones in stream bed. Larvae hatch August to September at ¾" (19 mm), transform when 2¼–3½" (5.7–8.9 cm). Sexually mature in 3½ years.

Habitat: Prefers sizable swift and boulder-strewn mountain streams, at elevations of 1,600–5,000' (488–1,524 m).

Range: S. West Virginia through mountains to ne. Georgia; scattered populations in South Carolina and Georgia Piedmont.

This husky dusky rarely ventures beyond a stream's splash zone. Though nocturnal, it sometimes basks by day on wet rocks. Agile in the water, it darts quickly to safety when disturbed. It used to be sold as fish bait in the Piedmont, which may explain its presence there. Sometimes plagued by leeches. Feeds on insects, snails, small salamanders.

92 Black Mountain Dusky Salamander
(*Desmognathus welteri*)

Description: 3–6⅝" (7.6–17 cm). Large, stocky, easily confused with Dusky and Seal salamanders. *Brown above,* with small dark-brown or black blotches; whitish below, with fine to heavy brown stippling. Back and belly colors blend at sides. End of tail keeled, compressed. Toe tips usually dark. Costal grooves, 14.

Breeding: About 24 eggs are laid March to November, under stones or forest litter at waterside. Female attends nest. Larvae hatch at about ¾" (19 mm), transform in 21–24 months.

Habitat: Mountain brooks and streams, springs, and seepage areas; 1,000–4,000' (305–1,219 m).

Range: Big Black Mountain in e. Kentucky and adjacent Virginia; also Warren County, Kentucky.

This species shares its range with Dusky and Seal salamanders but is more aquatic. It is found under stones in or along stream beds.

66 Pygmy Salamander
(*Desmognathus wrighti*)

Description: 1½–2" (3.7–5.1 cm). A terrestrial, high-altitude, midget species. Pale line runs from eye to angle of jaw. Wide tan to reddish-brown back strip with *dark herringbone pattern*. Belly flesh-colored. Tail rounded, less than half total body length. Rough skin atop head. Costal grooves, 13–14.

Breeding: Courts in spring and fall. Nests in pockets of wet gravel and mud near stream headwaters or springs; clutch of 3–8 eggs dangles from rocks by single stalk. Female attends nest. Eggs hatch in fall. No aquatic larval stage. Hatchlings are ½" (11 mm) long.

Habitat: Chiefly Fraser fir and red spruce forests; 2,750–6,500' (838–1,981 m).

Range: S. Appalachians from sw. Virginia into sw. North Carolina.

By day this diminutive creature hides under leaf litter, stones near seeps, or rotting logs; at night, during wet weather, it forages on ground surfaces or tree trunks for small insects. During fall, Pygmies congregate in underground seepage areas, spending the winter there.

104, 105, 107, 108 **Ensatina**
(*Ensatina eschscholtzi*)
Subspecies: Monterey, Yellow-
blotched, Large-blotched, Oregon,
Painted, Sierra Nevada, Yellow-eyed

Description: 3–5⅞" (7.6–14.9 cm). Only
salamander with *tail constricted at base*
and *5 toes on hind feet*. Displays
amazing variety of colors and patterns
—uniform brown or reddish-brown to
dark brown or black above, with cream,
yellow, or sometimes orange spots,
blotches, or mottling. Base of limbs
lighter in color than tips. Belly whitish
or flesh-colored. Costal grooves, 12–
13. Male's tail often longer than body;
female's shorter.

Breeding: Late spring to early summer. Cluster of
7–25 eggs, laid underground, is
brooded by female; young hatch at ¾"
(19 mm) in fall or early winter. No
aquatic larval stage. Sexually mature in
2½–3½ years.

Habitat: Douglas fir–vine maple forests in
northwestern areas. Redwood forest,
chaparral, and coast live oak–black
walnut woodlands along coast. Yellow
pine–black oak–incense cedar forests of
Sierra Nevadas. To 10,000′
(3,048 m).

Range: E. Vancouver Island; sw. British
Columbia, south along Coast Ranges to
Baja California and along w. slopes of
Cascade Range and Sierra Nevadas;
absent from Sacramento–San Joaquin
Valley.

Subspecies: Monterey (*E. e. eschscholtzi*), reddish-
brown above, whitish below, eyes
black; Coastal s. California into nw.
Baja California. Yellow-blotched (*E. e.
croceator*), black above with cream or
yellow blotches, large blotch behind
each eye; interior mountains of Kern
County, California. Large-blotched (*E.
e. klauberi*), black above with cream or
orange blotches and crossbars, U-
shaped blotch behind eyes; interior
mountains of extreme sc. California.

Oregon (*E. e. oregonensis*), plain brown
or blackish above, whitish or yellowish
below, with fine black dots; sw. British
Columbia south through w.
Washington, Oregon and into nw.
California. Painted (*E. e. picta*), brown
above, mottled with black, yellow and
orange, yellow-orange underside,
peppered with black dots; coastal sw.
Oregon and nw. California. Sierra
Nevada (*E. e. platensis*), brown above
with orange spots; Sierra Nevada
Mountains, e. of the Great Valley,
California. Yellow-eyed (*E. e.
xanthoptica*), brownish orange above,
belly orange, eye with yellow patch;
San Francisco Bay and opposite side of
Great Valley on w. slope of Sierra
Nevadas.

After the fall rains this wide-ranging
salamander remains active until the end
of May or sometimes through June in
northerly or high locations. It is found
under rocks and rotting logs in moist
forest areas and shaded canyons. In cold
or dry weather it retreats into caves,
animal burrows, and crevices among
rotted roots and logs. When
threatened, the Ensatina assumes a stiff-
legged, sway-backed stance, with tail
elevated and arched. If the tail is
seized, it easily snaps off, allowing
escape. Eats spiders, beetles,
springtails, crickets. Some may live
as long as 10–15 years in the wild.

88, 121 Two-lined Salamander
(*Eurycea bislineata*)
Subspecies: Northern, Southern,
Blue Ridge

Description: 2½–4¾" (6.4–12.1 cm). An
abundant brookside species. Broad,
basically yellow band above may be
tinged with brown, green, or orange-
bronze; often darkly speckled. *Band*

bordered by dark-brown or black stripe
running from each eye well out onto tail.
Tail oval, keeled, compressed. Costal
grooves, 13–16.

Breeding: A dozen to 100 eggs are laid on
undersides of submerged rocks, logs, or
aquatic plants; female may guard eggs.
Larvae hatch at ½" (13 mm) and
transform in 1–3 years at 1¾" (44
mm).

Habitat: Rock-bottomed brooks, springs,
seepages, river swamps, and floodplain
bottoms in coastal plain to damp forest
floors at high elevations; near sea level
to 6,000' (1,829 m).

Range: Mouth of St. Lawrence River, Quebec,
to n. Florida and west to se. Ontario, e.
Illinois, and Mississippi River.

Subspecies: Northern (*E. b. bislineata*), back stripes
tend to break up into dashes or dots on
tail, males do not develop downward
projections from nostrils; mouth of St.
Lawrence to Virginia and west to
Tennessee River. Southern (*E. b.
cirrigera*), back stripes narrow and
continue close to tail tip, males with
downward projections from nostrils; s.
Virginia to n. Florida west to
Mississippi River. Blue Ridge (*E. b.
wilderae*), back stripes broad and break
up into dots on tail, males with
downward projections from nostrils;
sw. Virginia to n. Georgia, in s. Blue
Ridge Mountains.

Its life history is largely unknown. In
some populations larvae retain gills and
do not transform. A stocky short-tailed
Two-lined Salamander, recognized as *E.
aquatica* by some authorities, lives in
scattered springs and small streams in
the South.

125 Junaluska Salamander
(*Eurycea junaluska*)

Description: 2½–3½" (6.4–9 cm). Resembles Two-lined. *Dull yellowish-brown above,* sprinkled with tiny black spots most numerous on either side of back, where they form an irregular stripe from behind eye to tail. Sides stippled with black. Belly greenish-yellow. Costal grooves, 14.

Breeding: Season probably runs through fall, winter, and early spring.

Habitat: Tulula and Santeetlah creeks and Cheoah River, and their feeders and tributaries; 1,200–2,000' (366–610 m).

Range: Cheoah River Valley, Graham County, North Carolina.

This species is named after a Cherokee chief who played an important role in the region's history. At night during rain storms the Junaluska can be seen crossing roads. Days are spent under stones or debris along or in streams. Its habitat is shared by Two-lined and Long-tailed salamanders.

Cascade Cavern Salamander
(*Eurycea latitans*)

Description: 2½–4⅛" (6.4–10.5 cm). Aquatic, with well-developed, lightly pigmented gills. *Eyes small and lidless.* Snout flattened; head slopes sharply upward from eyes. Light tan, with faint network of darker pigment, white flecks on sides. Legs short, stout. *Tail fin conspicuous, elevated.* Costal grooves, 14–15.

Breeding: Habits unknown.

Habitat: Subterranean streams and pools.

Range: Cascade Cavern, Kendall Co., Texas.

Very little is known about this cave-dwelling salamander.

113, 114, Long-tailed Salamander
123 (*Eurycea longicauda*)
Subspecies: Long-tailed, Three-lined,
Dark-sided

Description: 3⅛–7⅞" (9.8–20 cm). *Slender tail much longer than body.* Northern form: yellow to bright red-orange; scattered black spots, heaviest on sides, form bars or herringbone pattern on tail. Southern form: wide yellowish band above, with dark broken or continuous stripe down center; band flanked by wide dark stripes. Western form: broad yellowish band above, flanked by wide, grayish- to reddish-brown side bands; black spots on back; gray or yellowish flecks on sides. Costal grooves, 13–14.

Breeding: October to March. Eggs laid in underground crevices associated with springs, temporary pools, and streams. Larvae hatch in 6–8 weeks at ¾" (19 mm); transform in 3½–7 months at 1⅝" (41 mm). Sexually mature in 1–2 years.

Habitat: Stream sides, spring runs, seepages, cave mouths, forested floodplains in Deep South.

Range: S. New York to Florida panhandle, west to Mississippi River and beyond in s. Missouri, n. Arkansas, and adjacent Kansas and Oklahoma.

Subspecies: Long-tailed (*E. l. longicauda*), vertical black bars on tail, belly yellow; s. New York and n. New Jersey southwestward to n. Alabama and s. Illinois. Three-lined (*E. l. guttolineata*), narrow black back stripe, belly yellowish and heavily mottled with greenish gray; Virginia southwestward to Florida panhandle, west to Mississippi River. Dark-sided (*E. l. melanopleura*), dark sides contrast strongly with light back band, belly dull white; chiefly in portion of species' range west of the Mississippi River.

At night during warm, rainy weather this salamander ventures about on the forest floor in search of tiny invertebrate prey. The colors of the northern form often make it quite conspicuous.

122 Cave Salamander
(*Eurycea lucifuga*)

Description: 4⅞–7⅛" (12.5–18.1 cm). Slender, long-tailed brook salamander. Dull yellow to bright orange above; scattered black spots sometimes form 2 or 3 longitudinal rows. Belly yellowish and unspotted. *Tail prehensile, clearly longer than body.* Costal grooves, 14–15.

Breeding: Fall to spring. Lays 50–90 eggs that cling separately to undersides of stones in cave or spring waters. Larvae hatch at ⅜" (10 mm), transform during second year at about 2" (57 mm).

Habitat: Twilight zone of limestone caves or nearby springs and woodlands.

Range: Limited to limestone areas; w. Virginia and se. West Virginia south to c. Alabama, west to ne. Oklahoma.

Adults, assisted by a prehensile tail, are great climbers, exploring crevices of cave walls for insect prey. They also forage outside caves under stones and woodland debris. Larvae grow slowly because of low water temperatures and the general scarcity of food.

93 Many-ribbed Salamander
(*Eurycea multiplicata*)
Subspecies: Many-ribbed, Gray-bellied

Description: 2⅜–4⅛" (6–10.6 cm). A brook salamander; name refers to its *many costal grooves,* 19–20. Adults have yellowish-brown to chocolate-brown band above, often bordered by dark lines and sometimes with narrow line down band's center. Lower sides show silvery flecks. Belly pale gray to yellow.

Permanent larvae are pale, lack much of normal pattern; larger than transformed animals.

Breeding: September to April: season strongly influenced by water availability and temperature. Females lay row of 3–21 eggs on rock undersides. Nest left unguarded. Larvae hatch in 4–6 weeks at ½" (13 mm), transform in 5–8 months, mature soon thereafter.

Habitat: Cave springs and their runs, cold spring-fed brooks; 350–2,500' (107–762 m).

Range: C. Missouri and nc. Arkansas into Oklahoma; isolated population in extreme se. Kansas.

Subspecies: Many-ribbed (*E. m. multiplicata*), back bordered by indistinct lines or lines absent, belly uniformly yellow; nc. Arkansas southwestward to se. Oklahoma. Gray-bellied (*E. m. griseogaster*), back banded by well defined dark line, belly gray to yellow with tiny dark dots or dark patches; c. Missouri and ne. Arkansas west into ne. Oklahoma; also se. Arkansas.

Permanent larvae occur most often among cave-dwelling populations of this species. During wet weather, Many-ribbed adults leave their aquatic sanctuaries and forage in adjacent woodlands.

5 San Marcos Salamander
(*Eurycea nana*)

Description: 1⅝–2" (4.1–5.1 cm). Short, slender-legged aquatic salamander, with prominent, pigmented external gills. Light brown above, with pale-yellowish row of flecks or spots along sides. *Belly yellowish-white.* Eyes show dark ring around lens. Tail has well-developed back fin. Costal grooves, 16–17.

Breeding: Habits unknown.

Habitat: Spring pool at source of San Marcos River.
Range: Hays County, Texas.

This salamander is apparently restricted to the site where it was first collected in 1938. Neotenic. It lives among the algae mats covering the spring pool.

4 Texas Salamander
(*Eurycea neotenes*)

Description: 1⅞–3⅞″ (4.8–10 cm). Aquatic. *External gills have long, bright-red filaments.* Light brown to yellow above, with 1 or 2 rows of light flecks on sides; belly cream. Dark bar from eye to nostrils. Short stout legs; narrow tail fin. Costal grooves, 15–17.
Breeding: Larvae hatch at about ⅝″ (14 mm), grow to sexual maturity in 2 years.
Habitat: Small cave streams, springs, seeps, and headwaters of creeks.
Range: Edwards Plateau, wc. Texas.

Usually this animal spends its days under rocks or leaves, but sometimes it is seen walking along the bottom of a stream or pool. It rarely transforms to a terrestrial form. It is freely preyed on; few survive the first year.

109, 126 Dwarf Salamander
(*Eurycea quadridigitata*)

Description: 2⅛–3½″ (5.4–9 cm). Elongated miniature species, with *4 toes on hind feet.* Yellowish to dark-brown band runs along back, often with row of dark spots down center and bordered by a darker stripe running from eye to tip of tail. Costal grooves, 14–17.
Breeding: Courts in fall. Lays 1–4 dozen eggs in shallow water, late fall to early winter, singly or in small groups on undersides

of leaves and logs. Larvae hatch in 30–40 days, transform April to June at 1½" (38 mm). Sexually mature by first fall.

Habitat: Margins of pine savannah ponds, swampy areas in low flatwoods.

Range: Coastal plain, North Carolina to Lake Okeechobee, Florida, and west to e. Texas. Isolated populations in Arkansas and Missouri.

Dwarfs frequent soggy beds of pine needles or damp places under logs. Surface activity is greatest during fall and early winter breeding season.

2, 11 Comal Blind Salamander
(*Eurycea tridentifera*)

Description: 2–3⅜" (5.1–8.5 cm). Thoroughly adapted for subterranean aquatic life. External gills have pink-red filaments. *Large head,* with flattened snout, rises abruptly behind eyes. Eyes very small, often misshapen and covered with skin. Cream to pale yellow above, with traces of brown or gray pigment; faint line from eye to nostril. Legs slender. Costal grooves, 11–12.

Breeding: Lays 7–18 eggs; larvae hatch at ½" (13 mm).

Habitat: Underground waters of limestone caves.

Range: Honey Creek Cave and sinkhole caves on floodplain of Cibolo Creek, Comal County, and Elm Springs Cave, Bexar County, Texas.

At Honey Creek Cave, where the Comal Blind was discovered in 1965, it shares waters with the Texas Salamander. Its transparent underside allows quick determination of sex.

1 Valdina Farms Salamander
(Eurycea troglodytes)

Description: 2–3″ (5.1–7.8 cm). Eyes of this
aquatic cave-dweller are quite small and
partially or entirely covered by skin.
External gills retained throughout life.
Light gray or cream above, with white
specks and faint yellow stripes along
sides of body and top of tail. *Legs long
and skinny.* Costal grooves, 13–17.

Breeding: Habits unknown.

Habitat: Intermittent pools of subterranean
streams.

Range: Valdina Farms sinkhole, nw. Medina
County, Texas.

The Valdina was first collected in 1956.
Its natural history remains unknown.
Like other troglodytes that lack
pigmentation, its transparent belly
reveals the outlines of internal organs.

10 Oklahoma Salamander
(Eurycea tynerensis)

Description: 2–3⅛″ (5.1–8.1 cm). Aquatic; retains
short, dark external gills throughout
life. *Light to dark gray above,* with black
stippling or streaks; occasionally
uniformly gray. Usually shows one or
more rows of small light spots along
sides; belly pale. Tail has weakly
developed back fin. Costal grooves, 19–
21.

Breeding: Fall or late spring. Eggs laid on
undersides of stones. Hatchlings are
⅜–½″ (10–13 mm) long.

Habitat: Small gravelly-bottomed, spring-fed
streams below 1,000′ (305 m).

Range: Drainages of Neosho and Illinois rivers,
sw. Missouri to ne. Oklahoma.

Named for Tyner Creek in Oklahoma
where it was first collected, this species
does not transform to a terrestrial stage.
It has strong preferences for certain

types of bottom and thus tends to separate into isolated colonies along stream beds.

6, 8 Tennessee Cave Salamander
(Gyrinophilus palleucus)
Subspecies: Sinking Cove, Big Mouth, Berry Cave

Description: 4–8⅞" (10.2–22.7 cm). Cave-dwelling "gyro" that retains its larval characteristics. *Bright-red external gills; small, lidless eyes; large tail fin.* Snout broad, turned up at tip. Stout body ranges from nearly white to dark brown. Costal grooves, 17–19.

Breeding: Poorly known.

Habitat: Subterranean waters.

Range: Se. and c. Tennessee, n. Alabama, and extreme nw. Georgia.

Subspecies: Sinking Cove (*G. p. palleucus*), flesh colored, unspotted; se. edge of Cumberland Plateau, Tennessee. Big Mouth (*G. p. necturoides*), brown to purplish with heavy spotting; Big Mouth Cave, Grundy County, Tennessee. Berry Cave (*G. p. gulolineatus*), similar to Big Mouth, with dark stripe or blotch on throat; se. Tennessee.

The full range of this species is far from known. Although in nature the larvae never transform, they do so in the laboratory when treated with thyroxin. Food supply in caves is limited, hence the young grow slowly.

101, 127, 130 Spring Salamander
(Gyrinophilus porphyriticus)
Subspecies: Northern, Kentucky, Blue Ridge, Carolina

Description: .4¼–8⅜" (10.8–21.9 cm). One of the largest lungless salamanders. *Light bar runs from eye to nostril.* Sturdy body and keeled tail. Salmon, brownish-pink, yellowish-brown or orange, or reddish-brown; patterns vary. Often cloudy in appearance, with vague darker markings. Costal grooves, 17–19.

Breeding: During warmer months 11–100 eggs are singly attached to undersides of stones in cool water. Larvae hatch late summer or fall when ¾″ (19 mm) long, may reach 4″ (10.2 cm) before transforming in 2–3 years.

Habitat: Springs, cool and clear mountain brooks, shaded seepages, and wet caves; 300–6,600′ (91–2,012 m).

Range: S. Quebec and s. Maine to n. Georgia, Alabama, and Mississippi.

Subspecies: Northern (*G. p. porphyriticus*), mottled or netlike pattern on back; s. Maine southwestward to n. Georgia. Kentucky (*G. p. duryi*), back with scattered small black spots; sw. Ohio southeastward to extreme sw. Virginia. Blue Ridge (*G. p. danielsi*), distinct pale bar with black line below, back with scattered small black dots; high mountains along North Carolina–Tennessee border. Carolina (*G. p. dunni*), distinct pale bar with black line below, back profusely flecked with black; sw. North Carolina to ne. Alabama.

Ecology poorly known. At night during downpours, these salamanders may leave their aquatic homes and prowl about for food, including other salamanders.

West Virginia Spring Salamander
(*Gyrinophilus subterraneus*)

Description: 6½–7½″ (15.9–18.4 cm). Subterranean. Huge larvae with bright-red gills may reach 7″. Flesh-colored above, with gray netlike pattern; sides show 2–3 irregular rows of pale spots.

Indistinct bar from eye to nostril. Costal grooves, 17.

Breeding: Poorly known.

Habitat: Single limestone-cave stream and its steep mudbanks.

Range: General Davis Cave, Greenbrier County, West Virginia.

It shares its habitat with the Spring Salamander, which the West Virginia closely resembles. Larvae reach sexual maturity shortly before or during transformation. Adults eat worms, ground beetles, small crustaceans, and insects that fall into the stream.

7 Georgia Blind Salamander
(*Haideotriton wallacei*)

Description: 2–3″ (5.1–7.6 cm). Subterranean. Blind. A pinkish-white salamander, with *deep-red plumelike gills.* Vestiges of eyes visible under skin as minute black dots. Head not flattened, *snout very broad and rounded.* Costal grooves, 12–13.

Breeding: Habits unknown.

Habitat: Cave and other subterranean waters, deep wells.

Range: Limestone sink areas of sw. Georgia and adjacent Florida.

So few Georgia Blinds have been collected that little is known about them. Adults retain larval characteristics and are unable to transform to a terrestrial stage.

103 Four-toed Salamander
(*Hemidactylium scutatum*)

Description: 2–4″ (5.1–10.2 cm). A small species distinguished by *hind feet with 4 toes* and marked constriction at base of tail. Reddish-brown above, grayish sides;

white belly, with black spots. Costal grooves, 13–14.

Breeding: Late winter to spring. 2–3 dozen eggs, singly attached to sphagnum moss or other plants close to water; female guards eggs until hatching 6–8 weeks later. The ½" (13 mm) aquatic larvae transform in 1½ months at ⅞" (22 mm). Mature in 2½ years.

Habitat: Bogs, boggy streams, and floodplains; usually associated with sphagnum moss.

Range: Discontinuous. Chiefly east of Mississippi River; Nova Scotia to Wisconsin south to Gulf, but absent from Florida peninsula.

Adults live under stones and leaf litter in hardwood forests surrounding boggy areas; the need for this special habitat accounts for its spotty distribution. When a predator grabs the Four-toed's tail, it readily breaks off—a twitching morsel that distracts the enemy. A new tail is soon regenerated.

---

97 Limestone Salamander
(Hydromantes brunus)

Description: 2¾–4⅜" (7–11.1 cm). *Web-toed,* with short tail. Head and body flattened. Eyes prominent. Uniformly brown above, underside of tail yellow. Young are pale yellow to green above. Costal grooves, 13.

Breeding: Poorly known.

Habitat: Moss-covered limestone cliffs, ledges, and talus rubble in Digger pine and chaparral community; 1,200–2,500' (366–762 m).

Range: Lower Merced Canyon, Mariposa County, California.

Discovered in 1952, this species belongs to the only genus of salamanders found in both the New and Old World. Projecting upper teeth in males and long mushroomlike tongues are shared characteristics.

86 Mount Lyell Salamander
(*Hydromantes platycephalus*)

Description: 2¾–4½″ (7–11.4 cm). *Web-toed;* head and body flattened. Prominent eyes. Short blunt tail. Brown or blackish above, with lichenlike patches and flecks of gray or gold. Young are black, with dense gold stippling; appear greenish. Costal grooves, 13.

Breeding: Lays 6–14 eggs; young probably hatch in fall.

Habitat: Exposed granite outcrops, associated fissures, cave openings, and talus; favors north-facing slopes, 4,000–11,000′+ (1,219–3,353+ m).

Range: In the Sierra Nevadas from Sonora Pass south to Sequoia National Park, California.

This nocturnal species may be encountered under rock slabs near melting snowbanks or in seepage-moistened talus at bases of cliffs. Its broad-webbed feet allow it to scale smooth surfaces in search of insects and spiders. The tail is blunt, the tip used as a braking or anchoring device as it roves steep cliffs.

87 Shasta Salamander
(*Hydromantes shastae*)

Description: 3–4¼″ (7.6–10.8 cm). *Web-toed.* Head flattened, with prominent eyes. Mottled gray-green, tan, or reddish-brown above; tail short, usually yellowish. Young resemble adults. Costal grooves, 13.

Breeding: Summer. About 9 eggs are deposited in crevice of moist cave; female broods clutch. No aquatic larval stage.

Hatchlings are ⅞″ (22 mm).

Habitat: Fissures and mouths of caves in limestone outcroppings surrounded by Digger pine–Douglas fir–oak forest; 1,000–2,500′ (305–762 m).

Range: Vicinity of Shasta Reservoir headwaters, Shasta County, California.

This species was found in 1950 by herpetologist Joe Gorman, who also first described the Limestone Salamander. In wet weather the Shasta may be found away from caves under rock slabs.

---

135 Shovel-nosed Salamander
(*Leurognathus marmoratus*)

Description: 3⅛–5¾″ (7.9–14.6 cm). Aquatic; heavy-bodied. *Dark brown to blackish above,* with dull mottled pattern; usually shows 2 rows of irregular gray, olive, or yellowish blotches. Belly often dark gray with lighter central area. Snout U-shaped from above; usually darkest part of body. Tail topped by *prominent knife-edged keel,* sometimes ragged. Costal grooves, 14.

Breeding: June to July. Some 3 dozen eggs are attached singly or in small groups to undersides of rocks in stream's main current. Female stands guard during 2½-month development. Hatchlings approach ⅝″ (16 mm), transform at about 1¾″ (44 mm).

Habitat: Small to medium-sized, cold, fast-flowing streams; 1,000–5,500′ (305–1,676 m).

Range: White Top Mountain, sw. Virginia, southwest into Smokies and ne. Georgia.

It is often confused with the Black-bellied Salamander, with which it shares its habitat. The Shovel-nosed prefers fast-current areas and when disturbed swims a short distance, stopping in the open. The Black-bellied lives at the stream's periphery and vigorously retreats. If still in doubt, open its mouth and check internal nares. In the Shovel-nosed they are small slits; in the Black-bellied they are conspicuous round openings. The Shovel-nosed eats larvae of the mayfly, caddis fly, and stone fly.

68 Red Hills Salamander
(Phaeognathus hubrichti)

Description: 4–10″ (10.2–25.6 cm). Elongated, short-legged terrestrial salamander. Uniformly dark brown. *Males have many tiny glands on tail and body.* Costal grooves, 20–22.

Breeding: Uncertain. Believed to take place in spring, entirely on land.

Habitat: Slopes of moist, shaded, hardwood-dominated ravines.

Range: Red Hills of sc. Alabama between Alabama and Conecuh rivers.

Discovered in 1960, this salamander is extremely secretive and rarely seen. It lives in burrows near tree bases or under outcroppings of siltstone on steep slopes. At night it leaves the burrow to feed on spiders and insects. Because of its limited distribution and threatened environment, it is protected by the U.S. Endangered Species Act.

144 Caddo Mountain Salamander
(Plethodon caddoensis)

Description: 3½–4⅜″ (8.9–11.1 cm). Slender black woodland salamander with white spots and *numerous brassy flecks scattered on back.* White pigment on sides. Underside marked with white or yellow from chin to front limbs, remainder darkly pigmented. Costal grooves, 16.

Breeding: Habits poorly known.
Habitat: Mixed deciduous, second-growth woodlands with some pine; 900–2,150' (274–655 m).
Range: Caddo Mountains, Arkansas.

Terrestrial. During the day this salamander stays in its burrow beneath rotting logs. It emerges during evening hours and forages for small invertebrates amid moist forest litter. It is sometimes confused with the closely related Rich Mountain Salamander or the Slimy Salamander; the Rich Mountain has a dark chest and lives at higher elevations, the Slimy a dark throat.

71, 117 Red-backed Salamander
(*Plethodon cinereus*)

Description: 2½–5" (6.4–12.7 cm). Long and slender. Two color phases: "Red-backed" has broad, straight-edged, dark bordered red stripe extending along back from head onto tail and narrowing at base of tail. Stripe may be yellow, orange, pink, or gray. "Lead-backed" is light gray to almost black, without stripe. Intermediates are occasionally found. *Belly has black and white mottling.* Costal grooves, usually 19.

Breeding: Courts and mates October to April. In June or July suspends a grapelike cluster of about 6–12 eggs from roof of small cavity under stone or within rotten log. Female coils about and attends eggs until they hatch 2 months later. Larva hatch at ⅞" (22 mm); omit aquatic stage; mature in 2 years. Females lay eggs every other year.

Habitat: Cool moist coniferous, mixed, and hardwood forests; sea level to 5,600' (1,707 m).

Range: W. Ontario to s. Quebec and

Newfoundland, south to North Carolina and s. Indiana.

Completely terrestrial. Most abundant and commonly encountered salamander throughout much of its range. Because of its ability to tolerate cold the Red-backed has survived in glaciated areas of northeast United States and southeast Canada. During the day it hides under stones or woodland debris. At night it searches amid moist leaf litter for tiny invertebrates. During warm spells in winter it may be found at the surface. In dry weather it retreats underground and only surfaces after rainstorms.

96 Zigzag Salamander
(*Plethodon dorsalis*)
Subspecies: Eastern, Ozark

Description: 2½–4⅜" (6.4–11.1 cm). Small slender salamander. Back stripe red, orange, or yellow, with well-defined wavy border or straight-edged, poorly defined border. *Stripe widens on tail base. Belly mottled* with orange, white, and black. Unstriped dark-phase individuals may be found. Costal grooves, 18.

Breeding: Probably nests in an underground retreat during summer. Larvae skip aquatic stage and hatch in fall.

Habitat: Moist rocky retreats—ravines, canyons, escarpments, talus rubble, spring seepages, and caves; 400–2,500′ (122–762 m).

Range: From c. Indiana south to c. Alabama and s. Missouri, n. Arkansas, and ne. Oklahoma.

Subspecies: Eastern (*P.d. dorsalis*), broad back stripe with well-defined wavy borders; c. Indiana to Alabama, and satellite populations. Ozark (*P. d. angusticlavius*), narrow back stripe with poorly defined, often straight-edged borders; sw. Missouri, n. Arkansas, and ne. Oklahoma.

The Zigzag spends May through September underground. During fall, winter, and early spring it can be found in wet rocky areas or near fallen logs. Some southern populations are considered to be a new species, *P. websteri.*

110 Dunn's Salamander
(Plethodon dunni)

Description: 3⅞–5⅞″ (9.8–15 cm). Mottled *greenish-yellow back stripe does not reach tail tip;* sometimes all dark brown and unstriped. Sides dark brown or black with small light flecks. Small, scattered yellowish spots on belly. Costal grooves, usually 15.

Breeding: Deposits cluster of 6–18 eggs in late spring to early winter.

Habitat: Moist, shaded rocky areas, such as wet talus rubble, spring seepages, shale or sandstone outcrops along streams; from sea level to 2,500′ (762 m).

Range: Extreme sw. Washington south through w. Oregon to extreme nw. California.

This salamander is especially common under stones in saturated areas. It takes to water when disturbed. Oregon Ensatina and Larch Mountain salamanders share its habitat. It eats mites, springtails, flies, and small snails.

67 Del Norte Salamander
(Plethodon elongatus)

Description: 4½–6″ (11.4–15.2 cm). Long slender body. Dark brown or black, with or without straight-edged orange or reddish-brown *back stripe that often extends to tail tip.* Belly very dark with a few scattered white spots. Many tiny white or yellow flecks on sides. Toes short and slightly webbed. Costal grooves, usually 18.

Breeding: Probably nests late spring to early summer with fall or winter hatching.

Habitat: Old rock slides and outcrops under redwood or Douglas fir cover.

Range: Sw. Oregon and nw. California.

Unlike other western woodland salamanders, this species is not usually seen in seepages or other extremely moist environments. It is more common among moss-covered rock rubble, under slabs of bark, or in decaying logs. In coastal Oregon, it is found with Dunn's Salamander. It eats springtails and beetles.

Fourche Mountain Salamander
(*Plethodon fourchensis*)

Description: 4–6¼" (10.2–16 cm). Black with two rows of large brassy-flecked *white spots* down back and many smaller spots. Abundant yellowish-white spots on cheeks, legs, and sides; few on belly. Chin light. Costal grooves, 16.

Breeding: Habits unknown.

Habitat: Shaded, moist hardwood forests; 1,560–2,150' (503–655 m).

Range: Fourche and Irons Fork mountains, Arkansas.

Officially described as a species in 1979, the Fourche Mountain was formerly known as the "Buck Knob" phase of the Rich Mountain Salamander. The two species interbreed along a narrow zone on Fourche Mountain. Like its relatives, adult Fourche Mountain Salamanders are often infected with chiggers.

140, 141 Slimy Salamander
(*Plethodon glutinosus*)
Subspecies: Slimy, White-throated

Description: 4½–8⅛" (11.4–20.6 cm). Usually *shiny black with large white, gray, or yellow spots on sides* and scattered smaller silvery white spots and/or brassy flecks atop head, back, and tail. Belly slate colored. Costal grooves, 16.

Breeding: Courts and mates spring and fall in north, summer in south. Lays clutch of 6–36 eggs in underground retreat or in rotting log, late spring in north, August–September in south. Female guards nest. No aquatic larval stage. Larvae hatch late summer in north, October in south; mature in 3 years. Northern females lay every other year, southern females every year.

Habitat: Shaded ravine slopes, shale banks, wooded floodplains, cave entrances; near sea level to 5,500′ (1,676 m).

Range: C. New York south to Florida and west to Mississippi River and beyond in s. Missouri, nw. Arkansas, and e. Oklahoma. Isolated populations in s. New Hampshire, nc. Louisiana, e. and sc. Texas.

Subspecies: Slimy (*P. g. glutinosus*), dark throat; c. New York to c. Florida, west to e. Oklahoma and e. Texas. White-throated (*P. g. albagula*), light throat, large yellow spots on sides; Edwards Plateau, sc. Texas.

Nocturnal. Active all year in the south. In the north, appears at the surface in early spring and, except during summer dry spells, can be found under flat rocks and rotten logs until the onset of sub-freezing temperatures in fall. After showers, the Slimy moves about the forest floor in search of invertebrate prey. Slimy's skin glands secrete a gluey substance that is next to impossible to remove from the fingers —thus the species name *glutinosis*. Slimy is closely related to the

Appalachian Woodland Salamander and interbreeds with it in sw. North Carolina.

70 Valley and Ridge Salamander
(*Plethodon hoffmani*)

Description: 3¼–5⅜″ (8.3–13.7 cm). Small and slender; resembles Ravine Salamander. Deep brown above with abundant brassy flecking; sometimes a narrow red back stripe. *Belly dark with moderate white mottling;* chin heavily mottled. Costal grooves, 21.

Breeding: May to June. Female lays clutch of 3–8 eggs in moist shelter. Larvae hatch at ⅞″ (22 mm), August to September. Males reach sexual maturity in 2 years, females in 3.

Habitat: Valley and Ridge province, drier and with better drained soils than adjacent regions.

Range: West Branch of Susquehanna River Valley in c. Pennsylvania, southwest in Appalachians to New River, sw. Virginia.

Rarely seen in summer. Most surface activity occurs at night in early spring and fall. Although this species' range is encircled by that of closely related Red-backed Salamander, the two are not usually found together. Dry regional conditions apparently keep the Red-backed from advancing into Valley and Ridge's area.

133, 136, 138 Appalachian Woodland Salamander
(*Plethodon jordani*)

Description: 3¼–7¼″ (8.3–18.4 cm). *Highly variable throughout range.* Many are black above, gray below, unmarked or with small white spots on sides and cheeks. Others have heavy brassy flecking on upper surfaces. In Great Smokies most have bright red cheek patches; in the Nantahala Mountains, red legs. No red on back, except for an occasional spot in young; no back stripe. Chin light. Costal grooves, 16.

Breeding: Courts late July to August. Egg-laying probably occurs in late spring in underground cavities. No aquatic larval stage. Young hatch in fall, appear on surface following spring.

Habitat: Humid, heavily forested slopes, with moss-covered logs and slabs of rock; most populations live at higher elevations; 700–6,400' (213–1,951 m).

Range: S. Appalachians; sw. Virginia south through e. Tennessee and w. North Carolina into ne. Georgia and nw. South Carolina.

Well known for its extensive geographical variation. Early herpetologists thought the various color forms were different species or subspecies. Occasionally this salamander hybridizes with the closely related Slimy Salamander. Active April to November, it is usually seen well away from water, under rotting logs, flat stones, or at night at the entrance of small burrow holes. The cheek patches and leg markings are mimicked by the Imitator Salamander. When moved 500' from its retreat, this salamander can find its way home.

84 Larch Mountain Salamander
(*Plethodon larselli*)

Description: 3–4" (7.6–10.3 cm). Smallest western woodland salamander. *Ragged-edged back stripe* is red to yellow and heavily mottled with small dark flecks, often divided by black line from head to tail. Gold specks on upper sides; large white spots on lower sides. Short 5th rear toe. Costal grooves, 15.

Breeding: Habits unknown.

Habitat: Prefers lava talus slopes and outcrops in dense Douglas fir stands; 100–3,900′ (30–1,189 m).

Range: Lower Columbia River gorge, Skamania County, Washington; Multnomah and Hood River counties, Oregon.

Found under rocks at base of outcrops or under forest litter away from seepages and streams. When discovered, it rapidly coils and uncoils itself several times, perhaps mimicking the numerous unpleasant-tasting millipedes in its habitat. The Larch Mountain shares its habitat with the Western Red-backed Salamander in Washington, with Dunn's Salamander in Oregon. Mites and springtails are common in its diet.

143 Crevice Salamander
(*Plethodon longicrus*)

Description: 5–8⅝″ (12.7–22.1 cm). Largest of woodland salamanders. Black above with scattered white patches from head to tail; heavy white pigment on sides. Belly dark gray with light spotting. *Chestnut blotches or flecks may cover up to three-fourths of back.* Costal grooves, 16.

Breeding: Habits poorly known.

Habitat: Cool, damp vertical faces of granite outcrops, associated cracks, and deep crevices; occasionally shaded forest floor; 1,400–3,200′ (427–975 m).

Range: Henderson and Rutherford counties, North Carolina.

Discovered in 1961. Some authorities believe this rock crevice dweller is simply a slightly differentiated form of the Yonahlossee Salamander. Ants and spiders are favorite foods.

69 Jemez Mountains Salamander
(*Plethodon neomexicanus*)

Description: 3¾–5⅝" (9.5–14.3 cm). Long slender body. Back brown, finely dusted with gold. *Internal organs visible through transparent, sooty belly skin.* Legs short, 5th toe noticeably abbreviated. Young have faint gray or brassy back stripe. Costal grooves, 19.

Breeding: Courts July to August. Clutch of about 8 eggs is laid underground in spring. Young hatch in July. No aquatic larval stage. Females mature in 3 years and lay eggs every other year.

Habitat: Cool, damp north-facing slopes and canyons supporting mixed evergreen forest with Rocky Mountain maple and aspen; 7,200–9,200' (2,195–2,804 m).

Range: Jemez Mountains, New Mexico.

Active at night, June to September. This salamander is found in tunnels under rock talus and in or under rotting logs. It eats large numbers of ants, and beetle and moth larvae. At the end of last Ice Age this species was stranded on a high-altitude "island" as southwestern forests began to disappear. Its nearest relative lives almost 600 miles away.

102 Netting's Salamander
(*Plethodon nettingi*)
Subspecies: Cheat Mountain, Peaks of Otter

Description: 3–4¾" (7.6–12.2 cm). Long and slender. Black back boldly marked with small brassy flecks or with brassy blotches and spots that fuse and may almost form stripe. Belly dark gray to black. Costal grooves, 18–19.

Breeding: Egg clutches with attendant females have been found from June to July.

Habitat: Moist, shaded ravines in spruce forests

above 3,550' (1,082 m); cool, humid deciduous forests above 3,100' (845 m).

Range: Cheat Mountains, West Virginia; Peaks of Otter region, Virginia.

Subspecies: Cheat Mountain (*P. n. nettingi*), numerous small brassy flecks, 18 costal grooves; Cheat Mountains, West Virginia. Peaks of Otter (*P. n. hubrichti*), abundant brassy pigment forms blotches or back stripe, 19 costal grooves; Bedford and s. Rockbridge counties, Virginia.

The two subspecies of Netting's Salamander are now considered separate, full species. Both occupy very restricted ranges, which probably were larger in the past. It appears as if the more successful Red-backed Salamander prevents these species from extending their ranges.

115 Rich Mountain Salamander
(*Plethodon ouachitae*)

Description: 4–6¼" (10.2–15.9 cm). Color varies greatly; back usually black accented with wash of chestnut covering numerous small white spots and brassy flecks. *Dense white pigment on sides,* often forming band. Belly black with small white spots. Throat usually light; chest usually dark. Costal grooves, 16.

Breeding: Habits poorly known.

Habitat: Shaded ravines and northerly exposed slopes of cool, moist hardwood forest, chiefly 1,700–2,850' (518–869 m).

Range: Ouachita Mountains, Arkansas and Oklahoma.

Observed March to November—especially after rainstorms—under loose bark, in rotten logs, under sandstone boulders. Some populations completely lack chestnut markings; one striking form has paired white spots on back. This species is known to hybridize with

the Slimy Salamander on Kiamichi
Mountain south of Big Cedar,
Oklahoma.

---

139 White-spotted Salamander
(*Plethodon punctatus*)

Description: 4–6³⁄₁₆″ (10.2–15.7 cm). Large and
blackish with *many scattered small
yellowish-white spots on upper surfaces.*
Forward part of dark belly spotted with
yellow. Costal grooves, 17–18.

Breeding: Habits poorly known.

Habitat: High elevations of Valley and Ridge
province.

Range: Shenandoah and North mountains,
Virginia–West Virginia border.

White-spotted is closely related to
Wehrle's Salamander but lacks the
dorsal red spots and brassy flecking and
has more webbing on hind toes. It
shares habitat with Slimy Salamander
and strongly resembles it, but the two
are rarely found together.

---

72 Ravine Salamander
(*Plethodon richmondi*)

Description: 3⅛–5⅝″ (8–14.3 cm). An
inconspicuous woodland salamander
with *long slender body and narrow head;*
resembles "lead-backed" phase of Red-
backed Salamander. Back dark brown
to blackish, peppered with brassy and
silvery specks. Small white or yellow
spots on sides. Belly uniformly dark;
throat lightly mottled with white.
Costal grooves, 19–22.

Breeding: Lays cluster of about 6 eggs in
underground cavity in mid- to late
spring. Larvae skip aquatic stage and
hatch at ⅞″ (22 mm), August to
September, ready for terrestrial
life.

Habitat: Ravines and sloping hillsides in open to dense woodlands.

Range: W. Pennsylvania south to ne. Tennessee and nw. North Carolina, west to se. Indiana.

Completely terrestrial and may be seen well away from water under flat stones, leaf litter or rotting logs. Active at surface spring to fall, but they may retreat into deep moist crevices during dry summer months.

---

118 Southern Red-backed Salamander
(*Plethodon serratus*)

Description: 2½–4⅞" (6.4–12.5 cm). Two color phases: "red-backed" has red or orange back stripe, sometimes with saw-toothed edges; "lead-backed" is stripeless, may have red specks on sides. Costal grooves, 18–21.

Breeding: Habits presumably similar to Northern Red-backed Salamander.

Habitat: Moist woodlands, often well away from water.

Range: Widely scattered populations in Southeast. Both phases are common in Alabama and Georgia; lead-backed is rare elsewhere.

Most frequently encountered under rocks, logs, or leaf litter during the cooler months or after showers. Mental gland on chin of sexually active males is shelflike, that of related male Zigzag Salamander is rounded or oblong. The Southern Red-backed closely resembles the Red-backed Salamander; the two can be distinguished by range.

Shenandoah Salamander
(*Plethodon shenandoah*)

Description: 2¾–4" (7–10.2 cm). *Two color phases:* black with narrow, straight-edged red back stripe; or unstriped with some brassy flecking and often small red spots on middle of back. *Belly nearly black.* Costal grooves, 18.

Breeding: Habits poorly known.

Habitat: Steep north- and northwest-facing talus slopes of Hawksbill Mountain, Stony Man Mountain, and the Pinnacle; 3,000–3,750' (914–1,143 m).

Range: Shenandoah National Park, Virginia.

The Red-backed Salamander keeps the Shenandoah from colonizing the moist woodlands surrounding its rocky habitat by getting to the food first. The Shenandoah is more tolerant of dry conditions and so it survives where the Red-backed cannot penetrate.

79 Siskiyou Mountain Salamander
(*Plethodon stormi*)

Description: 3⅞–5½" (9.8–14 cm). Looks like Del Norte Salamander but lighter in color, no red back stripe, and longer legs. Light brown above; head and back peppered with small white flecks. Belly purplish-gray with white or yellowish speckling. Costal grooves, 17–18.

Breeding: Habits poorly known.

Habitat: North-facing slopes or heavily shaded areas associated with Douglas fir, big leaf maple, black oak, canyon oak, sword fern, and poison oak.

Range: Siskiyou County, California, and adjacent sw. Oregon.

Active aboveground only during spring and fall rains, it is seen under surface litter or rock rubble at base of slopes. It tolerates drier conditions than other western woodland species.

111 Van Dyke's Salamander
(*Plethodon vandykei*)
Subspecies: Washington, Coeur d'Alene

Description: 3¾–4⅞" (9.5–12.3 cm). Stocky and long-legged. *Distinct parotoid glands on head.* Black, tan, yellow, or reddish above, with even or scalloped-edged tan, yellowish, or reddish back stripe. Some lack stripe. *Throat pale yellow.* Costal grooves, 14–15.

Breeding: Habits poorly known, probably similar to those of other western lungless species.

Habitat: Moist coniferous forest; in seepages, streamsides, talus slopes, and forest litter; to 5,000' (1,524 m).

Range: Spotty distribution in Olympic Mountains, Willapa Hills, and Cascade Mountains, Washington; isolated populations east and west of Bitteroot Range between Idaho and Montana.

Subspecies: Washington (*P. v. vandykei*), straight-edged back stripe, top of legs light colored; w. Washington. Coeur d'Alene, (*P. v. idahoensis*), scalloped back stripe, legs black; n. Idaho and nw. Montana.

Most aquatic woodland salamander. Found under rocks, logs, and leaf litter in wet areas. It readily takes to fast-moving water when disturbed. The Tailed Frog shares Van Dyke's habitat.

---

119 Western Red-backed Salamander
(*Plethodon vehiculum*)

Description: 2¾–4½" (7–11.5 cm). Resembles Red-backed Salamander found in the East. *Back stripe well defined; even edged;* red, yellow, green or tan; extends to tail tip. Some dark individuals lack stripe; others lack dark pigment and are largely color of stripe. Upper sides dark brown or black; salt and pepper flecking on belly. Costal grooves, 16.

Breeding: Mates November to December; lays clutch of about 10 eggs April to May; hatchlings ⅞" (22 mm) long appear in fall, mature in 2½ years. Females lay eggs every other year.

Habitat: Variety of habitats, including moisture laden rock slides; damp, shaded ravines; Douglas fir forests; near sea level to 4,100′ (1,250 m).

Range: From sw. Oregon north to Vancouver Island, British Columbia.

Most numerous and wide-ranging western woodland salamander. It is found under stones, logs, moss, and woodland debris, usually in situations less damp than those preferred by Dunn's Salamander.

80 Wehrle's Salamander
(*Plethodon wehrlei*)

Description: 4¾–6¼" (12.1–16 cm). Dark brown to nearly black, with *irregular bluish-white or yellow spots and dashes on sides* and sometimes a few small white spots or brassy flecks on back. Southern populations frequently show *paired red spots on back.* Throat to chest area mottled with white, remainder of underside gray. Webbing on hind feet nearly reaches tips of first two toes. Costal grooves, usually 17.

Breeding: Winter in northern areas; late spring in southern. Female deposits about a dozen eggs in an inaccessible retreat; probably guards nest. Sexual maturity reached in 4–5 years.

Habitat: Second-growth woodlands to climax American beech–sugar maple–hemlock forests of unglaciated uplands.

Range: Appalachian Plateau, sw. New York, to nw. North Carolina.

Wehrle's Salamander is usually found in deep crevices in rocky outcrops or under nearby logs or flat rocks. It eats

weevils, centipedes, spiders, and insect larvae, in turn is preyed upon by ringneck snakes and shrews. Slimy and Red-backed salamanders may be found in the same area as Wehrle's, but they prefer the wetter portions of the habitat.

82 Weller's Salamander
(*Plethodon welleri*)

Description: 2½–3⅝" (6.4–9.2 cm). *Large, fused gold or brassy blotches* accent black back. Belly uniformly black or white-spotted. Costal grooves, 16.

Breeding: Grapelike cluster of 4–11 eggs, suspended by single stalk in well rotted evergreen log. No aquatic larval stage; eggs hatch mid-August to early September, hatchlings ¾–⅞" (19–22 mm) long. Female tends nest.

Habitat: Deciduous woodlands with some rhododendron and hemlock at lower elevations, 2,300–5,700' (701–1,737 m). Spruce–fir and hemlock–yellow birch forests at higher altitudes, 5,000–5,900' (1,524–1,798 m).

Range: Mt. Rogers and Whitetop Mountain, Virginia, southwestward along Unaka mountain ridges in Tennessee, into Yancy County, North Carolina. Isolated population on Grandfather Mountain, North Carolina.

This salamander lives amid or under forest floor litter well away from stream beds. It eats tiny invertebrates. Courting males locate other salamanders by sight and smell. A male either flees an approaching male or viciously bites it on legs and tail. Named for W. H. Weller, who first collected the species.

116 Yonahlossee Salamander
(*Plethodon yonahlossee*)

Description: 4¼–7½" (10.8–19 cm). A large and
attractive woodland species. *Back red or
chestnut from neck to well out onto tail;*
sides gray or white; belly black with
white spots. Young have light
undersides and paired red spots on
back. Costal grooves, 16.

Breeding: Habits poorly known.

Habitat: Favors rich forests, especially areas with
old windfalls; mixed deciduous second-
growth woodlands; 2,500–5,700'
(762–1,737 m).

Range: S. Blue Ridge Mountains, French
Broad River Valley northeast to sw.
Virginia and North Carolina.

First described in 1917 by E. R. Dunn,
noted for his classic salamander works.
Named after Yonahlossee Road, near
Dunn's collecting site on Grandfather
Mountain, North Carolina. Yonahlossee
lives in burrows beneath rotting logs
and stumps, venturing out on the forest
floor during early evening hours. It is
extremely agile and hastily retreats into
burrows when disturbed. Shares habitat
with Slimy, Red-backed, and
Appalachian Woodland salamanders.

98, 99, 128, 131 Mud Salamander
(*Pseudotriton montanus*)
Subspecies: Eastern, Midland,
Gulf Coast, Rusty

Description: 2⅞–7⅝" (7.3–19.5 cm). Stout and
strikingly colored, with short legs and
tail. Coral-pink, bright red, or
brownish-salmon above, generally with
well-scattered black spots. Older
animals reddish- to chocolate-brown;
spots obscured. Underside reddish or
yellowish. *Back and belly colors sharply
separated. Eyes brown.* Costal grooves,
16–17.

Breeding: Late fall to early winter. Female lays 75–190 eggs. Larvae hatch late winter at ¾" (19 mm), transform in 1½–2½ years when about 3" (7.6 cm) long. In North Carolina males mature in 2½ years, females in 4–5 years.

Habitat: Muddy springs and streams, wooded floodplains, and swampy pools; low elevations.

Range: Chiefly coastal plain and Piedmont areas, s. New Jersey to c. Florida; low elevations west of Appalachians from s. Ohio to sc. Tennessee.

Subspecies: Eastern (*P. m. montanus*), scattered spots on back and sides, some belly spotting; s. New Jersey to n. Georgia. Midland (*P. m. diastictus*), coral pink to bright red, large spots mostly on back, belly unspotted; west of Appalachians, s. Ohio to s. Tennessee. Gulf Coast (*P. m. flavissimus*), numerous spots on back and sides, belly unspotted; s. tip of South Carolina west to e. Louisiana. Rusty (*P. m. floridanus*), back purplish brown without spots, sides streaked, belly spotted; s. Georgia and n. Florida.

Aquatic larvae live in silt or decaying vegetation; adults in muck, under logs or stones, or in mud burrows along stream banks. On the coastal plain the Mud is often found with the Southern Dusky.

129, 132 Red Salamander
(*Pseudotriton ruber*)
Subspecies: Northern, Southern, Black-chinned, Blue Ridge

Description: 3⅞–7⅛" (9.7–18.1 cm). Robust red salamander with short legs and tail. Young coral-red to reddish-orange above, adults orange-brown to purple-brown; numerous irregularly shaped black spots. *Back and belly colors blend gradually* along sides. Eyes yellow. Costal grooves, 16–17.

Breeding: Females retreat to nesting sites in early fall, lay 50–100 eggs. Larvae hatch late fall to early winter at ⅞" (22 mm); transform in 2½ years when 2¾–4¼" (7–11 cm) long. Female spawn at 5 years.

Habitat: Springs, their seepages, cool clear brooks and surrounding woodlands, swamps, and meadows; elevations to 5,000+' (1,524+ m).

Range: S. New York west to se. Indiana, southward east of the Mississippi River to the Gulf. Absent from Atlantic coastal plain south of Virginia and from peninsular Florida.

Subspecies: Northern (*P. r. ruber*), edges of back spots diffuse and tending to fuse, small dark spots on belly; s. New York west to n. Indiana, south to Georgia and Alabama. Southern (*P. r. vioscai*), profusion of minute white flecks, especially on head, black spots usually distinct; s. Georgia to se. Louisiana, north to w. Tennessee. Black-chinned (*P. r. schencki*), heavy black pigment on jaws; s. Blue Ridge Mountains, south of French Broad River. Blue Ridge (*P. r. nitidus*), no black spots on latter part of tail, belly unspotted; s. Blue Ridge Mountains, north and east of French Broad River.

More terrestrial than the Mud Salamander, the Red may be encountered some distance from water, but usually is seen in the leaf litter of spring-fed brooks or under nearby forest debris or rocks. It is fond of earthworms.

90 Many-lined Salamander
(*Stereochilus marginatus*)

Description: 2½–4½" (6.4–11.4 cm). Small brown or dull-yellow aquatic species with *parallel lines on sides,* sometimes indistinct or reduced to a few dark spots. Belly yellow with scattered specks. Head small, narrow, and pointed; short tail; tiny legs. Costal grooves, 18.

Breeding: Mates in autumn. In winter female lays 6–100 eggs that adhere singly to aquatic vegetation or in a clump to a log next to water. Larvae hatch in spring at ⅝" (14 mm), transform in 13–28 months when 2½–2¾" (54–70 mm) long. Males first breed at 2¾ years, females at 4 years.

Habitat: Lowland ponds, swamps, large drainage ditches, and sluggish streams.

Range: Coastal plain from Virginia to se. Georgia.

Occasionally these salamanders are seen under logs or damp sphagnum moss on banks; most are found in water amid bottom debris or the roots of a floating sphagnum mat.

3, 9 Texas Blind Salamander
(*Typhlomolge rathbuni*)

Description: 3⅜–5⅜" (9.2–13.7 cm). A ghostly-white cave dweller with red plumelike gills and *long spindly legs*. Larvae do not transform to adults. Eyes reduced to black dots, buried under skin. Snout flattened, duckbilled in appearance. Costal grooves, 12.

Breeding: Habits unknown.

Habitat: Subterranean waters of the Purgatory Creek system.

Range: Balcones Escarpment of Edwards Plateau, vicinity of San Marcos, Texas.

An endangered species. Ezell's Cave, now a nature preserve, is the only entrance to the Texas Blind's habitat. There roosting bats supply nutrients (guano) to the cave's unique invertebrates. These in turn are eaten by the Texas Blind.

12 Grotto Salamander
(*Typhlotriton spelaeus*)

Description: 3–5¼″ (7.6–13.5 cm). Only blind cave-dwelling salamander that loses its external gills and transforms. Gilled larvae have functional eyes, high tail fin, are brownish- or purplish-gray, with yellowish flecks. *Adults slender; flesh-colored,* with tinges of orange on tail, sides, and feet; eyes partly overgrown by lids. Costal grooves, 16–19.

Breeding: Eggs laid in or near water on rocks; larvae hatch at about ¾″ (17 mm), transform in 2–3 years at about 3½″ (8.9 cm).

Habitat: Adults restricted to limestone caves; larvae to springs or streams in cave or cave vicinity.

Range: Ozark Plateau in s. Missouri, n. Arkansas, and adjacent Kansas and Oklahoma.

Found in 1891, the Grotto was the first subterranean salamander discovered in the New World. Unlike other cave species, which retain larval characteristics throughout life, Grottoes completely transform to a terrestrial stage. Larvae may develop in outside waters but return to caves before transformation and spend the rest of their lives there.

Frogs and Toads
(Order Salientia)

There are nearly 2,700 species of frogs
and toads known. They are divided into
16 families, with 9 occurring in North
America. The remainder are found in
South America, Europe, Africa, Asia,
Australia, and the South Pacific islands.
A total of 81 species occur north of
Mexico.

Adult frogs and toads lack tails. They
have well-developed forelimbs and even
larger hind legs. They lack a clear neck,
the head seeming to be attached
directly to the body. Most have a well-
developed ear, as evidenced externally
by a conspicuous tympanum, and a
voice used to attract mates, drive off
intruders, and to signal distress and
presence. All are carnivorous as adults.
With their moist skin, most frogs and
toads are prone to dessication, and
therefore are confined to wet or moist
habitats. However, some species have
adapted to more arid habitats by
burrowing into the soil or hiding
beneath rocks or logs to avoid the heat
of the day.

Most species return to water to breed.
Eggs laid in the water are fertilized by
the male while clasping the female.
The eggs hatch into tadpoles which
later transform into young frogs. A
number of frogs and toads lay their
eggs in shaded moist sites on land, or
in nests constructed over water. The
eggs may hatch into tadpoles that
either drop into the water, are swept
into the water by rains, or are carried to
water by a parent. Other species bypass
the tadpole stage, their eggs hatching
directly into miniature frogs. Finally,
the African live-bearing toad,
Nectophrynoides, and the North
American tailed frog, *Ascaphus,* are
unique among frogs and toads in
fertilizing the eggs within the cloaca of
the female.

TAILED FROG FAMILY
(Ascaphidae)

Two genera: *Leiopelma* with 3 species in New Zealand, and *Ascaphus* with 1 species in the northwestern United States and southwestern Canada. These are primitive frogs with 2 pairs of unattached ribs and 9 presacral vertebrae. The pupil is vertical. None of the species has a true tail, but all possess tail-wagging muscles. Male *Ascaphus* have a unique tail-like extension of the vent that serves as a copulatory organ for passing sperm directly into the body of the female, while clasping her around the waist.

All tailed frogs live in cool mountain habitats, but reproduction varies: *Leiopelma* lays its eggs on damp earth where they hatch 6 weeks later as miniature frogs; *Ascaphus* lays its eggs in cold mountain streams where they hatch into tadpoles, transforming into frogs more than 6 months later.

165 Tailed Frog
(*Ascaphus truei*)

Description: 1–2″ (2.5–5.1 cm). Usually olive or gray to almost black; many dark spots on back. Dark stripe often runs from snout through eye; snout bears yellowish triangle. Some small tubercles and warts on skin. Toes long and slender; outer toes of hind feet thicker than others. Eye pupil vertical. *No external eardrum.* Male has *pear-shaped "tail,"* actually a copulatory organ for internal fertilization of eggs.

Voice: No vocal sac.

Breeding: May to September. Eggs are attached to downstream side of rocks.

Habitat: Usually clear, cold swift-flowing mountain streams; sometimes found near water in damp forests or in more open areas in cold, wet weather.

Range:

From s. British Columbia south to nw
California; also w. Oregon and
Washington east to nw. Montana and
Idaho. Many separate populations.

Aquatic. It is a primitive frog with rib
and vestigial tail-wagging muscles.
Sucking mouthparts equip tadpoles for
clinging to rocks in strong currents.
They will hang on to anything solid,
including human flesh. They feed on
algae and invertebrates and transform i
1–3 years.

BURROWING TOAD FAMILY
(Rhinophrynidae)

Only 1 species, *Rhinophrynus dorsalis*,
found from Costa Rica to Texas.
This frog lacks a breastbone. The pupil
is vertical. In addition, its tongue is
attached in the back and free in the
front, unlike the typical toad or frog, so
that it protrudes like the tongue of
most vertebrates. *Rhinophrynus* is adept
at burrowing, making use of the spade
on each hindfoot to shuffle backward
into the soil. Breeding males clasp the
female around the waist. Eggs are laid
in water.

217, 220 Mexican Burrowing Toad
(*Rhinophrynus dorsalis*)

Description: 2–3½" (5.1–8.9 cm). Looks like a
giant Narrow-mouthed Toad. Egg-
shaped body, narrow pointed head, and
smooth moist skin; brown with *broad
yellow to reddish-orange stripe down middle
of back*. Hind limbs partly hidden
within skin of body. Large spade on
each hind foot.

Voice: Loud guttural moan. Males call from
burrow.

Breeding: After rainfall creates temporary pools.
Eggs are laid in clumps in water, then
quickly separate and float to surface.
Tadpoles have several long "whiskers."

Habitat: Forested and cultivated areas with soil
suitable for burrowing.

Range: Extreme s. Texas into Mexico.

Nocturnal. This toad is the only species
of the family Rhinophrynidae. When
frightened, it inflates its body and looks
like a walking balloon. It feeds on
termites.

SPADEFOOT TOAD FAMILY
(Pelobatidae)

Ten genera with 69 species known;
1 genus in North America: *Scaphiopus*,
with 5 species.

Spadefoots derive their names from the
"spade," a horny, dark, sharp-edged
tubercle on the inner surface of the hind
foot, used to dig their daytime burrow.
Scaphiopus differs from other North
American frogs and toads in having the
coccyx, the bony end of the vertebral
column, fused to the sacral vertebra.
The teeth on their upper jaw, the
vertical pupils, relatively smooth skin,
and absence of parotoid glands easily
separate them from true toads (Family
Bufonidae).

Breeding males clasp the female around
the waist. *Scaphiopus* breeds in
temporary rainpools which may dry up
soon after the rains end. As an
adaptation to this, spadefoots have an
accelerated development. The total
period of time from egg to tadpole to
metamorphosed toad can be as little as
2 weeks.

231 Plains Spadefoot
(*Scaphiopus bombifrons*)

Description: 1½–2½" (3.8–6.3 cm). A stout-
bodied toad with *round- to wedge-shaped
spade on hind feet and prominent bony hump
between eyes.* External eardrum apparent.
Skin relatively smooth with scattered
small tubercles; gray to brown, often
with overtones of green; tubercles
orange. Usually light stripes on back
are vaguely discernible. Belly white.
Male throat bluish-gray on sides.

Voice: A dissonant grating note given at 1-
second intervals; sometimes a hoarse
trill lasting 1 second.

Breeding: May to August, stimulated by rain.
Eggs in masses of 10 to 200 are

attached to submerged vegetation in
shallow ponds, hatching within 48
hours. Tadpoles are omnivorous;
transform within 2 months.

Habitat: Shortgrass prairie where soil is loose
and dry, rainfall low. Likes sandy and
gravelly soils.

Range: The Great Plains from s. Alberta and
Saskatchewan se. through Montana to
Missouri and c. Oklahoma, south
through w. Texas and e. Arizona and
into Mexico. Separate population in
extreme s. Texas.

Nocturnal. A single sharp-edged spade
on the inside of each hind foot pushes
aside soil as the Spadefoot backs into
the ground. Burrows may be a few
inches to several feet long. They remain
open but are difficult to locate in sandy
soil. Occasionally, sticky matter is seen
at the entrance, probably to cement soil
in place and prevent burrow collapse.

252 Couch's Spadefoot
(*Scaphiopus couchi*)

Description: 2¼–3½" (5.7–8.9 cm). A plump toad
with *elongated, sickle-shaped spade on each
hind foot* and no hump between eyes.
External eardrum apparent. Skin
smooth, but with many minute light-
colored tubercles. Bright greenish-
yellow to brown with variable dark
marbling. Belly mostly white.

Voice: Like the bleat of a lamb; lasts about 1
second. Very noisy chorus can be heard
for a long way.

Breeding: Mates after heavy rainfall, April to
September. Eggs are laid on plant
stems in temporary pools and, if the
spot is warm, may hatch in 36 hours.
Tadpoles transform in 2–6 weeks,
before pools dry up.

Habitat: Tolerant of dry terrain; likes shortgrass
prairie as well as mesquite savannah and
creosote bush desert.

Range: Extreme se. California, s. Arizona and New Mexico, and sw. Oklahoma south into Mexico.

Nocturnal. During dry periods this toad stays underground in the burrow of a small mammal or buried in loose soil.

229, 230, 253 Western Spadefoot
(Scaphiopus hammondi)
Subspecies: Western, New Mexico

Description: 1½–2½″ (3.8–6.4 cm). Stout-bodied toad with *wedge-shaped spade on each hind foot and no hump between eyes.* External eardrum apparent. Dusky-olive to brown or gray, with irregular light stripes and random darker blotches. Skin relatively smooth with scattered small tubercles, red- or orange-tipped in some specimens. Belly white.

Voice: A rolling trill like the purr of a cat. Males call while floating on surface of water.

Breeding: January to August, depending on rainfall. Eggs are laid in cylindrical masses attached to vegetation. Hatching occurs within 2 days, transformation in 4–6 weeks. Tadpoles are carnivorous and feed on mosquito larvae.

Habitat: Tolerates wide range of conditions from semiarid to arid. Prefers shortgrass plains and sandy, gravelly areas such as alkali flats, washes, and river floodplains.

Range: Arizona, New Mexico, parts of s. Colorado, and w. Oklahoma south into Mexico. Separate population in California south of San Francisco through the central valley and foothills into northern Baja California.

Subspecies: Western (*S. h. hammondi*), with trill lasting less than 1 second; California south into Baja. New Mexico (*S. h. multiplicata*), with trill lasting longer

than 1 second; s. Colorado through
Arizona and New Mexico into
Mexico.

Nocturnal. They are often numerous
where soil conditions are favorable for
burrowing. Deep burrows provide a
microhabitat with moderate
temperatures and humidity. When
handled, the Western Spadefoot
produces a secretion that smells like
peanuts and can inflame the skin or
cause hay-fever symptoms like runny
nose and watery eyes. The Western and
Plains Spadefoots hybridize in areas
where their ranges coincide.

233 Eastern, Spadefoot
(*Scaphiopus holbrooki*)
Subspecies: Eastern, Hurter's

Description: 1¾–3¼" (4.4–8.3 cm). Stout, with
sickle-shaped spade on each hind foot.
External eardrum apparent. Skin
relatively smooth, with scattered tiny
tubercles. Olive to brown to nearly
black, often with 2 irregular light lines
down back. Underside white to
grayish.

Voice: Sounds like the coarse low-pitched
complaint of a young crow, given at
about 2-second intervals. Chorus may
be heard for half a mile. Spadefoots may
call from burrow.

Breeding: March to September, after rains fill
temporary pools. Eggs laid in
gelatinous bands are attached to aquatic
vegetation; hatch within 2 days and
transform in 2–8 weeks.

Habitat: Forested, brushy, or cultivated areas of
sandy, gravelly, or loose loam.

Range: From s. New England through the
Florida Keys (except the high
Appalachians), west to c. Louisiana,
north to s. Illinois and Ohio. A
separate population inhabits sw.
Missouri, w. Arkansas and Louisiana,

extreme se. Nebraska, e. Oklahoma, and e. Texas into Mexico.

Subspecies: Eastern (*S. h. holbrooki*), lacks bony hump between the eyes; eastern part of range. Hurter's (*S. h. hurteri*), with a bony hump between the eyes; western part of range.

Nocturnal. It lives in shallow burrows, usually of own making, safe from weather on the surface. It is often found on damp summer nights at the mouth of its burrow. Eastern Spadefoots have been observed, unscathed, amid the smoldering ashes of a brush fire.

232 Great Basin Spadefoot
(*Scaphiopus intermontanus*)

Description: 1½–2″ (3.8–5.1 cm). Stout toad with *wedge-shaped spade on each hind foot* and *glandular hump between eyes.* External eardrum apparent. Skin relatively smooth, but many minute tubercles present. Olive to gray-green, with light stripes along flanks. Belly white.

Voice: A series of low-pitched throaty notes given rapidly.

Breeding: April to July, usually following heavy rains. Utilizes springs and slow-moving water as well as temporary pools.

Habitat: Forested areas and sagebrush flats. Digs burrow in loose soil or uses burrows of other animals.

Range: From s. British Columbia south to e. California, east to Colorado and nw. New Mexico.

Primarily nocturnal. It is occasionally found abroad in daylight foraging for insects and can sometimes be brought to the surface by loud stamping near its burrow.

TRUE FROG FAMILY
(Ranidae)

Forty-six genera with approximately
560 species worldwide. Only 1 genus
in North America: *Rana*, with 21
species.
True frogs have a bony breastbone and
horizontal pupils. North American
species are large frogs with slim waists,
long legs, pointed toes, and extensive
webbing on the hind feet. They are
excellent jumpers. The adults are truly
amphibious, typically living along the
edge of water and entering it daily to
catch prey, flee danger, or to mate.
They are all voracious carnivores,
feeding primarily on insects, spiders,
and crustaceans, but readily accepting
anything else that can be caught and
swallowed.
Mating usually is initiated in the spring
with aggregations of males calling in
chorus to attract females to the
breeding site. The breeding male has
swollen forearms and thumbs for
clasping the female behind her forelegs.
In the water, female *Rana* may lay
strings or rafts containing up to 20,000
eggs. Eggs hatch within a month, with
tadpoles metamorphosing into frogs 6
to 24 months later.

194, 195, 198 | **Crawfish Frog**
(*Rana areolata*)
Subspecies: Carolina, Florida, Dusky,
Southern, Northern

Description: 2¼–4½" (5.7–11.4 cm). Stout-bodied
frog. Cream to brown or black, with
irregular dark spots on back and sides.
Adjacent spots frequently are fused into
horizontal blotches. Dorsolateral folds
sometimes highlighted with yellow.
Web extends about half length of
longest toe on hind foot. *Skin smooth to
warty.*

Voice: A deep, resonating guttural snore. Heavy rains in late winter sometimes stimulate choruses.

Breeding: Year-round in southern areas; in spring and early summer in northern areas. Egg masses are laid in shallow water, sometimes attached to vegetation.

Habitat: Moist meadows, prairie woodlands, and pine scrub.

Range: Coastal plain from North Carolina to Florida and along the Gulf coast to extreme e. Louisiana. Separate populations in w. Louisiana and e. Texas, and drainage systems of Arkansas, Missouri, Mississippi, and Ohio rivers.

Subspecies: Carolina (*R. a. capito*), dark, chest with dark marbling; coastal plain from North Carolina to c. Georgia. Florida (*R. a. aesopus*), white to cream, chin spotted, belly white; c. Georgia through most of the Florida peninsula to the Florida panhandle. Dusky (*R. a. sevosa*), dark, chest spotted, back rough; from the Florida panhandle to e. Louisiana. Southern (*R. a. areolata*), dark spots with light borders, chest white, back smooth; Louisiana and e. Texas to sw. Arkansas and se. Oklahoma. Northern (*R. a. circulosa*), dark spots with light borders, chest white, back rough; ne. Oklahoma and e. Kansas to Indiana and south to Mississippi.
The break in distribution in Louisiana, coupled with differences in coloration between the eastern and western populations, leads some authorities to consider the two populations to be different species: Crawfish Frog (*Rana areolata*) in the West with 2 subspecies; Gopher Frog (*Rana capito*) in the East with 3 subspecies.

Nocturnal. These frogs will travel great distances to reach breeding ponds. Otherwise they seldom range far from the entrance of their daytime retreat—a stump hole, an abandoned burrow of a

crawfish or small rodent, or the active
burrow of a Gopher Tortoise. They will
swallow almost any prey they can catch,
including other frogs and toads.

208, 215 Red-legged Frog
(*Rana aurora*)
Subspecies: Northern, California

Description: 2–5⅜" (5.1–13.6 cm). Large;
reddish-brown to gray, with many
poorly defined dark specks and
blotches. Dorsolateral folds present.
Dark mask bordered by *light stripe on
jaw.* Eardrum smooth. Underside
yellow, with *wash of red on lower abdomen
and hind legs.* Toes not fully webbed.
Male has enlarged forearms and swollen
thumbs.

Voice: Series of weak throaty notes, rather
harsh, lasting 2–3 seconds.

Breeding: January to March. Egg masses laid in
permanent bodies of water.

Habitat: Usually found near ponds or other
permanent water with extensive
vegetation. Also likes damp woods.

Range: Vancouver Island, B.C., south along
the Pacific coast west of the Cascade
and Sierra Mts. to n. Baja California.
From sea level to 8,000′ (2,400 m).
Introduced populations in Nye County,
Nevada.

Subspecies: Northern (*R. a. aurora*), dark blotches
without light centers; British Columbia
to n. California. California (*R. a.
draytoni*), dark blotches with light
centers; California to Baja California.

Primarily diurnal. Breeding takes place
over a very few days.

196 **Rio Grande Leopard Frog**
(*Rana berlandieri*)

Description: 2¼–4½″ (6–11.4 cm). Pale green, with *large dark spots between russet dorsolateral ridges*—ridges broken near hind legs. Light jaw stripe poorly defined.

Voice: A short rapid trill, low in pitch.

Breeding: Year-round. Egg masses are attached to submerged vegetation.

Habitat: Any water or moist conditions, natural or artificial.

Range: Sw. Arizona and s. New Mexico to c. Texas, south into Mexico.

Subspecies: One in our range, *R. b. berlandieri.*

Primarily nocturnal. The Rio Grande Leopard Frog can tolerate fairly dry conditions by burrowing under rocks and shingle.

197, 202 **Plains Leopard Frog**
(*Rana blairi*)

Description: 2–4⅜″ (5.1–11.1 cm). Stout green to brown frog. *Large dark spots between yellow dorsolateral ridges*—ridges broken near hind legs. Prominent light-colored jaw stripe. Usually has light spot on eardrum. Groin and underside of thigh yellow.

Voice: Two or 3 guttural notes a second; almost a chuckle.

Breeding: Throughout the year, depending on adequate rainfall.

Habitat: Prairies and other grassy, moist areas, along margins of ponds, streams, marshes.

Range: From c. Nebraska to Illinois and extreme w. Indiana, south through Kansas to c. Texas, north to e. Colorado.

Primarily nocturnal. It can be found foraging along the water's edge on cloudy days.

209 Foothill Yellow-legged Frog
(*Rana boylei*)

Description: 1⅝–3" (3.9–7.5 cm). Gray to brown to olive, with *light-colored band across top of head*. Gray mottling may be present on back. Dorsolateral ridges indistinct. No dark mask. Eardrum granular. Lower abdomen and *underside of legs yellow*. Toes fully webbed. Male has swollen, darkened thumbs.

Voice: Rasping, unmusical note given 4–5 times in rapid series; rarely heard.

Breeding: March to May, when streams have slowed after winter runoff. Egg clusters are attached to downstream side of submerged rocks.

Habitat: Aquatic. Prefers gravelly or sandy streams with sunny banks and open woodlands nearby. From sea level to about 6,000' (1,800 m).

Range: From w. Oregon to s. California—Los Angeles County near coast and Kern County inland; absent from the central valley.

Primarily diurnal. When threatened, it dives quickly to the bottom and hides among the rocks.

199 Cascades Frog
(*Rana cascadae*)

Description: 1¾–2⁵⁄₁₆" (4.4–7.5 cm). Small, olive to brown, with *black spots on back and legs*. Dorsolateral folds present. Dark mask bordered by light jaw stripe. Eardrum smooth. Underside yellow, becoming more intense posteriorly. Toes not fully webbed.

Voice: A low-pitched, raspy series of 4–6 notes a second.

Breeding: May to August.

Habitat: Mountain meadows, streams, ponds, and lakes above 3,000' (900 m), in the water and vegetation around it.

Range: In the Olympic Mts. of Washington

and Cascade Mts. of Washington, Oregon, and n. California. To 9,000′ (2,700 m) near timberline.

Diurnal. It is frequently seen preying on insects or basking on a rock in or near the water. When threatened, it swims away instead of diving and hiding.

187, 190 Bullfrog
(*Rana catesbeiana*)

Description: 3½–8″ (9–20.3 cm). The largest frog in North America. Green to yellow above with random mottling of darker gray. *Large external eardrum; hind feet fully webbed except for last joint of longest toe. No dorsolateral ridges.* Belly cream to white, may be mottled with gray.

Voice: Deep-pitched *jug o'rum* call can be heard for more than a quarter mile on quiet mornings.

Breeding: Northern areas, May to July; southern, February to October. Egg masses are attached to submerged vegetation. Tadpoles are large, 4–6¾″ (10.2–17.1 cm), olive-green, and may take almost 2 years to transform.

Habitat: Aquatic. Prefers ponds, lakes, and slow-moving streams large enough to avoid crowding and with sufficient vegetation to provide easy cover.

Range: Eastern and central United States; also New Brunswick and parts of Nova Scotia. Extensively introduced in the West.

Nocturnal. Less aquatic than the Pig Frog, it is usually found on the bank at water's edge. When frightened, it will as soon flee into nearby vegetation as take to the water. Large specimens have been known to catch and swallow small birds and young snakes; its usual diet includes insects, crayfish, other frogs, and minnows. Attempts to commercially harvest frogs' legs have prompted many introductions of the Bullfrog outside its natural range.

189, 213 Green Frog
(*Rana clamitans*)
Subspecies: Bronze, Green

Description: 2⅛–4" (5.4–10.2 cm). *Green, bronze, or brown frog;* large external eardrum and prominent *dorsolateral ridges that do not reach groin.* Typically green on upper lip. Belly white with darker pattern of lines or spots. Male has yellow throat and swollen thumbs.

Voice: Like the twang of a loose banjo string, usually given as a single note, but sometimes repeated rapidly several times.

Breeding: March to August. Eggs are usually laid in 3–4 small clutches attached to submerged vegetation.

Habitat: Lives close to shallow water, springs, swamps, brooks, and edges of ponds and lakes. May be found among rotting debris of fallen trees.

Range: Widespread throughout eastern North America.

Subspecies: Bronze (*R. c. clamitans*), brown or bronze; Carolinas to c. Florida and through the gulf coast states to e. Texas and s. Arkansas. Green (*R. c. melanota*), green or greenish-brown; s. Ontario east to Newfoundland, south to North Carolina, west to Oklahoma, and introduced into Canada, the West, and Hawaii.

Primarily nocturnal. Green Frogs are not as wary as many other species of frog. They seldom scream in alarm when caught.

Las Vegas Leopard Frog
(*Rana fisheri*)

Description: 1¾–3″ (4.4–7.6 cm). Plump olive-green or brown frog with *a few dark spots between light-olive dorsolateral ridges —ridges broken near hind legs. Dark spots lack light-colored edges,* are less numerous on head. No light stripe on upper jaw. Eardrum larger than eye. Underside of hind legs golden yellow. Male has swollen thumbs.

Voice: Similar to that of Northern Leopard Frog.

Breeding: In spring.

Habitat: Springs, seepage areas, and their grassy margins.

Range: Clark County, Nevada.

Probably extinct. It was last seen in 1942, long before the public began to worry about endangered species. As the city of Las Vegas grew, *Rana fisheri*'s habitat was destroyed: Groundwater was pumped out, springs capped, and streams and pools cemented up. Discovery of a remnant population would be a herpetological event.

188 Pig Frog
(*Rana grylio*)

Description: 3¼–6⅜″ (8–16.2 cm). Large bullfrog; olive to grayish-green to dark brown, with numerous dark spots and distinct band on thighs. *Large eardrum, and fully webbed hind feet. No dorsolateral ridges.* Belly cream, often with heavy mottling.

Voice: A short, explosive piglike grunt, from which it gets its common name. Chorus sound is a steady roar. Calls year-round, usually from a floating position.

Breeding: March through September. Tadpoles are large, to 5″ (12.7 cm), greenish, and transform within 2 years.

Habitat: Aquatic. Marshes, shores of lakes,

ponds, or other waters with a dense cover of emergent or floating vegetation.

Range: From s. South Carolina through Florida west along Gulf coast to se. Texas.

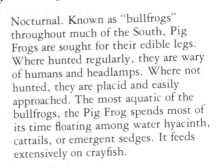

Nocturnal. Known as "bullfrogs" throughout much of the South, Pig Frogs are sought for their edible legs. Where hunted regularly, they are wary of humans and headlamps. Where not hunted, they are placid and easily approached. The most aquatic of the bullfrogs, the Pig Frog spends most of its time floating among water hyacinth, cattails, or emergent sedges. It feeds extensively on crayfish.

210 River Frog
(Rana heckscheri)

Description: 3¼–5⁵⁄₁₆" (8–13.5 cm). Large and rough-skinned; dark olive to almost black; light spots along edge of jaws. *Belly dark*, from gray to almost black, with irregular light spots and lines. *Large external eardrum; hind feet webbed to last joint of largest toe. No dorsolateral ridges.* Male has a swollen thumb and darker throat than female.

Voice: Sharp grunt or low-pitched snore.

Breeding: April to August. Tadpoles are large, to 5" (12.7 cm), black with black borders on tail, and transform in about a year.

Habitat: Swamps bordering slow-moving rivers and creeks throughout range.

Range: From se. South Carolina to c. Florida.

Nocturnal. Newly metamorphosed River Swamp Frogs apparently have a toxic skin secretion, for water snakes and indigo snakes become violently ill after ingesting them. Even after regurgitating, the snakes will continue to wipe their mouths on the ground.

207 Mountain Yellow-legged Frog
(*Rana muscosa*)

Description: 2–3¼" (5.1–8 cm). Brown with black or dark-brown spots or lichenlike markings. Dorsolateral ridges present, but may be indistinct. No dark mask. Eardrum smooth. *Belly yellow to pale orange.* Toes fully webbed, and *toe tips dark.* Male has swollen thumbs.

Voice: No vocal sacs; mating call unknown.

Breeding: Season dependent on altitude and weather conditions: March to May in lower regions; June to August at higher elevations. Egg masses are attached to vegetation.

Habitat: Sunny stream banks and undisturbed ponds and lakes, usually with sloping gravel banks.

Range: The Sierra Nevada Mountains of California and extreme w. Nevada. Separate population in the San Gabriel, San Bernardino, San Jacinto and Palomar mountains of s. California.

This is the only frog in the high Sierras, from 6,000–12,000′ (1,800–3,600 m). In the south it occurs from 1,200–7,500′ (365–2,300 m). It has a pungent, musky odor. Primarily diurnal.

Relict Leopard Frog
(*Rana onca*)

Description: 1¾–3⅜" (4.4–8.4 cm). Slender, brown frog with *large gray-edged dark spots between light-colored dorsolateral ridges*—ridges broken near hind legs. Light stripe on upper jaw. Eardrum may have light center. Male has swollen thumbs.

Voice: A low throaty snore.

Breeding: March to May.

Habitat: Spring seeps and stands of bulrush along the edges of marshes and pools.

Range: The Virgin River drainage in se.
Nevada, nw. Arizona, and sw. Utah.

Primarily nocturnal. Like other leopard
frogs, when frightened this species will
flee into the water or to the security of
an overgrown grassy bank.

---

201 **Pickerel Frog**
(*Rana palustris*)

Description: 1¾–3⁷⁄₁₆″ (4.4–8.7 cm). Smooth-
skinned, tan, with *parallel rows of dark
squarish blotches running down back*. Jaw
has light stripe. Dorsolateral folds
yellow. Belly and under surfaces of hind
legs bright yellow to orange.
Voice: A steady low croak. May call in a
rolling snore while under water.
Breeding: March to May. Egg masses are attached
to submerged vegetation.
Habitat: Slow-moving water and other damp
areas, preferably with low, dense
vegetation; streams, swamps, and
meadows.
Range: Throughout the eastern states except
the extreme Southeast.

Nocturnal. An irritating skin secretion
makes this frog unappetizing to some
predators. The secretion will kill other
frogs kept in the same collecting
container or terrarium. Pickerel frogs
hibernate from October until March or
April.

192, 203 **Northern Leopard Frog**
(*Rana pipiens*)

Description: 2–5″ (5.1–12.8 cm). Slender brown or
green frog with *large, light-edged dark
spots between light-colored dorsolateral
ridges—ridges continuous to groin*. Light
stripe on upper jaw. Eardrum without
light center.

Voice: A low guttural snore lasting about 3 seconds, followed by several clucking notes.

Breeding: March to June. Egg masses are attached to submerged vegetation or laid on bottom.

Habitat: From freshwater sites with profuse vegetation to brackish marshes and moist fields; from desert to mountain meadow.

Range: Throughout northern North America, except West Coast.

Primarily nocturnal. When pursued on land, it flees in zigzag leaps to the security of water.

206 Spotted Frog
(*Rana pretiosa*)

Description: 2–4″ (5.1–10.2 cm). Large and brown, with *ill-defined dark spots, sometimes with light centers.* Dorsolateral ridges present. Dark mask is bordered by *light stripe on upper jaw.* Underside varies from yellow to orange to red, usually with dark mottling on throat. Toes fully webbed. Eyes turned slightly upward. Male smaller than female and has swollen thumbs.

Voice: A series of short rapid croaks without much carrying power.

Breeding: March to June. Egg clusters free.

Habitat: Mountainous areas near cold streams and lakes. Reported to move overland in spring and summer.

Range: Extreme se. Alaska and nc. British Columbia, south to w. Montana, Wyoming, Idaho, and Oregon. Scattered populations in Utah, Nevada, and w. Oregon and Washington.

Diurnal. It does not regularly frequent ponds and lakes where water is warm enough to allow for extensive growth of emergent vegetation, such as cattails.

200 Mink Frog
(*Rana septentrionalis*)

Description: 1⅞–3" (4.8–7.6 cm). Uniformly olive to brown, or with *dark spots or mottling on sides and hind legs*. Large eardrum. Has pungent, musky, minklike odor. Belly yellowish. Dorsolateral ridges vary from prominent to absent. Webbing on hind foot reaches last joint of longest toe. Eyes turned slightly upward. Male has swollen thumbs.

Voice: Low-pitched, metallic croak. Males call while floating or seated on lily pads.

Breeding: June to August. Egg masses are attached to submerged vegetation.

Habitat: At home in cold waters where vegetation is abundant; especially partial to water lilies. Also found near edges of ponds or lakes with emergent vegetation or peaty bottoms.

Range: In Canada from Manitoba to Labrador and the Maritime Provinces, south along the St. Lawrence River to n. New York, west to n. Minnesota.

Primarily nocturnal. At night males can be found squatting on lily pads far from shore. During the day they usually hide in the vegetation on shore.

191, 204 Southern Leopard Frog
(*Rana sphenocephala*)

Description: 2–5" (5.1–12.7 cm). Slender and narrow-headed; green to brown, with *large dark spots between light-colored dorsolateral ridges*—ridges continuous to groin. Light stripe along upper jaw; typically, *a light spot in center of eardrum.*

Voice: Series of short throaty croaks. Males call while afloat or from land.

Breeding: Year-round in southern areas, March to June in northern areas. Egg masses are laid in shallow water.

Habitat: Any freshwater location. Wanders among moist vegetation in the summer, returns to freshwater ponds and streams and brackish marshes rest of year.

Range: From s. New York to the Florida Keys, west to Texas and e. Oklahoma, north to ec. Kansas.

The most ubiquitous frog of the eastern states. Primarily nocturnal. During the day, it hides in grass or sedges of sunny banks. To elude a predator—such as a raccoon or water bird—this frog dives into the water, makes a sharp turn while still submerged, and surfaces amid vegetation at the water's edge; meanwhile the predator continues to search in the direction of the original dive. This is the species most frequently hunted for frogs' legs by youngsters in the Southeast.

211, 212, 214, **Wood Frog**
216 (*Rana sylvatica*)

Description: 1⅜–3¼″ (3.5–8.3 cm). Pink, tan, or dark brown, with *prominent dark mask ending abruptly behind eardrum.* Light stripe on upper jaw; sometimes light line down middle of back. Dorsolateral ridges prominent. Dark blotch on chest near base of each front leg. Belly white, may have dark mottling. Toes not fully webbed; male has swollen thumbs.

Voice: A series of short raspy quacks.

Breeding: Early spring, before ice has completely melted from water. Egg masses are attached to submerged vegetation.

Habitat: Moist woodlands in eastern areas; open grasslands in western; tundra in the far north.

Range: Widespread throughout northern North America.

The only North American frog found north of the Arctic Circle. Primarily diurnal. In the colder parts of its range, the Wood Frog is an explosive breeder.

Swarms of pairs lay fertilized eggs
within 1 or 2 days, then disappear into
the surrounding country. It may
venture far from water during summer,
and hibernates in forest debris during
winter.

193 Tarahumara Frog
(*Rana tarahumarae*)

Description: 2⅜–4½" (5.8–11.3 cm). Large and
chunky; olive to rust or brown, with
dark spots on back and prominent
crossbanding on hind legs. *No dark
mask* or light stripe on jaw. Belly white
to cream, often thinly overcast with
gray on throat and chest. Groin may be
yellow. Dorsolateral folds absent or
indistinct. Eardrum indistinct and
granular. Toes fully webbed. Male has
swollen, darkened thumbs.

Voice: No vocal sacs; no mating call reported.

Breeding: During the summer rainy season, July
and August.

Habitat: Canyon streams between 1,500' and
6,000' (450–1,800 m). During dry
season may be encountered in isolated
pools.

Range: From extreme sc. Arizona near Nogales
into e. Sonora, Mexico, and south.

Nocturnal. It is closely related to the
Rough-skinned Frog (*Rana pustulosa*)
found farther south in Mexico.

205 Carpenter Frog
(*Rana virgatipes*)

Description: 1⅝–2⅝" (4.1–6.7 cm). Small, brown;
*4 distinct yellowish stripes down back but
no dorsolateral folds.* Underside cream to
yellow, with random dark spotting.
Webbing does not reach tip of longest
toe.

Voice: A rhythmic hammering sound; chorus

sounds like several carpenters at work.

Breeding: April to August. Eggs laid in clusters are attached to vegetation. Tadpoles transform during the following spring.

Habitat: Sphagnum bogs and sphagnum fringes of lakes and ponds. Also found in tea-colored, slow-moving water with abundant emergent vegetation.

Range: The coastal plain from the Pine Barrens of New Jersey to s. Georgia.

Nocturnal. This frog is often seen wholly out of water but is never far from the edge. When frightened, it leaps into water and hides beneath vegetation, swimming only a short distance before it breaks the surface for a quick look at the pursuer.

NARROW-MOUTHED FROG FAMILY
(Microhylidae)

Sixty-one genera with approximately 270 species known. Only 2 genera occur in our range: *Gastrophryne* with 2 species, and *Hypopachus* with 1 species. Narrow-mouthed frogs have a reduced shoulder girdle. The North American species lack teeth, and have a fold of skin across the back of the narrow, pointed head. The body is plump and the skin smooth and moist. The legs are short and the hind feet have enlarged tubercles used in digging. They are secretive creatures active only at night; they feed almost exclusively on ants.

Males give a bleating call to attract females to the breeding ponds. In addition to being clasped firmly behind the forelimbs by the male, the female narrow-mouthed's sticky skin secretion also holds the breeding pair together. Eggs are laid in a thin, floating film and hatch in a few days. Tadpoles metamorphose in about 30 days.

---

219 Eastern Narrow-mouthed Frog
(*Gastrophryne carolinensis*)

Description: ⅞–1½" (2.2–3.8 cm). Small, plump, and smooth-skinned, with egg-shaped body, pointed snout, and *fold of skin across back of head*. Reddish-brown to dark gray, often with light stripes along sides. *Belly heavily mottled*. Single spade on each hind foot. Male has dark throat.

Voice: Like the weak bleat of a sheep.

Breeding: April to October; initiated by rains, especially in the deep South. Males call while sitting on vegetation floating in the water, usually concealed in grass. Eggs float on surface as a thin sheet.

Habitat: Near water, especially along the edge of ponds or ditches and under moist debris and decaying vegetative matter.

Range: From s. Missouri east through s. Kentucky and Tennessee, and from s. Maryland south through the Florida Keys, and west to e. Texas and Oklahoma. Isolated populations reported from areas near western and northern borders.

Nocturnal. It eats a variety of insects but prefers ants. A prodigious burrower, it can disappear into leaf litter or loose soil in a minute or two.

221 Great Plains Narrow-mouthed Frog
(*Gastrophryne olivacea*)

Description: ⅞–1⅝" (2.2–4.1 cm). Small, plump, smooth-skinned, with egg-shaped body, pointed snout, and *fold of skin across back of head.* Gray to olive, with scattered black flecks. *Abdomen lacks mottling.* Single spade on each hind foot. Male has dark throat.

Voice: High-pitched buzzing bleat.

Breeding: March to September, when heavy rains create favorable water levels. Eggs are laid as a surface film.

Habitat: Montane woodlands, grasslands, and desert from sea level to 4,000′ (1,200 m). Moist or damp areas from marshes to leaf litter and rodent burrows.

Range: In a band from e. Nebraska and w. Missouri through Oklahoma and Texas into Mexico, west through n. Mexico into sc. Arizona.

Nocturnal. This frog often shares the burrow of a tarantula (where the two apparently live in harmony), a lizard, or a mole. It feeds primarily on ants.

218 Sheep Frog
(*Hypopachus variolosus*)

Description: 1–1¾" (2.5–4.4 cm). Small, plump, smooth-skinned frog with pointed snout and *fold of skin across back of head.* Olive to brown, with *light yellow line down middle of back.* Belly has dark mottling and a light line down middle. Other pale lines may be present across chest and along rear surface of limbs. Two prominent spades on each hind foot. Throat of male dark.

Voice: Sounds like the bleating of sheep, hence the common name.

Breeding: Prompted by sufficient rain or artificial irrigation. Eggs are laid as floating rafts and hatch within 24 hours. Tadpoles transform in about 4 weeks.

Habitat: Moist places in arid areas. Margins of ponds, marshes, under leaf litter, or in subterranean burrows.

Range: Southeastern Texas into Mexico.

Nocturnal. It is almost never seen during the day unless disturbed. It hides under rocks, logs, debris, or cow dung, or in the burrow or nest of a rodent. It feeds on ants and termites.

TOAD FAMILY
(Bufonidae)

Nineteen genera with approximately 300 species known. Only 1 genus occurs in our range: *Bufo*, with 18 species.

Toads are squat and plump with rough warty skin. They have horizontal pupils, no teeth on the upper jaw, and lack an anterior breastbone. Enlarged parotoid glands are located on each side of the neck over or behind the tympanum. These glands secrete a viscous white poison, which gets smeared in the mouth of any would-be predator. The poison inflames the mouth and throat, causes nausea, irregular heart beat, and, in extreme cases, death. Survivors of such a poisoning seldom ever again attack toads.

Bufo breeds in spring and summer. Males congregate at the breeding ponds and sing in order to attract females. Males clasp the willing females around the body behind the forelimbs. Males also have a rudimentary ovary, which can become functional if the testes are damaged or removed. Eggs are usually laid in strings attached to vegetation; they hatch into tiny black tadpoles, which weeks later metamorphose.

225 Colorado River Toad
(*Bufo alvarius*)

Description: 3–7" (7.7–17.9 cm). Largest native toad in the United States. Olive to dark brown, with a *relatively smooth, shiny skin*. Elongate parotoid glands touch prominent cranial crests. One or 2 white warts at corner of mouth. Other large warty glands on hind legs. Belly is cream-colored.

Voice: A weak low-pitched toot, lasting less than a second.

Breeding: May to July.
  Habitat: Desert. Prefers damp areas near
           permanent springs or man-made
           watering holes but may be found in
           arid grasslands and woodlands. From
           sea level to 5,300' (1,600 m).
    Range: Extreme se. California to extreme sw.
           New Mexico, south into Mexico.

Nocturnal. It sometimes appears before
seasonal rains fill breeding pools. When
the rains finally arrive, breeding
commences. It eats insects, spiders,
lizards, and other toads.

237 American Toad
   (*Bufo americanus*)
   Subspecies: American, Hudson Bay,
   Dwarf American

Description: 2–4⅜" (5.1–11.1 cm). Large, with
             *elongate parotoid glands not touching
             prominent cranial crests* or connected by
             spur. Brown to brick-red to olive, with
             various patterns in lighter colors. Spots
             brownish, warts brown to orange-red.
             Light stripe down middle of back may
             be present. Belly usually spotted. Male
             has dark throat.
      Voice: A pleasant musical trill lasting up to 30
             seconds.
  Breeding: March to July. Egg strings are attached
             to vegetation.
    Habitat: Common in a variety of habitats from
             mowed grassy yards to heavily forested
             mountains; wherever there are abundant
             insects and moisture.
      Range: In Canada from se. Manitoba to James
             Bay and Labrador, south in the east
             through Maritime Provinces, New
             England, and the Appalachian
             Mountains; west from c. Georgia to e.
             Oklahoma and Kansas; north through
             Wisconsin into Canada.
  Subspecies: American (*B. a. americanus*), with 1 or
             2 warts in each dark spot on back;
             Manitoba to the Maritime Provinces,

south in the East to Louisiana, north through Tennessee and Kentucky, west through Indiana to ne. Kansas, and north to Canada. Hudson Bay (*B. a. copei*), with heavily spotted belly; Manitoba to James Bay and Labrador. Dwarf American (*B. a. charlesmithi*), small, with 1 wart in each dark spot on back; se. Kansas through c. Missouri to sw. Illinois and Indiana, south in the East to ne. Louisiana and west to e. Oklahoma.

Primarily nocturnal. It is a prodigious insect eater. The American Toad can be distinguished from Woodhouse's Toad by the separation of the parotoid glands from the cranial crests. However, occasionally hybrids are found.

226, 240, 243 Western Toad
(*Bufo boreas*)
Subspecies: Boreal, California, Amargosa

Description: 2½–5″ (6.4–12.8 cm). Large; lacks cranial crests but has oval parotoid glands. Gray to green, with *light-colored stripe down middle of back.* Warts tinged with red and surrounded by black blotches. Male has pale throat.

Voice: Like the weak peeping of baby chicks. No vocal sacs.

Breeding: January to September, depending on weather. Egg strings are attached to vegetation in shallow, usually still water.

Habitat: Near springs, streams, meadows, woodlands.

Range: Pacific Coast from s. Alaska to Baja California, east to wc. Alberta, Montana, Wyoming, Utah, Colorado, and Nevada.

Subspecies: Boreal (*B. b. boreas*), dark blotches on belly; se. Alaska south through the Rocky Mountains to c. Colorado, and south into n. California in the West.

California (*B. b. halophilus*), with fewer dark blotches on belly, and wide head; nc. California south into n. Baja California. Amargosa (*B. b. nelsoni*), with narrow head, and fewer warts; Nye and Lincoln counties, Nevada.

Active at twilight. At higher elevations, where nighttime temperatures are low, it is often active during the day. It lives in burrows of its own construction or those of small rodents.

227, 244 Yosemite Toad
(*Bufo canorus*)

Description: 1¾–3" (4.5–7.7 cm). Moderate-sized; no cranial crests, but has large oval parotoid glands. *Male uniformly pale olive to yellow-green,* with small scattered flecks of darker colors. *Female has many large dark blotches* on a yellowish ground. Both have pale throats and bellies.

Voice: A long musical trill, given at frequent intervals.

Breeding: May to July, in available shallow water.

Habitat: The high Sierra Nevadas from 6,400 to 11,000' (2,000–3,000 m) in damp meadows and forest margins.

Range: Alpine County to Fresno County, California.

Diurnal. Solitary, except at breeding choruses. It seeks refuge from the cold by burrowing or hiding in rodent holes.

247 Great Plains Toad
(*Bufo cognatus*)

Description: 2–4½" (5.1–11.4 cm). Large, with prominent cranial crests that converge to form bony hump on snout. Behind eyes, cranial crests meet elongate parotoid glands. Gray to olive to brown, with *large, symmetrical, light-bordered dark blotches.* Sharp-edged tubercle on each hind foot. Flap of skin conceals the deflated male vocal sac.

Voice: A high-pitched, almost metallic trill.

Breeding: April to September, usually during or after heavy rainfall. Egg strings are attached to debris on bottom of pool.

Habitat: Grasslands of the prairie and drier bushy areas.

Range: From se. Alberta to w. Wisconsin in the north, south through the Great Plains to nw. Texas and into Mexico, west to s. New Mexico, Arizona and se. California, and north to parts of se. Nevada and c. Utah.

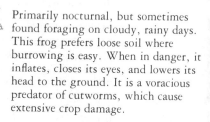

Primarily nocturnal, but sometimes found foraging on cloudy, rainy days. This frog prefers loose soil where burrowing is easy. When in danger, it inflates, closes its eyes, and lowers its head to the ground. It is a voracious predator of cutworms, which cause extensive crop damage.

254, 255 Green Toad
(*Bufo debilis*)
Subspecies: Eastern, Western

Description: 1¼–2⅛" (3.2–5.4 cm). Small flat *bright-green* toad with many small warts and black spots. Large parotoids extend onto sides; cranial crests absent. Male has dark throat.

Voice: A piercing cricketlike trill. Males call while floating head-up in the water.

Breeding: March to September, but only when rains are adequate to fill pools. If conditions are not favorable, breeding season may be skipped. Egg strings are attached to vegetation.

Habitat: The shelter of rocks in semiarid regions. Also found in prairies.

Range: From sw. Kansas south through Texas to the Gulf coast and into Mexico, north to se. areas of Arizona, New Mexico, and Colorado.

Subspecies: Eastern (*B. d. debilis*), black spots usually discrete; sc. Kansas through c. Oklahoma and e. Texas into Mexico. Western (*B. d. insidior*), black spots usually interconnected; se. Colorado and w. Kansas through w. Texas and e. New Mexico and Arizona to Mexico.

Active at twilight, but frequently will forage during the day following heavy rains. When threatened, it frequently flattens itself against the ground.

245 Black Toad
(*Bufo exsul*)

Description: 1¾–2⅜" (4.4–6.2 cm). Small; lacks cranial crests but has oval parotoid glands. *Heavy dark-olive to black mottling separated by light wavy marks.* Usually has a light stripe down middle of back. Heavy black blotches on belly.

Voice: A weak chirp.

Breeding: May to July.

Habitat: Quiet streams or springs; sometimes found in grass or open woods.

Range: Deep Springs Valley, Inyo County, California.

Active at twilight. An aquatic species that forages among the grassy tussocks bordering the spring runs at Antelope and Deep Springs. Now on the California endangered list.

239 Canadian Toad
(*Bufo hemiophrys*)
Subspecies: Canadian, Wyoming

Description: 2–3¼" (5.1–8.3 cm). Large, with cranial crests fused into *bony hump between eyes.* Green to brownish-red, with brownish-red warts; light line down middle of back. Parotoid glands oval, somewhat indistinct.

Voice: Short, weak low-pitched trill, repeated every 15–20 seconds.

Breeding: Through summer, March to September. Eggs are laid in shallow water at edge of ponds and lakes.

Habitat: Margins of ponds, lakes, and potholes.

Range: In Canada from extreme s. Mackenzie and e. Alberta southeast to ne. S. Dakota and n. Montana. Separate population in se. Wyoming.

Subspecies: Canadian (*B. h. hemiophrys*), with wide light line on back; s. District of Mackenzie south and east to Minnesota and ne. South Dakota, west in the south to ec. Alberta. Wyoming (*B. h. baxteri*), with narrow light line on back; se. Wyoming.

Primarily nocturnal. It readily takes to water to avoid capture. It is also an adept burrower, utilizing two tubercles on the hind feet in the same manner as a spadefoot toad.

---

228 Houston Toad
(*Bufo houstonensis*)

Description: 2–3¼" (5.1–8.0 cm). Medium-sized, with prominent cranial crests thickened behind the eyes. Spur of crests contacts elongate parotoids. Light brown to gray, with *dark mottling on back,* sometimes producing *vague oblique lines.* Light stripe down middle of back often present. Male has dark throat.

Voice: Long, high-pitched musical trill. Males call when in or near grass-fringed pools.

Breeding: February to June, once rain creates shallow pools. Rain permitting, breeds twice a year, in early spring and again in early summer.

Habitat: Pine forests and coastal prairies along sand ridges.

Range: C. and se. Texas.

Primarily nocturnal. Now an endangered species, its survival is threatened by urbanization, which has destroyed its habitat, and by hybridization with other species where the habitat has been partially modified. Now protected by federal order in Bastrop and Burleson counties (Texas).

242 Giant Toad
(*Bufo marinus*)

Description: 4–9½" (10–23.8 cm). A true giant, with *enormous parotoids extending down sides of body*. Brown to yellow-brown. Cranial crests prominent; body outline somewhat round and flattened.

Voice: Low-pitched trill. A distant chorus sounds like an idling diesel engine.

Breeding: Year-round with suitable temperature and rainfall. Eggs are laid in long-standing water—ditches, canals, streams, ponds, and fishponds.

Habitat: Various humid sites with adequate hiding places.

Range: Extreme s. Texas and into Mexico. Introduced in s. Florida.

Primarily nocturnal. It is probably the most widely introduced amphibian in the world. Agriculturalists released it in attempts to control the beetles that damage sugarcane. It readily adapts to backyard gardens and will accept food from fingertips. A milky secretion from the Giant Toad's parotoid glands is highly toxic; it will burn the eyes and may inflame the skin. A dog or cat that bites a Giant Toad will sicken and may die.

223, 235   Southwestern Toad
(*Bufo microscaphus*)
Subspecies: Arizona, Arroyo

Description: 2–3″ (5.1–7.7 cm). Medium-sized plump toad, olive to brown to pink, with or without dark spots. Usually ha *light stripe or patch on head and back.* Parotoid glands oval, widely separated; forward ends light-colored. No cranial crests. Male throat not dark.

Voice: A pleasing musical trill; ends abruptly after 10 seconds.

Breeding: March to July; not dependent on rainfall. Egg strings are laid on bottom of pools.

Habitat: Loose gravelly areas of streams and arroyos in drier portion of range; often on the sandy banks of quiet water in other areas.

Range: Coastal s. California and n. Baja California. Scattered localities in Utah, Nevada, Arizona, and New Mexico.

Subspecies: Arizona (*B. m. microscaphus*), with dark spots on back; scattered populations along the headwaters and tributaries of the Colorado River from sw. Utah, s. Nevada, c. Arizona, and sw. New Mexico, into Mexico. Arroyo (*B. m. californicus*), with fewer spots on back, back warty; sw. California into nw. Baja California.

Primarily nocturnal, but also found foraging by day. It hops instead of walking.

234   Red-spotted Toad
(*Bufo punctatus*)

Description: 1½–3″ (3.8–7.6 cm). Small flat toad with *round parotoids.* Olive to grayish-brown, usually with *reddish warts.* Cranial crests weak or absent.

Voice: A high-pitched musical trill. Males call while sitting near water's edge.

Breeding: April to September, initiated by

rainfall. The only North American toad that lays eggs one at a time, not in long strings, on bottoms of pools.

Habitat: Desert and rocky regions and prairie grasslands, usually near source of permanent water or dampness, natural or man-made, from sea level to 6,000' (1,800 m).

Range: From c. Texas west into se. California and south into Mexico.

Active at twilight. Red-spotted Toads are most often collected at breeding choruses, but animals have been encountered over a mile from water and even in prairie dog burrows.

250 Oak Toad
(Bufo quercicus)

Description: ¾–1¼" (1.9–3.3 cm). The smallest toad in North America, with an obvious *white to orange stripe down middle of back*. Has 4–5 pairs of dark blotches on back. Cranial crests not apparent. Parotoid glands elongate. Male's throat dark.

Voice: A single high-pitched whistle, reminiscent of a bird or chick.

Breeding: April to October, with the advent of violent, warm thundershowers.

Habitat: Loose and sandy soil of pine flatwoods and oak scrub.

Range: Coastal plain from se. Virginia to Louisiana.

Primarily diurnal, the Oak Toad spends the day insect-hunting among the debris of brushy undergrowth. Camouflage coloration makes resting toads difficult to spot. They are seldom seen at night, except in breeding choruses.

251 Sonoran Green Toad
(*Bufo retiformis*)

Description: 1½–2¼" (3.8–5.7 cm). Small flat toad with *green to greenish-yellow spots* surrounded by *striking broad patterns of black.* Large, elongate parotoid glands extend onto sides. Cranial crests weak or absent. Male has dusky throat.

Voice: A buzz. Males call from edge of grass-bordered pools.

Breeding: July, with advent of rains.

Habitat: Semiarid grassland and creosote bush desert.

Range: A narrow band from sc. Arizona to wc. Sonora in Mexico.

Nocturnal. It is a shy toad, rarely seen in numbers except at breeding ponds. As a consequence, most specimens collected are males. Occasionally single individuals are found abroad at night among the mesquite.

238 Texas Toad
(*Bufo speciosus*)

Description: 2–3⅝" (5.1–9.2 cm). Medium-sized plump toad with indistinct cranial crests and widely separated oval parotoids. *Olive to grayish-brown, with greenish warts in darker spots.* Two sharp-edged tubercles on each hind foot; inner tubercle is sickle-shaped. Flap of skin covers deflated vocal sac.

Voice: An abrupt high-pitched trill.

Breeding: April to September, after heavy rains. Uses temporary pools or man-made waterholes and ditches.

Habitat: Prairie grasslands and open woodlands; adapted for dry conditions.

Range: Extreme sc. Kansas through Oklahoma, Texas, and s. New Mexico into n. Mexico.

Nocturnal. An effective burrower, the Texas Toad disappears rapidly in loose soil. When threatened, it often flattens itself on the ground.

236, 241 Southern Toad
(Bufo terrestris)

Description: 1⅝–4½" (4.1–11.3 cm). A large plump toad with high, conspicuously *knobby cranial crests* and prominent parotoids. Brown, reddish, or black; some dark spotting may surround warts. Occasional light stripe down middle of back. Male has dark throat.

Voice: A high-pitched, musical trill, piercing at close range. Males call when in or near the water.

Breeding: March to October, typically in temporary pools and flooded meadows.

Habitat: Widely distributed. Abundant in open scrub oak where the soil is sandy and easily burrowed.

Range: Coastal plain from se. Virginia to Louisiana.

Nocturnal. Spends the day inside its burrow. It is often found in suburban areas, near houses and mowed lawns, where it feeds on insects drawn to night-lights.

246 Gulf Coast Toad
(Bufo valliceps)

Description: 2–5⅛" (5–13.0 cm). Medium-sized, with *light-bordered dark band along side*. Brown to black, with orange highlights and white spots. Usually a light stripe down middle of back. Cranial crests prominent, creating a depression between them on top of the skull. Crests connected to triangular parotoid glands. Male has yellow-green throat.

Voice: A short trill, repeated often.

Breeding: March to September.

Habitat: Various humid locations, from roadside

ditches to the barrier beaches of the
Gulf of Mexico.

Range: Gulf coast from s. Mississippi, west
through e. Texas and south into
Mexico. Also, a small section of sc.
Arkansas.

Active at twilight. Common in
gardens. Frequently seen catching
insects under streetlights. It even turns
up in city storm sewers.

224, 248, 249   Woodhouse's Toad
(*Bufo woodhousei*)
Subspecies: Woodhouse's,
Southwestern, East Texas, Fowler's

Description: 2½–5″ (6–12.7 cm). Large toad with
light stripe down middle of back.
*Prominent cranial crests contact elongate
parotoids.* Yellow to green to brown.

Voice: Like the bleat of a sheep with a cold.
Males call while sitting in quiet water.

Breeding: March to August. Egg strings are
attached to vegetation in shallow water.

Habitat: Sandy areas near marshes, irrigation
ditches, backyards, and temporary rain
pools.

Range: Widespread throughout most of the
United States.

Subspecies: Woodhouse's (*B. w. woodhousei*),

spotted back with light stripe to snout,
chest unspotted; e. Montana and North
Dakota, south through the plains states
to c. Texas, and west of the Rocky
Mountains from Idaho south to
Colorado and Arizona, with isolated
populations in w. Texas, se. California,
and the Oregon–Washington border.
Southwestern (*B. w. australis*), spotted
back with light stripe that does not
reach snout, chest also spotted; c.
Colorado through New Mexico and
Arizona, to Sonora, Mexico, and along
the Rio Grande drainage into sw. Texas
and adjacent Mexico. East Texas (*B. w.
velatus*), back dark, chest spotted; ne.

Texas. Fowler's (*B. w. fowleri*), back blotched, chest unspotted; Lake Michigan east through most of Pennsylvania to se. New York and s. New England, south to the Gulf coast (excluding coastal South Carolina, Georgia, and most of Florida), west to e. Texas and north to Missouri and s. Illinois.

Primarily nocturnal; it is the toad commonly seen at night catching insects beneath lights. Occasionally it is active during the day, but more frequently remains in its burrow or hides in vegetation.

TREEFROG FAMILY
(Hylidae)

Thirty-four genera with approximately 600 species known. Seven genera with 26 species occur in our range. Treefrogs are small and have slender legs; their pupils are horizontal.

Arboreal treefrogs are typically walkers and climbers—they are reluctant jumpers. Their toe tips are expanded into sticky adhesive pads used in climbing. Climbing is further aided by the presence of a cartilage between the last 2 bones of each toe. The cartilage allows the tip of the toe to swivel backward and sideways while keeping the sticky toe pad flat against the climbing surface. A few treefrogs, such as the North American *Acris,* have returned to a terrestrial existence, lack large toe pads, and are active leapers.

Male treefrogs in our range typically call while perched on vegetation in, over, or near water. Males clasp females just behind the forelimbs; masses of eggs are laid in the water.

---

153 Northern Cricket Frog
(*Acris crepitans*)
Subspecies: Northern, Blanchard's

Description: ⅝–1½" (1.6–3.8 cm). Skin rough; greenish-brown, yellow, red, or black. *Dark triangle between eyes* and longitudinal dark stripes on back of thigh. Snout rounded. *Legs relatively short.* Webbing of hind foot quite extensive, reaching tip of first toe and next-to-last joint of longest toe.

Voice: A shrill, measured clicking.

Breeding: April to August in northern areas, earlier in western areas.

Habitat: Sunny ponds of shallow water with good growth of vegetation in the water or on shore; slow-moving streams with sunny banks.

Range: Southern New York to Florida panhandle west to Texas and se. New Mexico, north to South Dakota, Wisconsin, and Michigan.

Subspecies: Northern (*A. c. crepitans*), with sharply defined stripe along rear of thigh; se. New York and Long Island through the coastal states and s. Tennessee to w. Florida panhandle and e. Texas. Blanchard's (*A. c. blanchardi*), with poorly defined dark stripe along rear of thigh, rougher skin; Ohio to n. Tennessee and Arkansas to s. Texas and just into Mexico, se. New Mexico north to se. South Dakota, to n. Michigan.

Diurnal. These frogs are often abundant but are difficult to catch as they hop among the grass at the water's edge.

161, 162 Southern Cricket Frog
(*Acris gryllus*)
Subspecies: Southern, Florida, Coastal

Description: ⅝–1¼″ (1.6–3.2 cm). Tiny, with rough skin; red, brown, black, or green. *Dark triangle between the eyes,* and longitudinal dark stripes on back of thigh. Snout pointed. *Legs relatively long.* Webbing of hind foot extensive, but not reaching tip of first toe or next-to-last joint of longest toe.

Voice: A rapid cricketlike click of 1 or 2 syllables. Calls while sitting on the ground, usually near edge of water.

Breeding: Throughout the year; infrequent during winter.

Habitat: Margins of swamps, marshes, lakes, streams, and roadside ditches.

Range: The coastal plain from se. Virginia to Mississippi and e. Louisiana.

Subspecies: Southern (*A. g. gryllus*), se. Virginia through inland Georgia to Louisiana. Florida (*A. g. dorsalis*), Georgia coast through Florida peninsula to se.

Alabama. Coastal (*A. g. paludicola*), ne. Texas coast.

It is active throughout the day and a strong jumper. A quick walk along the water's edge will usually flush cricket frogs from cover. Most will stop jumping after a series of erratic leaps.

149 Pine Barrens Treefrog
(Hyla andersoni)

Description: 1⅛–2" (2.9–5.1 cm). Bright green with *white-edged lavender stripes on side.* Hidden surface of thighs orange; toe pads large.

Voice: A nasal honk without much carrying power, repeated about once a second.

Breeding: April to August.

Habitat: Swamps, streams, and acid bog areas.

Range: Population centers are scattered in mid-Atlantic and South.

Nocturnal. A rare species, seldom encountered except at breeding time, when it may be seen at a favorite mating site. Its survival is threatened by habitat destruction, especially the draining of wetlands.

159 Canyon Treefrog
(Hyla arenicolor)

Description: 1¼–2¼" (3.2–5.7 cm). Plump and warty, with a toadlike appearance. Olive to brownish-gray, with darker blotches present in most populations. *Dark-edged light spot beneath eye.* Under surfaces of thigh yellow to orange. Large toe pads; webbing does not extend to tip of fifth toe of hind foot. Male has gray to black throat.

Voice: An explosive, rather hollow and nasal

series of notes, all of one tone and lasting 1–3 seconds.

Breeding: Usually March to July; may be delayed until adequate rain falls.

Habitat: Arid areas close to rocky washes, streams, and permanent pools.

Range: From s. Utah to c. Colorado, south into Mexico. Isolated populations in ne. New Mexico and the Big Bend region of Texas.

Primarily nocturnal; often seen along water courses. During the day it can be found hiding among rocks or in stony crevices near streams, camouflaged by its color.

156 Bird-voiced Treefrog
(*Hyla avivoca*)
Subspecies: Western, Eastern

Description: 1⅛–2⅛″ (2.9–5.2 cm). Greenish, brownish, or shades of gray, with darker blotches on back; *dark-edged light spot beneath eye. Hidden surfaces of thigh green to greenish-yellow.* Toe pads large.

Voice: The most beautiful frog call in North America: a resonant, birdlike whistle. Males call while perched on shrubs 3–5′ (1–1.5 m) above water.

Breeding: March to September.

Habitat: Wooded river swamps and bayous. Especially frequent in cypress, tupelo, and other trees that grow in standing water, and creepers that grow among the trees.

Range: Extreme s. Illinois southwest to Florida panhandle and south to Louisiana. Isolated populations in ne. Georgia and on the Georgia–South Carolina border.

Subspecies: Western (*H. a. avivoca*), smaller size, white or pale yellow light spot beneath eye; Illinois to Louisiana and w. Florida. Eastern (*H. a. ogechiensis*), larger size, yellow or green light spot beneath eye; c. Georgia and sw. South Carolina.

Nocturnal. This species is almost never seen except during nighttime choruses; apparently it remains in the treetops during the day. Its skin secretions may cause runny nose and watery eyes in people who handle it.

---

158 California Treefrog
(*Hyla cadaverina*)

Description: 1–2″ (2.5–5.1 cm). *Skin rough, gray, with dark blotches.* Dark stripe through eye usually absent. Expanded toe pads; webbing extends to tip of fifth toe of hind foot. Male has gray throat.

Voice: An abrupt low-pitched quack, given during the day as well as at night. Males usually call while sitting in the water, often at the base of a rock.

Breeding: March to May.

Habitat: Near slow streams and rocky washes with permanent pools. Deserts to mountains, sea level to over 5,000′ (1,500 m).

Range: From sw. California into n. Baja California.

Primarily nocturnal. It seeks shade during the day among the rock crevices near water. Protective coloration helps it avoid daytime predators. When disturbed, it leaps into the water but returns almost immediately to shore.

152, 157, 160 Cope's Gray Treefrog and Common Gray Treefrog
(*Hyla chrysoscelis* and *Hyla versicolor*)

Description: 1¼–2⅜″ (3.2–6 cm). Skin rough; greenish or brownish to gray, with several large dark blotches on back. *Dark-edged light spot beneath eye. Under surfaces of thighs bright yellow-orange.* Large toe pads.

Voice:	A hearty, resonating trill, usually heard in spring and early summer.
Breeding:	April to August. Also breeds in winter in southern parts of range.
Habitat:	Trees and shrubs growing in or near permanent water.
Range:	From s. Ontario and Maine to n. Florida west to c. Texas, north through Oklahoma to Manitoba.

The two species of Gray Treefrog are identical in appearance, and since their ranges overlap extensively, they cannot be distinguished in the field. However, Cope's Treefrog has a faster trill and only half as many chromosomes as the Common Gray. Nocturnal; they live high in trees and descend only at night, usually just to chorus and to breed.

146 Green Treefrog
(*Hyla cinerea*)

Description:	1¼–2½" (3.2–6.4 cm). Bright green, yellow, or greenish-gray. Has *sharply defined light stripe along upper jaw and side of body;* side stripe occasionally absent. Sometimes has tiny, black-edged gold spots on back. Large toe pads.
Voice:	Cowbell-like when heard at a distance. Nearer, sound is *quank, quank.* Males call while clinging to vertical stems 1–2′ (.3–.5 m) above water.
Breeding:	March to October in southern areas, April to September in northern areas.
Habitat:	Vegetation near permanent water. During the day frequently found asleep on underside of large leaves or in other moist, shady places.
Range:	Delaware south along the coastal plain into Florida and the Keys, west to s. Texas, and north through c. Arkansas and w. Tennessee to Illinois.

Green Treefrogs congregate in large choruses of several hundred. A typical treefrog, this species prefers to walk rather than jump. When fleeing a predator in the trees it takes gangly leaps into space.

173 Spring Peeper
(*Hyla crucifer*)
Subspecies: Northern, Southern

Description: ¾–1⅜″ (1.9–3.5 cm). Tan to brown to gray, with *characteristic dark X on back*. Large toe pads.

Voice: A high-pitched ascending whistle, sometimes with a short trill. Chorus sounds like the jingle of bells. Males call from shrubs and trees standing in or overhanging water.

Breeding: In southern areas, November to March; in northern areas, March to June, with the start of warm rains.

Habitat: Wooded areas in or near permanent or temporarily flooded ponds and swamps.

Range: Manitoba to the Maritime Provinces, south through c. Florida, west to e. Texas, and north into c. Wisconsin.

Subspecies: Northern (*H. c. crucifer*), unmarked belly; throughout species' range except s. Georgia and Florida. Southern (*H. c. bartramiana*), spotted belly; se. Georgia and adjacent n. Florida.

Nocturnal. The Spring Peeper is one of the most familiar frogs in the East. Its chorus is among the first signs of spring. Peepers hibernate under logs and loose bark.

150 Mountain Treefrog
(*Hyla eximia*)

Description: ¾–2¼″ (1.9–5.7 cm). Skin smooth; green, sometimes tan or black, with a *light-edged purple to black stripe from snout through eye* and along side, sometimes to hind leg. Occasionally has longitudinal dark bars on lower back. Toe pads small. Male has olive throat.

Voice: A harsh, low, slightly metallic note repeated several times per second. Males call while sitting in shallow water.

Breeding: June to August, with the summer rains. Egg masses are attached to vegetation just below the water.

Habitat: Slow water and shallow grassy pools in oak and conifer forests over 5,000' (1,500 m). Rarely seen high in treetops.

Range: The mountains of c. Arizona and New Mexico.

Nocturnal. It was formerly known as *Hyla wrightorum*.

164 Pine Woods Treefrog
(*Hyla femoralis*)

Description: 1–1¾" (2.5–4.4 cm). Gray to reddish-brown, usually with dark blotches; *yellow to white spots on dark rear surface of thigh* visible only when leg is extended. Has large toe pads.

Voice: A series of measured notes, like the tapping together of wooden dowels. Chorus sounds like an office of industrious typists. Solitary males frequently call from high in the treetops.

Breeding: March to October in southern areas; June to September in northern areas.

Habitat: Pine flatwoods near ponds and marshes. Most common in treetops.

Range: Coastal plain from se. Virginia to s. Florida (except the Everglades) and west along the Gulf coast to Louisiana. Some isolated populations in c. Alabama.

Nocturnal. It is difficult to observe because of its treetop habitat and its mottled appearance, which blends well with pine tree bark. It is most easily located when it calls in early evening.

145, 151   Barking Treefrog
(*Hyla gratiosa*)

Description: 2–2¾" (5.1–7 cm). Plump, with
granular skin. Bright green, yellow,
gray, or dark brown; dark spots usually
present in lighter phases. Yellowish
stripes run from upper jaw along sides
of body. Large toe pads. Male has a
green or yellow throat.

Voice: Mating call is a single bell-like note
given near water. Loud, barking rain
call is given from high in trees.

Breeding: March to August. Choruses gather at
permanent water, streams, cypress
ponds, and bayheads.

Habitat: Spends warm months in treetops.
During winter and dry periods, seeks
moisture by burrowing among tree
roots, clumps of vegetation and in the
ground.

Range: The coastal plain from se. Virginia to s.
Florida and Louisiana. Also, isolated
populations in the northern parts of the
Gulf states and Tennessee and
Kentucky. Introduced into s. New
Jersey.

Nocturnal. An excellent terrarium pet,
it readily accepts insects from fingertips
at night, growing plump through the
years. During the day it sleeps in a
secluded spot.

148, 170, 182   Pacific Treefrog
(*Hyla regilla*)

Description: ¾–2" (1.9–5.1 cm). Skin rough; varies
greatly from green to light tan to
black, often with dark spots. *Black
stripe through eye* and usually a dark
triangle between the eyes. Large toe
pads. Male has gray throat.

Voice: A high-pitched, 2-part musical note.

Breeding: January to August.

Habitat: On the ground among shrubs and
grass, close to water.

Range: From s. British Columbia to Baja California east to Montana, Idaho, and Nevada. Also, Channel Islands off s. California.

This commonly heard frog of the Pacific coast, active both day and night, is found from sea level to over 10,000' (3,000 m). When Hollywood moviemakers need an authentic outdoor nighttime sound, they often record its call. As a consequence, the Pacific Treefrog has been heard around the world.

147, 174 Squirrel Treefrog
(*Hyla squirella*)

Description: ⅞–1⅝" (2.2–4.1 cm). Poorly defined *white stripe along upper jaw and side of body*, sometimes absent. Body green to brown, plain to spotted; may have yellow flecks on back; sometimes a dark bar between eyes. Large toe pads.

Voice: Mating call a nasal trill, usually given from perch on vertical stem about 3' (1 m) above water. Rain call a quack or grating like a scolding squirrel.

Breeding: March to October in southern areas, April to August in northern areas. Usually breeds among emergent vegetation.

Habitat: Any habitat with moisture and insects. Adults hide under loose bark, on the underside of palm leaves, or in hollow tree holes.

Range: The coastal plain from se. Virginia to Florida and the Keys west along the Gulf coast to c. Texas. Isolated populations in n. Mississippi, Louisiana, and southeastern Oklahoma.

The rain frog commonly heard before and after summer showers. Nocturnal and very active. An efficient predator, it can frequently be seen catching insects around patio lights. During the day it hides under roof flashing or in garden shrubs; sometimes dozens are found huddled together.

---

172 Little Grass Frog
(*Limnaoedus ocularis*)

Description: ½–⅝" (1.1–1.7 cm). The smallest frog in North America. Tan to gray-green, with a *dark line through eye and along side.* Sometimes has line down middle of back starting as triangle between the eyes. Round toe tips.

Voice: A high-pitched note or two like the tinkle of glass; may be mistaken for an insect and is inaudible to some people.

Breeding: January to September in most of range; year-round in Florida.

Habitat: Climbs among grass and sedges near roadside ditches, pond margins, and cypress bays.

Range: Southeastern Virginia along the coastal plain through s. Florida and west to se. Alabama.

Primarily nocturnal. It is a true treefrog despite its small size and preference for perches within 1–2 feet (.3–.6 m) of the ground. It is usually found clinging to upright stems, with head cocked to one side, looking for prey.

155, 178 Cuban Treefrog
(*Osteopilus septentrionalis*)

Description: 1½–5½" (3.8–14.0 cm). Largest treefrog in North America. Skin is warty; green, bronze, or gray. *Enormous toe pads* as large as external ear. Skin on top of head is fused to skull.

Voice: A variably pitched, slightly rasping or grating snore.

Breeding: May to October. Eggs deposited in lakes and ponds; also drainage ditches, swimming pools, cisterns.

Habitat: Moist and shady places in trees and shrubs, or around houses.

Range: Introduced into s. Florida from Cuba.

Nocturnal. Highly predaceous, it will eat anything it can catch and swallow, including insects, spiders, and other frogs. The Cuban Treefrog is most abundant around ornamental fishponds and well-lighted patios and can be found on highway billboards, hiding among the timbers during the day or feeding on insects attracted to the night lights. This species was probably introduced by accident into Key West on vegetable produce brought from Cuba early in the century. It continues to spread on the Florida mainland by hitchhiking on crates and transplanted shrubs.

163 Mountain Chorus Frog
(*Pseudacris brachyphona*)

Description: 1–1½" (2.5–3.8 cm). Small, brown to green, with 2 *back-to-back, crescent-shaped dark stripes*, which may touch to form an irregular X. Broad lateral stripes pass through eye. Belly yellowish. Small round toe tips.

Voice: A high-pitched squeak, given in a rapid series.

Breeding: Usually February to April, occasionally into summer months. Mates in shallow water near forest margins or in wooded ponds.

Habitat: Prefers woodlands, especially in hilly areas.

Range: From se. Ohio and sw. Pennsylvania in a band to c. Alabama. Isolated populations in West Virginia and Mississippi.

Primarily nocturnal. It behaves more like a wood frog than a treefrog in that it tends to leap more than walk.

175 Brimley's Chorus Frog
(*Pseudacris brimleyi*)

Description: 1–1¼" (2.5–3.2 cm). Slender, tan with *black stripe from snout through eye to groin*. Usually 3 longitudinal dark stripes and dark spots on chest. No transverse dark mark between eyes. Small round toe tips.

Voice: A rasping trill, of about 1 second, given often.

Breeding: February to April.

Habitat: Marshes and sunny wooded areas where vegetation is not too thick.

Range: The coastal plain from e. Virginia to e. Georgia.

Primarily nocturnal. It can change color rapidly, depending on temperature and activity.

180 Spotted Chorus Frog
(*Pseudacris clarki*)

Description: ¾–1¼" (1.9–3.2 cm). Small and gray with *black-edged bright-green blotches*. Blotches sometimes fuse to form longitudinal stripes. Green stripe through eye. Small round toe tips.

Voice: A rasping trill, like the sound of a saw.

Breeding: All year in southern part of range when stimulated by rainfall, late spring in northern part.

Habitat: Shortgrass prairie.

Range: From c. Kansas south through c. Texas to Gulf of Mexico; introduced into Montana.

Nocturnal. The Spotted Chorus Frog is inactive when conditions are dry. It is most frequently found around marshy breeding pools. It feeds on insects and spiders.

183 Southern Chorus Frog
(*Pseudacris nigrita*)
Subspecies: Southern, Northern

Description: ¾–1¼" (1.9–3.2 cm). Rough warty skin; light tan or gray, with 3 *longitudinal rows of black spots*, which may be fused into stripes. Black stripe runs through eye. Small round toe tips.

Voice: A rasping trill. Males call from floating vegetation or grassy tussocks.

Breeding: All year when rains are heavy in southern areas. November to April in northern areas.

Habitat: Wet or moist grassy meadows, ponds, and sinkholes; among damp leaf litter in woodlands.

Range: The coastal plain from e. North Carolina through Florida to s. Mississippi.

Subspecies: Southern (*P. n. nigrita*), with white stripe on upper lip; coastal plain from e. North Carolina to n. Florida and s. Mississippi. Northern (*P. n. verrucosa*), with spotted upper lip; throughout Florida peninsula.

Primarily nocturnal, but occasionally found foraging during daylight. The tiny toe pads limit its climbing ability. It burrows into the banks of ponds and ditches.

176, 177 Ornate Chorus Frog
(*Pseudacris ornata*)

Description: 1–1⅞₁₆" (2.5–3.7 cm). Small and plump, with *dark stripe through eye* and large *dark blotches on side and low on back*. Usually reddish-brown but may be ash-white, green, or nearly black. Small round toe tips.

Voice: A single high-pitched, metallic peep, uttered once a second.

Breeding: November to April, in shallow ponds, roadside ditches, and rain-flooded pastures.

Habitat: Areas of shallow water either temporary or permanent where vegetation is not too dense.

Range: The coastal plain from North Carolina to central Florida and eastern Louisiana.

Nocturnal. This species burrows, which probably explains why it is almost never found except in a breeding chorus. It has been dug up in cultivated fields and around the margins of ponds.

---

185 Strecker's Chorus Frog
(*Pseudacris streckeri*)
Subspecies: Strecker's, Illinois

Description: 1–1⅛" (2.5–4.8 cm). Plump toad-shaped body. Green to reddish-brown to gray, with *dark stripe through eye and a dark spot below it;* eye stripe continues onto side as series of spots. Small round toe tips.

Voice: A single clear, bell-like note.

Breeding: November to June. Eggs are attached to submerged aquatic vegetation. Tadpoles transform in about 2 months.

Habitat: Moist areas, including wooded and open fields, swamps, and streams.

Range: From nc. Oklahoma and w. Arkansas south through Texas to the Gulf. Separate populations in c. Illinois and se. Missouri and adjacent Arkansas.

Subspecies: Strecker's (*P. s. streckeri*), with yellow groin and prominent dark spots on sides; nc. Oklahoma and w. Arkansas south to the Gulf of Mexico. Illinois (*P. s. illinoensis*), groin not yellow, spots on side faded; wc. Illinois, with a separate population in adjacent se. Missouri and ne. Arkansas.

Nocturnal. This is the largest chorus frog. It burrows with its front feet, unlike other burrowing frogs and toads, which use the hind feet.

179 Chorus Frog
(*Pseudacris triseriata*)
Subspecies: Western, Upland,
New Jersey, Boreal

Description: ¾–1½" (1.9–3.8 cm). Skin smooth,
greenish-gray to brown. *Three dark
stripes down back;* may be broken,
reduced, or absent. Dark stripe through
eye and white stripe along upper lip.
Small round toe tips.

Voice: A rasping, rising trill lasting 1–2
seconds, like the sound of a fingernail
running over the teeth of a comb.
Males call while sitting upright on
floating vegetation.

Breeding: All winter in warmer areas of range,
late winter to summer in northern
areas.

Habitat: Grassy areas from dry to swampy to
agricultural; also suburbs where
pollution and pesticides are not a
problem; woodlands; and river swamps.

Range: Widespread. Alberta to n. New York
(except New England, the n.
Appalachians, and the southern coast)
south to Georgia, west to Arizona.

Subspecies: Western (*P. t. triseriata*), with dark
stripes; Wisconsin to extreme s.
Quebec, south through w. New York
and north of the Ohio River to c.
Oklahoma, west to Nebraska and South
Dakota, and ne. Lake Superior; a
disjunct population occurs in c. Arizona
and New Mexico. Upland (*P. t.
feriarum*), with thin dark stripes or rows
of small spots; e. Pennsylvania south to
the Florida panhandle and west to e.
Texas and Oklahoma, and north to
Kentucky. New Jersey (*P. t. kalmi*),
larger size, with prominent dark
stripes; extreme s. New York along the
coastal plain through the Delmarva
Peninsula. Boreal (*P. t. maculata*), with
shorter hind legs, and dark stripes or
spots; nw. Canada near Great Bear Lake
to n. Ontario, south through n.
Michigan to n. New Mexico, west to c.
Utah, e. Idaho and along the e. slopes

of the Rocky Mountains to British Columbia.

Nocturnal. Chorus frogs may be heard calling on warm nights in early spring even before all the ice has disappeared from the water. At the slightest threat they disappear beneath the surface.

186 Burrowing Treefrog
(*Pternohyla fodiens*)

Description: 1–2″ (2.5–5.1 cm). Light yellow to brown, with large black-edged brown spots or longitudinal stripes. *Skin is fused to bony skull; skin fold at back of head.* Toe pads small. Large spadelike tubercle on hind feet. Male has gray throat.

Voice: A series of loud deep squawks.

Breeding: July to August, with summer rains.

Habitat: Arid grassy areas or open mesquite woodlands. Sea level to about 5,000′ (1,500 m).

Range: Extreme sc. Arizona south along the Pacific coast of Mexico.

Nocturnal. This Mexican treefrog is adapted to living in burrows where humidity is high. It is easily caught when breeding or chorusing but at other times is extremely wary. It has been suggested that the bony skull is used like a stopper, to plug the burrow against predators and to slow the loss of moisture.

181, 184 Mexican Treefrog
(*Smilisca baudini*)

Description: 2–3⅝″ (5.1–9 cm). Large, with dark-edged *light spot under eye* and *dark line from snout through eye to shoulder area.* Color varies from dark to light and greenish to gray or yellow. Large toe pads.

Voice: A series of short explosive notes.

Breeding: In conjunction with adequate rainfall.

Habitat: Arid areas in the northern part of its range; moist forests in southern parts. Found in fairly humid places, along streams, in canyons, in trees and shrubs. Sea level to 3,300' (1,000 m).

Range: Extreme se. Texas, south into Mexico.

Nocturnal. It seeks refuge from the heat under loose bark, in tree holes, and in damp earth.

LEPTODACTYLID FROG FAMILY
(Leptodactylidae)

Fifty genera of about 650 species occur in Australia, southern Africa, and the tropical Americas; 4 genera with 7 species occur in our range.
Leptodactylids are an extremely diverse group adapted to a variety of habitats and reproductive strategies. The 7 species that occur in our section of North America have horizontal pupils, and a T-shaped bone in the tip of each toe.
Males of some species congregate together to call in loud choruses to attract females to the breeding ponds, but males of most species call singly, often while hidden in vegetation or burrows. Breeding males clasp the females behind the forelimbs. Some species, like *Leptodactylus,* lay numerous eggs in foam nests in the water. On hatching, the tadpoles escape into the water where they live until metamorphosing into frogs. Other species, like *Eleutherodactylus,* lay fewer than two dozen eggs in moist leaf litter or damp earth. They hatch 2 to 3 weeks later, releasing fully developed miniature frogs.

---

169 **Puerto Rican Coqui**
(*Eleutherodactylus coqui*)

Description: 1⁵⁄₁₆–2¼" (3.3–5.8 cm). Gray to gray-brown, usually with an *ill-defined dark W between shoulders* and scattered dark mottling. Light stripes down back in some specimens. Belly pale yellow, stippled with brown. Large toe pads and no webbing. Eye brown to golden. Males smaller than females.

Voice: A 2-note *co-qui,* of less than a second, like high-pitched chirp of a bird. Males call from 3 to 6 feet (1–2 m) above the ground, usually while sitting on leaves.

Breeding: Year-round. Eggs are laid on land and hatch into miniature frogs with tiny tails.

Habitat: Arboreal. Trees, epiphytes, and leaf litter in moist tropical hammocks and in greenhouses and garden trash.

Range: Introduced in s. Florida at Fairchild Tropical Gardens in Coral Gables.

Nocturnal. This species is usually seen only at night or on rainy, overcast days. The Florida population may not have survived the cold winter of 1977–78.

171 Greenhouse Frog
(*Eleutherodactylus planirostris*)

Description: ⅝–1¼" (1.6–3.2 cm). A tiny agile frog, with truncated toe pads and no webs. Brown to reddish-brown, with either *dark mottling or light dorsolateral stripes*. Eyes red to scarlet. Belly white with fine brown stippling.

Voice: A musical chirp repeated several times.

Breeding: With warm summer rains. Eggs are laid among vegetation. Tadpole develops inside the egg; young frog emerges with a tiny tail that is quickly lost.

Habitat: On the ground whenever there is shelter and moisture. Likes greenhouses, gardens, and hardwood hammocks with damp leaf litter in which to burrow.

Range: Widely introduced throughout the Florida peninsula.

Nocturnal. On humid nights it is frequently seen foraging on lawns. It can be brought out and encouraged to call by sprinkling.

154 **Barking Frog**
(*Hylactophryne augusti*)
Subspecies: Eastern, Western

Description: 2½–3¾″ (6.4–9.5 cm). Toad-shaped body, large head, and small truncated toe pads. Skin smooth; greenish to tan, with scattered dark spots. *Dorsolateral skin folds on back; one fold across back of head,* and one disk-shaped fold on belly.

Voice: A sharp throaty sound, like a dog's bark.

Breeding: February to May, when rainfall is sufficient. Eggs are laid under rocks and logs in moist soil. Tadpole stages occur within eggs, which hatch as miniature frogs.

Habitat: Damp limestone caves and crevices, especially where rain is frequent.

Range: From se. New Mexico and c. Texas south to Tehuantepec, Mexico, thence north to extreme s. Arizona.

Subspecies: Four; 2 in our range. Eastern (*H. a. latrans*), brown, with vague dark markings on hind legs; se. New Mexico to c. Texas and into adjacent Mexico. Western (*H. a. cactorum*), gray to olive, with prominent markings on hind legs. Extreme s. Arizona south into Mexico.

Nocturnal. It walks with the body held high off the ground. When threatened it inflates to several times its normal size. The skin fold on the belly may be useful in helping the frog cling to the sides of caves.

---

166 **White-lipped Frog**
(*Leptodactylus labialis*)

Description: 1⅜–2″ (3.5–5.1 cm). Gray to brown, with variable dark spots on back. White line along upper lip; no toe pads. Has *dorsolateral folds and prominent disk on belly*

Voice: A throaty 2-note call; the second note is a little higher in pitch.

Breeding: May and June with heavy rains. Lays eggs in a foamy meringuelike nest whipped from body secretion. Nests are frequently located in rain-filled excavations dug by the male.

Habitat: Grasslands, cultivated fields, irrigation ditches, under rocks, wherever moisture is plentiful.

Range: Extreme s. Texas through Mexico to Panama.

Nocturnal; burrowing. Males can be found hiding in burrows beneath grass clumps during the day or foraging on the surface at night.

---

167 Rio Grande Chirping Frog
(*Syrrhophus cystignathoides*)

Description: ⅝–1″ (1.6–2.5 cm). Grayish-yellow, with darker mottling; fine granular skin. *Vague eye stripe.* Truncated toe pads more prominent on front feet than on rear.

Voice: A short, cricketlike chirp with little carrying power. Calls whenever it rains.

Breeding: April and May.

Habitat: Moist and shaded vegetation; under leaf litter, rocks, and rubbish in ditches and palm groves; well-watered lawns and gardens.

Range: Extreme s. Texas and e. Mexico.

Nocturnal; sometimes burrows. It is easily discovered during the day under loose boards or discarded cardboard but is extremely difficult to catch.

Spotted Chirping Frog
(*Syrrhophus guttilatus*)

Description: ¾–1¼″ (1.9–3.2 cm). Yellowish-brown with a *wormlike pattern of dark lines;* smooth skin on back. Truncated toe pads more prominent on front feet.

Voice: Single short notes or chirps.

Breeding: Late winter and spring. Fewer than 15 eggs are laid.

Habitat: Near springs and moist areas of caves and ravines.

Range: Big Bend area of west Texas and south into Mexico.

Nocturnal; sometimes burrows. It hide under rocks and leaf litter during the day and emerges at night to feed on insects, spiders, and small crustaceans. A walking frog, it jumps only when frightened.

168. Cliff Chirping Frog
(*Syrrhophus marnocki*)

Description: ¾–1½″ (1.9–3.8 cm). Greenish with *brown mottling;* smooth-skinned. Flattened head and body. Truncated toe pads more prominent on front feet.

Voice: A trill or chirp like a cricket's.

Breeding: February to December. Female may lay eggs 3 times a year.

Habitat: Crevices and caves of limestone hills.

Range: From c. to w. Texas—San Antonio to Big Bend.

Nocturnal. The flattened shape of the Cliff Chirping Frog allows it to slip into narrow cracks in limestone to retain moisture and to elude predators.

TONGUELESS FROG FAMILY
(Pipidae)

Four genera: *Pipa* with 6 species in
South America, and *Xenopus,
Hymenochirus,* and *Pseudhymenochirus*
with 14 species in Africa. One species,
Xenopus laevis, has been introduced to
North America.
Tongueless frogs have attached ribs and
eight presacral vertebrae. Their pupils
are round. The South American species
have starlike projections on the tips of
the toes of the front feet; the African
species have simple pointed toes on the
front feet. The latter attach their eggs
singly to submerged vegetation, logs,
or stones; thousands of eggs are laid.
During breeding, males clasp the
female around the waist.

222 African Clawed Frog
(*Xenopus laevis*)

Description: 2–3¾" (5–9.5 cm). Olive-brown or
gray, with darker marks and mottling.
No external eardrums, tongue, or
teeth. *Horny black claws on outer 3 toes of
hind feet.* Hind toes fully webbed. Male
smaller than female; has dark nuptial
pads on front limbs.

Voice: Despite absence of vocal sacs, males
give a loud, rattling croak while
swimming.

Breeding: Follows warm, heavy rains (February to
May in California). Several hundred
eggs are deposited singly or in small
clusters on submerged vegetation and
stones. Tadpoles hatch within 36 hours;
they have catfishlike whiskers at mouth
corners and swim in a head-down
position.

Habitat: Standing water, from slow-moving
permanent streams to ponds and
marshes.

Range: Introduced from Africa into California
—Orange and San Diego counties.

Nocturnal. Although mainly aquatic, African Clawed Frogs sometimes turn up in rainpools, indicating land passage. When not actively foraging or mating, they rest quietly on the bottom or hide under rocks. They are highly carnivorous and eat anything they can catch.

REPTILES
(class Reptilia)

Living reptiles comprise 5 major groups, 4 of which occur in North America; the crocodiles (order Crocodylia), the turtles (order Testudines), the lizards (suborder Lacertilia), and the snakes (suborder Serpentes).

Derived from the ancient amphibians, the first reptiles appeared some 300 million years ago. The Age of Reptiles that ensued saw a proliferation of reptilian forms from small tree-dwellers to monstrous dinosaurs more than 100' (30 m) long and perhaps weighing in excess of 50 tons. Except for the crocodilians, all of the ruling reptiles are extinct; only 4 of 16 ancient orders are alive today.

The reptiles' protective dry scaly skin and tough-shelled eggs, as well as enlarged and improved lungs, enabled them to colonize areas where amphibian existence was impossible. The ancient reptile, in turn, gave rise to birds and mammals whose subsequent proliferation coincided with the reptiles' decline.

Crocodiles
(order Crocodylia)

Three families, 8 genera, and 21 species distributed worldwide in tropical and subtropical regions. The American Alligator and American Crocodile are native to the southeastern United States. A third species, the Spectacled Caiman, has been introduced. Crocodilians first appeared about 160 million years ago; they stem from the archosaurs, the dominant reptiles of the time. Crocodilians are large and well-armored with sculptured heads, protruding eyes and nostrils, and well-muscled, compressed tails. Front feet have 5 toes; webbed hind feet have 4 toes. The ear is covered with a movable flap.

All crocodilians are aquatic, carnivorous, and fond of basking. Females lay about 20–80 chinalike, elliptical eggs in a cavity dug in a sandbank or in a nest mound constructed of vegetation.

Sex of crocodilians is difficult to determine visually. Males tend to grow larger than females.

256, 259 American Alligator
(*Alligator mississippiensis*)

Description: 6'–19'2" (1.8–5.84 m). Largest reptile in North America. Distinguished from American Crocodile by *broad and rounded snout*. Generally black with yellowish or cream crossbands that become less apparent with age. *Large 4th tooth on bottom jaw* fits into a socket in upper jaw, *is not visible when mouth is closed*. No curved bony ridge in front of eyes, as seen in Spectacled Caiman.

Voice: During the breeding season adults produce a throaty, bellowing roar heard over considerable distance. Young give a high-pitched call: *y-eonk, y-eonk, y-eonk*.

Breeding: Mates April to May after emerging from hibernation. In June, female builds a mound-shaped nest about 5–7' (1.5–2.1 m) in diameter and 1½–3' (46–91 cm) high, of mud, leaves, and rotting organic material; deposits about 25–60 hard-shelled eggs, 3" (76 mm) long, in cavity scooped from center of nest. During 9-week incubation period, she remains near nest. The calling of hatching young prompts the female to scratch open the nest to free them. Hatchlings are 9–10" (22.8–25.4 cm) long and remain with the female for 1–3 years.

Habitat: Fresh and brackish marshes, ponds, lakes, rivers, swamps, bayous, and big spring runs.

Range: Coastal se. North Carolina to the Florida Keys and west along the coastal plain to s. Texas; north to extreme se. Oklahoma and s. Arkansas.

Alligators are important to the ecology of their habitat. During droughts they dig deep holes, or "dens," which provide water for the wildlife community. They hibernate in dens during the winter. Diet consists of rough fishes, small mammals, birds, turtles, snakes, frogs, and invertebrates. Alligators have been relentlessly hunted for their hides and are much reduced in numbers. Under state and federal protection they are beginning to make a comeback in some areas.

257, 260 Spectacled Caiman
(Caiman crocodilus)

Description: 4'–8'8" (1.2–2.64 m). Light brown to olive-brown or light yellow, with distinct or indistinct crossbanding on back and tail. *Curved bony ridge, or "spectacle," in front of eyes.* Enlarged 4th tooth on lower jaw not visible when jaws are closed.

Breeding: August to September. Female builds mound-shaped nest of soil and vegetation, somewhat smaller than alligators'. Clutch averages 28 eggs— each hard-shelled, elliptical, about 2½" (64 mm) long. Hatchlings emerge late October to November; average 8½" (21.6 cm) long.

Habitat: Ponds, streams, marshes, rivers, and drainage canals.

Range: Introduced from Central and South America; liberated in many areas. A small reproducing population inhabits South Florida.

Subspecies: Four; 1 in our range, *C. c. crocodilus.*

In the past 20 years, as a result of protective legislation for alligators, hundreds of thousands of caimans were imported for the pet trade. Most died quickly. Others, which escaped or were released, survived and have been sighted in the wild, especially in Florida drainage canals. They feed on fish, amphibians, small birds and mammals, and insects.

258, 261 American Crocodile
(*Crocodylus acutus*)

Description: 7–15′ (2.1–4.6 m). Long *slender snout*
distinguishes it from American
Alligator. Gray-green, dark olive-
green, or gray-brown with dark
crossbands on back and tail; crossbands
obscure in old adults. *Large 4th tooth on
bottom jaw visible when mouth is closed.*
No curved bony ridge in front of eyes,
as seen in caimans.

Breeding: Female builds mound-shaped nest of
soil, sand, and mangrove peat; lays 35–
50 eggs, late April to early May.
Hatchlings emerge July to early
August, are about 9″ (22.9 cm) long.

Habitat: Florida Bay in Everglades National
Park, Biscayne Bay, and Florida Keys;
bogs and mangrove swamps.

Range: Extreme coastal s. Florida and the
Keys.

May have reached 23′ (7 m) in South
America. The Florida population, fewer
than 500 in number, was declared
endangered in 1975. Poaching and
construction of highways, beachfront
homes, and mobile home parks have
reduced their numbers. The adult diet
includes crabs, fish (especially mullet),
raccoons, and water birds. Drawn to
the sounds of hatchlings, the female
opens the nest cavity and carefully picks
up the young in her mouth and, in a
series of trips, carries them to water.

Turtles
(order Testudines)

Of the 12 families of turtles living today, 7 are represented in the United States and Canada. These 7 families comprise 18 genera with 48 species.

The first turtles appeared about 200 million years ago, considerably before the dinosaurs. Today, these popular and easily recognized creatures are found in almost every environment—aquatic, oceanic, and terrestrial—throughout the tropical and temperate zones. Large land-dwelling turtles are often called tortoises, while those that are hard-shelled, edible, and aquatic are terrapins.

Structurally, turtles are a bizarre group of animals. With their expanded ribs incorporated into a protective shell, the unique placement of limb girdles inside the rib cage, and horny beak instead of teeth, turtles do not appear closely related to other reptiles. Indeed, turtle origins have puzzled paleontologists. However, like other reptiles, turtles have dry scaly skin and body temperature that is controlled behaviorally. Whether aquatic or terrestrial, all turtles enjoy basking in the sun.

Without exception, turtles lay eggs. Some small terrestrial species lay as few as 2 or 3 eggs per clutch, whereas sea turtles come ashore to lay several clutches of a hundred or more. Typically, the female digs a hole, deposits her hard-shelled eggs in it, then fills the hole with earth before departing. Depending on species, incubation may take as little as 6 weeks to more than 7 months. The hatchlings must then dig to the surface and, if aquatic, find their way to the water.

SNAPPING TURTLE FAMILY
(Chelydridae)

Two genera: *Chelydra* and *Macroclemys*, both restricted to the Americas; 2 species within the range of this guide. Snappers are among the largest of living freshwater species. They are characterized by massive heads with powerful hooked jaws, long tails, relatively small cross-shaped plastrons, and carapaces with 12 marginal scutes on each side.
These primitive-looking turtles occupy aquatic habitats of many descriptions, are opportunistic feeders, and lay large clutches of flexible-shelled, spherical-shaped eggs. Males grow larger than females.

322, 323, 324 **Snapping Turtle**
(*Chelydra serpentina*)
Subspecies: Common, Florida

Description: 8–18½" (20–47 cm). The familiar "snapper," with *massive head and powerful jaws*. Carapace tan to dark brown, often masked with algae or mud, bearing 3 rows of weak to prominent keels, and serrated toward the back. Plastron yellow to tan, unpatterned, relatively small, and cross-shaped in outline. Tail as long as carapace; with saw-toothed keels. Tubercles on neck. Wild specimens range to 45 lbs. (20.5 kg). Some fattened captives exceed 75 lbs. (34 kg).

Breeding: Mates April to November; peak laying season is June. Lays as many as 83 (typically 25–50) spherical, 1⅛" (29 mm) eggs in 4–7" (10–18 cm) deep, flask-shaped cavity. Each egg is directed into place by alternating movements of hind feet. Incubation, depending on weather, takes 9–18 weeks. In temperate localities, hatchlings overwinter in nest. Females may retain sperm for several years. Females often travel to a nesting site some distance from water.

Habitat: Freshwater. Likes soft mud bottoms and abundant vegetation. Also enters brackish waters.

Range: S. Alberta to Nova Scotia, south to the Gulf.

Subspecies: Four; 2 in our range. Common (*C. s. serpentina*), blunt tubercles on neck; throughout our range, except Florida. Florida (*C. s. osceola*), pointed tubercles on neck; throughout the Florida peninsula.

Highly aquatic, it likes to rest in warm shallows, often buried in mud, with only its eyes and nostrils exposed. It emerges in April from a winter retreat beneath an overhanging mudbank, under vegetative debris, or inside a muskrat lodge. The snapper eats invertebrates, carrion, aquatic plants, fish, birds, and small mammals. It is an excellent swimmer: Individuals displaced 2 miles have returned to their capture sites within several hours. Snappers strike viciously when lifted from water or teased and can inflict a serious bite. Some consider snapper meat a delicacy, and excellent soups are prepared from it.

325, 326, 327 Alligator Snapping Turtle
(*Macroclemys temmincki*)

Description: 13⅜–26+″ (34–66+ cm). Largest freshwater turtle in the world. Record weight is 219 lbs. (99.5 kg). Massive head with *strongly hooked beak;* very long tail. Carapace brown or gray, serrated, with *3 prominent keels* and an *extra row of small scutes* between marginal and costal scutes. Plastron comparatively small and gray.

Breeding: Mates underwater, February to April.

Lays one clutch, April to June, of 10–52 spherical, 1½" (38 mm) eggs, in a flask-shaped, earthen cavity a short distance from water's edge. Incubation takes 11½–16+ weeks.

Habitat: Deepwater rivers, lakes, oxbows, sloughs; occasionally enters brackish water.

Range: Coastal plain from se. Georgia and Florida panhandle to e. Texas, north to Iowa and Indiana.

Only nesting females are known to leave the water. The Alligator Snapper has a unique pink wormlike structure on its tongue. Resting quietly on the bottom, with mouth agape, it moves this "fishing lure" to attract prey. It also stalks anything it can capture and swallow, including other turtles. One in captivity has lived for more than 60 years.

MUSK AND MUD TURTLE FAMILY
(Kinosternidae)

A New World Family of 4 genera and 23 species. There are 2 genera north of Mexico: *Sternotherus* with 4 species of musk turtles, and *Kinosternon* with 5 species of mud turtles.

Musk and mud turtles can be distinguished by a small, smooth oval-shaped carapace bearing 11 marginal scutes on each side. Rear margin of carapace is not serrated; plastron is single or double hinged with 10 or 11 scutes. Females have short tails; those of males extend well beyond the carapace margin and end in a blunt or spinelike horny nail. The inner surface of males' hind legs have two small patches of tilted scales.

All musk and mud turtles have 2 pairs of musk glands beneath the border of the carapace. The secretions are very offensive, thus the common names "stinkpot," "stinking jim," "musk turtle."

Although occasionally encountered out of water, these turtles are strongly aquatic and are usually seen crawling along the bottom. They sun themselves in shallow water or amid floating vegetation with only the central portion of the shell exposed.

Each year females may lay one or more clutches of elliptical eggs with brittle, porcelainlike shells, translucent pink, bluish-white or banded with stark white.

317 **Striped Mud Turtle**
(*Kinosternon bauri*)

Description: 3–4¾" (7.5–12.2 cm). *Three long light stripes* on smooth, keelless carapace. Stripes may be obscure in older turtles. Scutes often translucent, with underlying bone sutures visible. Broad plastron has 2 well-developed hinges and 11 scutes. Male has spine-tipped tail and rough scale patches on insides of hind legs.

Breeding: Nests September to June. Lays 3 or more clutches yearly of 1–4 eggs, 1⅛" (28 mm) in length, in nest dug in sand or decaying vegetation. Incubation takes 13 to 19 weeks.

Habitat: Cypress swamps, sloughs, ponds, drainage canals, and wet meadows.

Range: S. Georgia south through the Florida Keys.

Our most terrestrial mud turtle; it is often observed crossing roads and visiting puddles after downpours. An opportunistic feeder, its diet includes insects, snails, dead fish, algae, cabbage palm seeds, and fishermen's bait. It is sometimes called the cow-dung cooter, because it forages in manure. The upper shell of turtles from the lower Florida Keys and Gulf hammock region is darkly pigmented, with stripes masked.

313 Yellow Mud Turtle
(*Kinosternon flavescens*)
Subspecies: Yellow, Illinois, Southwestern

Description: 3½–6⅜" (9–16.2 cm). Carapace olive to brown, smooth, keelless, and usually flattened; scutes dark-bordered. *Elevated 9th and 10th marginal scutes* (lacking in young). Plastron yellow to brown, with dark pigment along seams; *double-hinged,* with 11 scutes. Jaw and throat white or yellow, often spotted. Male has concave plastron; long, thick spine-tipped tail; and rough scale patches on insides of hind legs. Juveniles have a dark spot at rear edge of each carapace scute.

Breeding: Nests in June in New Mexico. 1 clutch of 1–6 (usually 4) hard-shelled elliptical eggs. Sexual maturity is reached in 6–7 years.

Habitat: Prefers quiet or slow-moving bodies of freshwater with mud or sandy bottoms.

Range: N. Nebraska south to Texas, e. and s. New Mexico, and se. Arizona into Mexico. Separate populations in nw. Illinois and on Illinois–Iowa border.

Subspecies: Yellow (*K. f. flavescens*), yellow lower jaw, throat, and carapace; n. Nebraska to Texas, New Mexico, and se. Arizona. Illinois (*K. f. spooneri*), carapace brown, skin gray to black, only front of lower jaw and barbels yellow; nw. Illinois and adjacent Iowa and Missouri. Southwestern (*K. f. arizonense*), lower chin and throat yellow, carapace olive; s. Arizona.

At dawn or twilight the Yellow Mud Turtle may be encountered foraging on land. It feeds on worms and arthropods as well as snails and tadpoles. These turtles spend the cooler months under brush piles or leaf litter, in stump holes or muskrat dens, or under water buried in mud. Normally shy, they usually do not attempt to bite.

315 Mexican Mud Turtle
(*Kinosternon hirtipes*)

Description: 3¾–6⅝″ (9.5–17 cm). Smooth oval carapace, olive to brown, with black-bordered seams; characterized by strong *middorsal keel; 10th marginal scute* extends well above other marginals. Plastron tan to brown, with dark-bordered seams; 11 scutes; 2 well-developed hinges.

Breeding: Habits little known. Lays 4–7 elliptical eggs, 1⅛″ (28 mm), in June.

Habitat: Mesquite and grassland ponds, lakes, rivers, and marshes.

Range: Big Bend region of Texas south through the Mexican plateau.

Subspecies: Two; 1 in our range, *K. h. murrayi*.

Primarily nocturnal, it is most active between 9 P.M. and midnight; there are some reports of activity during daylight hours.

314 Sonora Mud Turtle
(*Kinosternon sonoriense*)

Description: 3⅛–6½" (8–16.5 cm). Smooth elongated carapace, sometimes with 1 or 3 low keels. *Tenth marginal scute extends higher* than other marginals. Underside of marginals and bridge yellowish-brown.

Breeding: Nests May–September and lays 2–9 elliptical, 1½" (3 mm) eggs.

Habitat: Springs, water holes, ponds, and creeks. Desert, foothills, or in oak and piñon woodland and Ponderosa pine and Douglas fir forests at elevations up to 6,700' (2,042 m).

Range: Sw. New Mexico, Arizona, and adjacent extreme se. California south into Mexico.

During dry periods, Sonora Mud Turtles sometimes congregate in water holes. They feed on snails. One lived nearly 28 years in captivity.

318, 320, 321 Mud Turtle
(*Kinosternon subrubrum*)
Subspecies: Eastern, Florida, Mississippi

Description: 3–4⅞" (7.6–12.4 cm). Carapace olive to dark brown, patternless, smooth, keelless. *No enlarged marginal scutes.* Plastron yellow to brown, *double-hinged*, with *11 scutes*. Males have well-developed blunt spine at end of tail and rough scale patches on inside of hind legs.

Breeding: Sexually mature at 5–7 years. Breeds mid-March to May; usually nests in

June, but October through June
nestings have occurred. 1–6 elliptical
eggs—hard-shelled, pinkish or bluish-
white, 1″ (25 mm)—are deposited in a
3–5″ (7.6–12.7 cm) cavity dug in
vegetative debris or sandy loam soil.
Several clutches laid annually in
southern populations. Muskrat or
beaver lodges and alligator nests are
occasionally used.

Habitat: Fresh or brackish water. Prefers
shallow, soft-bottomed, slow-moving
water with abundant vegetation. Often
occupies muskrat lodges.

Range: Sw. Connecticut and Long Island south
to s. Florida, west to c. Texas, and
north in the Mississippi Valley to s.
Illinois and sw. Indiana; an isolated
population occurs in nw. Indiana.

Subspecies: Eastern (*K. s. subrubrum*), spotted or
mottled head; sw. Connecticut and
Long Island to Gulf, northwest to s.
Illinois and sw. Indiana, isolated
population in nw. Indiana. Florida (*K.
s. steindachneri*), plain or mottled head;
peninsular Florida. Mississippi (*K. s.
hippocrepis*), two light lines on side of
head; se. Missouri, Arkansas, e.
Oklahoma, e. Texas, Louisiana, w. and
s. Mississippi, s. Alabama, and extreme
w. Florida.

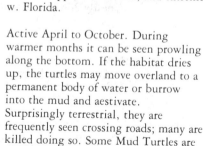

Active April to October. During
warmer months it can be seen prowling
along the bottom. If the habitat dries
up, the turtles may move overland to a
permanent body of water or burrow
into the mud and aestivate.
Surprisingly terrestrial, they are
frequently seen crossing roads; many are
killed doing so. Some Mud Turtles are
mild-tempered, while others are feisty
and do not hesitate to bite.

310 Razor-backed Musk Turtle
(*Sternotherus carinatus*)

Description: 4–5⅞" (10–14.9 cm). *Steeply sloped*
carapace with *prominent vertebral keel.*
Scutes light brown to orangish,
accented with small dark spots or
radiating streaks and dark borders.
Pattern may be lost in older turtles.
Plastron yellow; *lacks a gular scute;* has
only 10 scutes. Single hinge barely
discernible between pectoral and
abdominal scutes. Snout somewhat
tubular. Barbels on *chin only.*

Breeding: Reproductive cycle is poorly known.
Probably lays 2 clutches a season.

Habitat: Slow-moving streams and rivers with
soft bottoms and abundant aquatic
vegetation; swamplands.

Range: S. Mississippi west to Texas.

Quite shy, unlike other musk turtles; it
rarely bites or expels musk. It is active
from March to November and basks
frequently. Population density may
exceed 100 turtles an acre in good
habitats; in such locales bait fishermen
may catch more turtles than fish.

316 Flattened Musk Turtle
(*Sternotherus depressus*)

Description: 3–4½" (7.6–11.4 cm). *Carapace
extremely flattened.* Head and neck
greenish, covered with dark, netlike
pattern. *Barbels on chin only.*

Breeding: Little is known.

Habitat: Rock-bottomed streams and their
impoundments.

Range: Black Warrior River system in nw.
Alabama.

A shy little turtle, it is most active
during early morning. It interbreeds
with the Stripe-necked in parts of its
range and some consider it a subspecies
of the Stripe-necked.

311, 312 Loggerhead Musk Turtle
(*Sternotherus minor*)
Subspecies: Loggerhead, Stripe-necked

Description: 3⅛–5¼" (8–13.5 cm). Carapace keeled (prominent in juveniles), with overlapping vertebral scutes; brown or orange, with dark-bordered seams; may be patterned with dark spots or radiating streaks. Plastron small, pink or yellow, *with single indistinct hinge* and single gular scute. Barbels on chin only. Head has dark spots or stripes. Male has enlarged, spine-tipped tail; tip of female's tail barely reaches edge of carapace.

Breeding: Lays 1–4 clutches of 2–3 elliptical, 1⅛" (29 mm) eggs. 17 Loggerhead nests have been found in one 5' high pile of sand along a spring run. Brittle porcelainlike shells are translucent pink when deposited, turn opaque white as embryo development proceeds. Incubation takes 13–16 weeks.

Habitat: Large spring runs, creeks, rivers, oxbows, swamps, and sinkhole ponds.

Range: C. Georgia to c. Florida and panhandle, west to e. Mississippi and extreme e. Louisiana, north through e. Tennessee.

Subspecies: Loggerhead (*S. m. minor*), 3 dorsal keels, numerous spots on head; c. Georgia and se. Alabama to c. Florida. Stripe-necked (*S. m. peltifer*), 1 dorsal keel, stripes on neck; sw. Virginia and e. Tennessee south to Gulf and Pearl River, Mississippi and Louisiana.

Highly aquatic, and frequently observed crawling along bottom among rocks, submerged logs, and vegetation. Juveniles feed primarily on insects, adults on snails and clams, which they crush with the wide surfaces of their jaws. Loggerheads are as pugnacious as Stinkpots; the young are capable of expelling musk even before they hatch.

319 Stinkpot
(*Sternotherus odoratus*)

Description: 3–5⅜″ (7.6–13.7 cm). A feisty little
turtle with *2 light stripes on head, barbels
on chin and throat.* Carapace smooth or
with 3 keels, unserrated, highly
domed, and elongated; olive-brown to
dark gray and often obscured by a layer
of algae. Juveniles have keeled and
patterned carapace with irregular dark
streaks or spots. *Plastron small, with 11
scutes* and a single inconspicuous hinge.
Male's tail ends in a blunt horny nail;
inner surface of hind legs bears 2 small
patches of tilted scales. Female's tail
very short, may end in a sharp horny
tip.

Breeding: Nests February to June, depending on
latitude. Mates underwater. Lays 1–9
eggs—off-white with stark white band,
thick-shelled, elliptical, 1⅛″ (27 mm)
—in shallow nest under rotting stump
or in wall of muskrat lodge. Incubation
takes 9 to 12 weeks.

Habitat: Freshwater; prefers quiet or slow-
moving shallow, muddy-bottomed
waters.

Range: S. Ontario and coastal Maine to
Florida, west to c. Texas, north to s.
Wisconsin.

Also called Musk Turtle or Stinking
Jim. When disturbed, it secretes a foul-
smelling, yellowish fluid from 2 pairs
of musk glands under the border of the
carapace. Males are aggressive and bite
readily. Stinkpot's long neck can bring
its jaws as far back as its hind limbs. In
early spring it likes to bask in shallows
or amid floating vegetation with the
center of its carapace exposed to the
sun. Highly aquatic, Stinkpots rarely
leave the water, but they will
occasionally climb trees to bask.
Stinkpots annoy fishermen by grabbing
bait. One lived in captivity for almost
55 years.

POND, MARSH, AND BOX TURTLE FAMILY
(Emydidae)

Largest of living turtle families, with about 30 genera and 82 species worldwide; 7 genera with 26 species occur within our range—the majority in the eastern United States.

Emydids north of Mexico are small- to medium-sized turtles with horny scute-covered shells, 12 marginal scutes on each side of carapace, and 6 pairs of scutes on plastron. Plastron and bridge are well-developed. Unlike tortoises, hind feet are flattened and elongated with some webbing. With the exception of the high-domed *Terrapene*, emydid shells are low in profile. Blanding's and box turtles have a single plastral hinge.

Most emydids are semiaquatic; *Graptemys* and *Malaclemys* the most highly aquatic, *Clemmys* and *Terrapene* the most terrestrial. Diet is varied; while generally omnivorous, some species are carnivorous as juveniles and essentially herbivorous as adults. Basking behavior is well-developed. The courtship pattern of emydids is often elaborate. Male *Chrysemys* may face female and stroke her head and neck with his elongated foreclaws. Egg and clutch size vary with species, subspecies, size of female, and latitude. From 2–25 elliptical eggs are typically deposited in a flask-shaped nest. Northern forms usually nest once a season; southern 2, 3, or more times a season.

292 Alabama Red-bellied Turtle
(*Chrysemys alabamensis*)

Description: 8–13¼" (20.3–33.5 cm). *Prominent notch at tip of upper jaw flanked by toothlike cusps;* arrow-shaped stripe atop head (shaft between eyes, point at snout). Carapace olive-black, highly arched, with longitudinal wrinkles; vertebral scutes convex; light-centered dark rings on undersides of marginal scutes. *Plastron reddish;* elaborate pattern of light-centered marks spreads along scute seams in young, fades with age. Male has elongated straight claws on front feet.

Breeding: 3–6 eggs laid May to July.

Habitat: Prefers fresh to slightly brackish waterways with dense aquatic vegetation.

Range: Lower portion of Mobile Bay drainage in Alabama.

Alabama Red-bellies may bask with River Cooters, Southern Black-knobbed Sawbacks, and Pond Slider.

287 River Cooter
(*Chrysemys concinna*)
Subspecies: Eastern, Texas

Description: 5¾–16⅜″ (14.6–41.6 cm). *Light C-shaped figure on 2nd costal scute.* Carapace brown with cream-yellow markings, often in the form of concentric circles. Underside of each marginal scute has dark doughnut-shaped mark. *Plastron yellow with a dark pattern (fades with age) that follows scute seams.* An X-shaped design may be present on front lobe of plastron. Male has elongated claws on front feet.

Breeding: Nests late May to July. Lays up to 19 oval, pinkish-white eggs, 1½″ (38 mm) long. Hatchlings emerge August to September, may overwinter in the nest.

Habitat: Streams and rivers with moderate currents; large lakes, spring runs, and occasionally brackish tidal marshes.

Range: Coastal plain from se. Virginia to Georgia, southeast into Florida, west into Texas and New Mexico, and north in the Mississippi Valley to s. Illinois.

Subspecies: Eastern (*C. c. concinna*), see species description; se. Virginia south to wc. Florida, west to e. Texas. Texas (*C. c. texana*), pattern on plastron reduced to narrow lines that follow scute seams, upper jaw notched with cusp on each side; c. Texas and e. New Mexico south into Mexico.

A gregarious basker, it is often seen sunning with other turtle species. But it is quick to dive into the water when disturbed. The meat is favored locally in the South. One captive River Cooter lived more than 40 years.

288 Cooter
(*Chrysemys floridana*)
Subspecies: Florida, Peninsula, Missouri Slider

Description: 7½–15⅞" (19.1–40.3 cm). *Wide vertical stripe, often forked at one or both ends, on 2nd costal scute.* Carapace brownish with yellow markings; marginals have central bars on upper surface, dark doughnut-shaped marks on lower surface. *Plastron yellow, patternless.* Male has elongated claws on front feet.

Breeding: Nests late May to July; earlier in Florida. Lays 2 or more clutches of 4–22 elliptical eggs, 1⅜" (35 mm) long. Females sometimes dig one or more nest cavities several inches away from the main nest chamber and lay a single egg in each. Most hatchlings are seen in spring. Incubation takes 80–150 days. Males mature in 3 years, females in 6–7.

Habitat: Large ponds, lakes, spring runs, canals, and sluggish rivers with abundant aquatic vegetation and basking sites.

Range: Coastal plain from Virginia to e. Texas, north in Mississippi Valley to s. Illinois and e. Oklahoma.

Subspecies: Florida (*C. f. floridana*), with numerous

light head stripes; Atlantic coastal plain from Virginia to n. Florida, west to c. Alabama. Peninsula (*C. f. peninsularis*), with pair of "hairpin" markings on top of head; Florida peninsula. Missouri Slider (*C. f. hoyi*), with numerous broken and twisted head stripes, and notch at front of upper jaw; s. Illinois and Mississippi west to se. Kansas and e. Texas.

A gregarious basker, it is often seen with River Cooters and Florida Red-bellied Turtles. Sometimes the Cooter is encountered wandering on land.

---

296 Florida Red-bellied Turtle
(*Chrysemys nelsoni*)

Description: 8–13⅜" (20.3–34 cm). *Prominent notch at tip of upper jaw flanked by toothlike cusps; arrow-shaped stripe atop head* (shaft between eyes, point at snout). Carapace blackish, highly arched; vertebral scutes convex; red bar on each marginal scute, with dark blotch on underside. *Plastron reddish;* patternless or with dark semicircles along scute seams that fade with age. Male has elongated, slightly curved claws on front feet.

Breeding: Nests early June through August, but most clutches are laid mid-June to mid-July.

Habitat: Ponds, lakes, sloughs, marshes, and mangrove-bordered creeks.

Range: Florida peninsula and Apalachicola area of panhandle.

Active year-round, the Florida Red-belly is often seen basking with Florida Cooters and River Cooters on logs or floating mats of vegetation. Because of its thick shell, it can bask for long periods. Adults prefer a diet of aquatic plants.

293, 294, 297   Painted Turtle
(*Chrysemys picta*)
Subspecies: Eastern, Midland,
Southern, Western

Description: 4–9⅞" (10.2–25.1 cm). Carapace oli
or black; oval, smooth, flattened, and
unkeeled; scute seams bordered with
olive, yellow, or red. *Red bars or crescen
on marginal scutes.* Plastron yellow,
unpatterned or intricately marked.
Yellow and red stripes on neck, legs,
and tail. Notched upper jaw.

Breeding: Nests May to July. In north lays 1–2
clutches a year, in south 2–4, of 2–20
elliptical eggs, 1¼" (32 mm) long.
Flask-shaped nest cavity is 4" (10 cm)
deep. Hatchlings in north may
overwinter in the nest. Incubation
averages 10–11 weeks. Males reach
maturity in 2–5 years; females in 4–8.

Habitat: Slow-moving shallow streams, rivers,
and lakes. Likes soft bottoms with
vegetation and half-submerged logs.

Range: British Columbia to Nova Scotia, south
to Georgia, west to Louisiana, north to
Oklahoma, and northwest to Oregon.
Isolated populations in the Southwest.

Subspecies: Eastern (*C. p. picta*), vertebral and
costal scute seams aligned, plastron
yellow, not patterned; se. Canada
through New England and the Atlantic
coastal states to n. Georgia and e.
Alabama. Midland (*C. p. marginata*),
vertebral and costal scutes not aligned,
plastron yellow with dark blotch in
center; s. Quebec and s. Ontario to
Tennessee. Southern (*C. p. dorsalis*), red
or yellow stripe down carapace, plastron
yellow, not patterned; s. Illinois to
Gulf, se. Oklahoma to c. Alabama.
Western (*C. p. belli*), largest ssp., with
light netlike lines on carapace, bars on
marginals, and intricate branching
pattern on plastron; sw. Ontario south
to Missouri and west to Oregon and
British Columbia, isolated populations
in the Southwest. Specimens from areas
where ranges of subspecies overlap

display an intergradation of characteristics.

The most widespread turtle in North America. It is fond of basking and often dozens can be observed on a single log. Young turtles are basically carnivorous, but become herbivorous as they mature.

295 Red-bellied Turtle
(*Chrysemys rubriventris*)
Subspecies: Red-bellied, Plymouth

Description: 10–15¾" (25.4–40 cm). Carapace brown to black with flattened or slightly concave vertebral scutes; red bar on each marginal scute. *Prominent notch at tip of upper jaw flanked by toothlike cusps; arrow-shaped stripe runs atop head,* between eyes, to snout. *Plastron reddish;* dark markings along scute seams fade with age. Male has elongated, straight claws on front feet.

Breeding: Nests June to July. Lays 8–20 elliptical eggs, 1⅜" (35 mm) long, in 4" (10.2 cm) cavity. Hatchlings emerge in 10–15 weeks, may overwinter in nest.

Habitat: Deep ponds, lakes, streams and rivers and brackish marshes. In New England, restricted to deep ponds and lakes.

Range: Mid-Atlantic coastal plain from s. New Jersey to ne. North Carolina, and west in the Potomac River. Isolated population in se. Massachusetts.

Subspecies: Red-bellied (*C. r. rubriventris*), s. New Jersey and e. West Virginia to ne. North Carolina. Plymouth (*C. r. bangsi*), Plymouth County and Naushon Island, Massachusetts.

Shy and difficult to approach, Red-bellies prefer basking sites near deep water. They feed on snails, crayfish, tadpoles, and aquatic vegetation.

Draining of wetlands and overcollection
threaten their survival in many areas.

286, 289 Pond Slider
(*Chrysemys scripta*)
Subspecies: Yellow-bellied, Red-eared,
Cumberland, Big Bend

Description: 5–11⅜" (12.7–28.9 cm). The "dime
store" turtle. *Prominent yellow, orange, or
red blotch or stripe behind eyes.* Carapace
oval, weakly keeled, olive to brown,
with pattern ranging from yellow bars
and stripes to reticulations and eyelike
spots. Plastron yellow, plain to
intricately patterned. *Undersurface of
chin rounded. V-shaped notch at front of
upper jaw not flanked by cusps.* With age,
pattern and head blotch may become
masked by black pigment, making
identification difficult.

Breeding: Mates March to June. Nests June to
July. Lays 1–3 clutches of 4–23 oval
eggs, 1⅜" (37 mm) long, in nest cavity
1–4" (2.5–10.2 cm) deep, which may
be located some distance from water.
Hatchlings emerge in 2–2½ months,
but often overwinter in nest. Males
mature in 2–5 years.

Habitat: Sluggish rivers, shallow streams,
swamps, ponds, and lakes with soft
bottoms and dense vegetation.

Range: Se. Virginia to n. Florida west to New
Mexico, south to Brazil.

Subspecies: Yellow-bellied (*C. s. scripta*), with
conspicuous vertical yellow blotch
behind eye, vertical yellowish bar on
each costal scute, and dark round
smudges on forward part of plastron;
se. Virginia to n. Florida. Red-eared
(*C. s. elegans*), with wide red stripe
behind eye, dark smudge on each
plastron scute; Mississippi Valley from
n. Illinois to Gulf. Cumberland (*C. s.
troosti*), upper portions of Cumberland
and Tennessee River valleys from se.
Kentucky and sw. Virginia to ne.

Alabama. Big Bend (*C. s. gaigeae*), with large black-bordered orange spot on side of head, small orange spot behind eye, carapace with netlike pattern; Big Bend region of Texas and adjacent Mexico, also Rio Grande Valley in sc. New Mexico.

Fond of basking, Pond Sliders are often seen stacked one upon another on a favorite log. The young eat water insects, crustaceans, molluscs, and tadpoles, then turn to a plant diet as they mature. Millions have been raised on turtle farms and sold as pets. Few have survived to adulthood.

290 Spotted Turtle
(*Clemmys guttata*)

Description: 3½–5″ (8–12.7 cm). A small, attractive turtle. Carapace black, keelless, unserrated; usually *sprinkled with round yellow spots*. Spotting on head, neck, and limbs. Plastron creamy yellow with large black blotches along border. Male has brown eyes, tan chin, long thick tail; female has orange eyes and yellow chin.

Breeding: Courts March to May. In June, female digs shallow flask-shaped nest in sun-drenched areas and deposits up to 8 (typically 3–5) flexible-shelled, elliptical, 1¼″ (33 mm) eggs. Hatchlings emerge in late August and September or overwinter in the nest until the following spring.

Habitat: Marshy meadows, wet woodlands, boggy areas, beaver ponds, and shallow, muddy-bottomed streams.

Range: Southern Maine south along the Atlantic coastal plain to n. Florida, west through Maryland, Pennsylvania, and s. New York into n. Ohio and Indiana, extreme ne. Illinois and adjacent se. Wisconsin, s. Michigan, and Ontario.

Frequently seen basking in the cooler spring months, it is difficult to find during the summer, when dense vegetation obscures its movements. It is often found in association with Painted, Wood, and Bog turtles. It winters underwater in soft mud, accumulated debris, or muskrat burrows.

302 Wood Turtle
(*Clemmys insculpta*)

Description: 5–9″ (12.7–23 cm). Formed by concentric growth ridges, *each large carapace scute looks like an irregular pyramid.* Upper shell brown and keeled, appears sculptured and rough. Plastron yellow, with black blotches usually present along outer margins of scutes; hingeless. *Skin of neck and forelegs often reddish orange.* Male has concave plastron and thick tail, with anal opening beyond margin of carapace.

Breeding: One clutch of 6–8 (maximum 18) elliptical, flexible-shelled, 1⅝″ (41 mm) eggs; deposited May to June, hatch September to October. In north hatchlings may overwinter in the nest.

Habitat: Cool streams in diciduous woodlands, red maple swamps, marshy meadows, and farm country.

Range: Nova Scotia south to n. Virginia and discontinuously west through s. Quebec and the Great Lakes region to e. Minnesota and ne. Iowa.

"Ole redlegs" is reputedly an intelligent turtle. An excellent climber, it can surmount 6-foot (1.8 m) chain-link fences. After spring downpours, it is often seen searching for worms in freshly plowed fields. It also likes slugs, insects, tadpoles, and wild fruits. The Wood Turtle was once taken for food and now suffers from overcollection and habitat loss. It is currently protected in

some states. One lived 58 years in
captivity.

303 Western Pond Turtle
(*Clemmys marmorata*)
Subspecies: Northwestern,
Southwestern

Description: 3½–7″ (8.9–17.8 cm). Smooth, broad,
low carapace is olive to dark brown;
often marked with *network of dark flecks
and lines radiating from center of scutes.*
Plastron pale yellow, hingeless; may
have dark-brown or black blotches
along scute margins. Male plastron
concave.

Breeding: Lays 1 clutch of 3–11 oval, 1½″ (38
mm), hard-shelled eggs, April to
August (depending on latitude). Makes
an earthen chamber in a sunny spot
near water's edge or some distance away
in field. Hatchlings emerge in about 12
weeks.

Habitat: Ponds and small lakes with abundant
vegetation. Also seen in marshes, slow-
moving streams, reservoirs, and
occasionaly in brackish water.

Range: West of Cascade–Sierra Nevada crest
from extreme sw. British Columbia
south to Baja California. Isolated
population in Carson and Truckee rivers
in extreme w. Nevada.

Subspecies: Northwestern (*C. m. marmorata*), well-
developed triangular inguinal scutes on
bridge; British Columbia south to San
Francisco Bay, w. Nevada.
Southwestern (*C. m. pallida*), inguinal
scutes small or absent; San Francisco
Bay south into nw. Baja California.

It is often observed basking alone.
When disturbed, it will quickly dive
into water. One turtle may challenge
another for a favored basking site by
extending its neck, opening its mouth,
and exposing its yellow-edged jaws and
reddish interior.

301 Bog Turtle
(Clemmys muhlenbergi)

Description: 3–4½" (7.6–11.4 cm). A small brown turtle with *conspicuous yellow, orange, or reddish blotch* on each side of head. Carapace light brown to mahogany (a light-brown or orange sunburst pattern may be present on large scutes), weakly keeled, and rough or smooth depending on age. Plastron brownish-black with varying amounts of yellow along midline; hingeless, with 12 scutes. Male has concave plastron and thick tail, with anal opening beyond margin of carapace.

Breeding: Reaches sexual maturity in 5–7 years. Mates during first warm days of spring; nests in June. Lays single clutch of 1–6 (typically 3–4) elliptical, flexible-shelled eggs, averaging 1⅛" (29 mm) in length, in a 2" (51 mm) nest cavity. Hatchlings emerge in August and September after brief incubation of 6½ to 9 weeks, or may overwinter in the nest in northern localities.

Habitat: Sunlit marshy meadows, spring seepages, wet cow pastures, and bogs. Prefers narrow, shallow, slow-moving rivulets.

Range: E. New York and adjacent Massachusetts and Connecticut, south through New Jersey and parts of Pennsylvania, Delaware, and Maryland. Other populations in Finger Lakes region (New York) and parts of Pennsylvania, Virginia, and North Carolina.

Until recently, the Bog Turtle, or Muhlenberg's Turtle, was thought to be an endangered species, but it is now known to be more secretive than rare. Typically active from April to mid-October, the Bog Turtle searches out a wide variety of prey, including tadpoles, slugs, snails, worms, and insects. In spring it often basks in full sunlight atop grassy tussocks.

During hot periods it buries itself in
mud or vegetative debris, exposing only
a small portion of its shell to the sun.
Winter is spent buried deep in mud
flooded by subterranean waters. It is
now protected in most states where it is
found.

285 Chicken Turtle
(*Deirochelys reticularia*)
Subspecies: Eastern, Florida, Western

Description: 4–10" (10.2–25.4 cm). Easily
identified by *extraordinarily long, striped
neck, almost as long as shell.* Carapace
finely wrinkled, has *netlike pattern.
Vertical light stripes on rump; wide stripe
on each foreleg.* Male smaller than female
and has long thick tail.

Breeding: Nesting season varies with latitude;
March in South Carolina, throughout
the year in Florida. Lays several
clutches each year of 5–15
elliptical eggs, 1⅜" (37 mm) long, in
cavity about 4" (10.2 cm) deep.
Males mature in 2–4 years or at 4"
(10.2 cm) long; females 6–8 years at
7" (18 cm).

Habitat: Shallow ponds and lakes with dense
vegetation; also ditches and cypress
swamps.

Range: The coastal plain from se. Virginia to
Florida, west to e. Texas, north to
Oklahoma and Missouri.

Subspecies: Eastern (*D. r. reticularia*), carapace with
narrow greenish or brownish netlike
lines, and narrow yellow rim; coastal
plain from se. Virginia to Mississippi
River. Florida (*D. r. chrysea*), carapace
with bold orange or gold netlike lines,
and wide orange rim, plastron orange
or bright yellow; Florida peninsula.
Western (*D. r. miaria*), carapace with
broad but faint netlike lines, plastron
with dark seam; west of Mississippi
River from se. Oklahoma to Louisiana
and e. Texas.

It is often observed basking or encountered wandering on land. Some are struck down crossing highways. It is generally shy, but most bite. Chicken Turtle meat was once favored in southern markets.

291 Blanding's Turtle
(*Emydoidea blandingi*)

Description: 5–10½″ (12.7–26.8 cm). *Bright-yellow chin and throat* quickly identify basking specimens. Smooth helmet-shaped carapace is black with profusion of irregularly shaped radiating spots and vermiculations. *Plastron hinged;* yellow with large black blotches symmetrically arranged. *Neck long;* head flat with protruding eyes. Male has slightly concave plastron.

Breeding: Nests June to July. Lays about 8 oval, dull-white, 1½″ (38 mm) eggs. Hatchlings appear August to September.

Habitat: Lake shallows, ponds, marshes, and creeks with soft bottoms and dense aquatic vegetation.

Range: In the Great Lakes region and west to Nebraska. Separate population in s. New Hampshire, Massachusetts, and e. New York.

This turtle is very tolerant of cold temperatures. It is primarily aquatic but frequently comes on land to bask or search for insects and snails. It is shy, and when disturbed, abandons its basking spot atop a muskrat lodge or log.

283 Barbour's Map Turtle
(*Graptemys barbouri*)

Description: Males 3½–5″ (8.9–12.7 cm); females, 7–12¾″ (17.8–32.4 cm). Carapace olive to dark brown, with *black spinelike projections on keel* of juveniles and males; costal and marginal scutes have yellow oval markings. Plastron cream to greenish-yellow, with black-bordered scute seams; spines present on pectoral and abdominal scutes. *Large yellow or greenish-yellow blotches behind eyes* and patch atop snout. *Curved or transverse light bar under chin.* Adult females develop greatly enlarged head.

Breeding: Lays several clutches of 6–11 thick-shelled ellipsoidal eggs, about 1½" (40 mm) long, in nest cavities 3–6" (7.6–15 cm) deep near water's edge on sandbars. Hatchlings emerge mid-August through September.

Habitat: Streams and rivers with numerous stumps and logjams and an abundance of molluscs.

Range: Apalachicola River drainage of sw. Georgia, se. Alabama, and Florida panhandle.

Barbour's Map Turtle is vanishing. The species has been heavily exploited by collectors. Males and juveniles eat insects, small snails, and crayfish; the female's diet is mussels and large snails.

279 Cagle's Map Turtle
(Graptemys caglei)

Description: Males, 2½–3⅝" (6.4–9.2 cm); females larger, to at least 6⅜" (16.2 cm). May be distinguished from the Texas Map Turtle by green carapace with brown or black-tipped keel spines; cream-colored transverse chin bar; and light V-shaped mark atop head, the arms of which extend to form a crescent behind each eye. Adult males often have black flecks on plastron.

Breeding: Season is apparently late spring or early summer, as hatchlings have been found from September to November.

Habitat: Pools, impoundments, and slow-

moving stretches of rivers and
tributaries with exposed rocks, cypress
knees, and logs.

Range: Guadalupe–San Antonio river system of
Texas.

A recently described species, its natural
history is little known. It shuns
basking sites along the riverbank. It
eats caddis fly larvae and other small
aquatic invertebrates.

276 Yellow-blotched Sawback
(*Graptemys flavimaculata*)

Description: Males 3–4″ (7.6–10.2 cm); females 4–
6⅞″ (10.2–17.5 cm). Carapace olive to
brown with *conspicuous black spiny
projections on keel;* large scutes with
yellow blotch. Plastron cream-colored
with black pattern along scute seams.
Head markings similar to those of
Ringed Sawback. Male has elongated
claws on front feet.

Breeding: Nesting time uncertain. Males reach
sexual maturity in 3–4 years, females
later. Courtship pattern is similar to
that of the Painted Turtle.

Habitat: Slow-moving rivers with sand and rock
bottoms. Prefers sections with flood-
stranded debris and exposed tree root
tangles.

Range: Pascagoula River system of Mississippi.

This is the dominant turtle species in
the Pascagoula River. The habitat is
shared with the Alabama Map Turtle,
the Red-eared Turtle, the River Cooter;
and the Razor-backed Musk Turtle.
Favorite foods are insects and snails.

280 Map Turtle
(*Graptemys geographica*)

Description: Males 4–6¼" (10.2–15.9 cm); females, 7–10¾" (17.8–27.3 cm). Carapace greenish to olive-brown, with reticulated pattern of thin yellow-orange lines (obscure in adult females); somewhat flattened and with a low keel (with small spines in juveniles). Plastron yellowish; patternless in adults, with black-bordered scute seams in juveniles. Skin greenish with narrow yellow stripes; an *isolated yellow spot, often triangular in shape, found behind the eye.* Adult females have enlarged heads.

Breeding: Nests May to mid-July. Southern females lay 2 or more clutches a season (northern females, 1) of 12–14 ellipsoidal, 1¼" (33 mm) eggs. Hatchlings emerge mid-August through September or late May or June of following year.

Habitat: Slow-moving rivers and lakes with mud bottoms, abundant aquatic vegetation, and logjams.

Range: Lake George and Lake Champlain through St. Lawrence and Great Lakes drainage, south to Tennessee and Alabama; also Arkansas and Missouri river drainages. Isolated populations in Delaware River and Susquehanna River drainage.

Where common, this gregarious species can be observed stacked one upon another on a favorite basking log. Shy like other map turtles, they quickly slide into the water when disturbed. The female's large crushing jaws can break open freshwater clams and large snails. Males and juveniles eat insects, crayfish, and smaller molluscs.

278 Mississippi Map Turtle
(*Graptemys kohni*)

Description: Males, 3½–5″ (8.9–12.7 cm); females,
6–10″ (15.2–25.4 cm). Carapace olive
to brown, with dark-brown keel and
interconnected pattern of circular
markings. Plastron greenish-yellow
with highly variable and intricate
pattern of dark lines. *Yellow crescent-
shaped mark behind eyes prevents neck stripes
from reaching eye.* Round spot on chin.
Conspicuous white eye with black
pupil. Male has elongated claws on
front feet.

Breeding: Habits poorly known. Nests in early
June in Mississippi.

Habitat: Rivers, lakes, and sloughs with mud
bottoms and abundant aquatic
vegetation and basking sites.

Range: W. Mississippi River basin from sw.
Iowa to c. Illinois, south to the Gulf.

Although formerly sold as pets, they
did not fare well in captivity. They are
shy and difficult to approach when
basking. They feed on insects,
freshwater clams, snails, and aquatic
plants. Recently recognized as a
subspecies of False Map Turtle.

281 Black-knobbed Sawback
(*Graptemys nigrinoda*)
Subspecies: Black-knobbed, Southern

Description: Males, 3–4″ (7.6–10.2 cm); females,
4–7½″ (10.2–19.1 cm). Carapace dark
olive-brown; *black-bordered, narrow-
margined, circular or semicircular yellow or
orange mark on each costal and marginal
scute.* Plastron yellow, often with red
tint; variable black branching
pattern.

Breeding: Probably nests in July. Lays about 6
elliptical eggs, 1½″ (38 mm) long, in
nest 4″ (10 cm) deep in sandbank.

Habitat: Sand- and clay-bottomed streams and

rivers with moderate currents and many basking sites.

Range: Below the fall line in rivers of the Mobile Bay drainage, Alabama and Mississippi.

Subspecies: Black-knobbed (*G. n. nigrinoda*), plastron pattern covers a third or less of shell, broad yellow stripes on skin; Coosa, Tallapoosa, Cahaba, and Alabama rivers. Southern (*G. n. delticola*), plastron pattern covers two-thirds or more of shell, skin stripes narrow; Mobile and Tensaw rivers and interconnecting streams and lakes.

Wary and hard to approach. It sleeps on logs and brush piles, and eats insects.

275 Ringed Sawback
(*Graptemys oculifera*)

Description: Males, 3–4" (7.6–10.2 cm); females, 5–8½" (12.7–21.6 cm). Carapace dark olive-green with conspicuous black spinelike projections on keel. *Costal scutes marked with a single black-bordered, yellow or orange circle;* marginals bear wide bar or semicircular markings. Plastron yellow-orange with olive pattern following scute seams. *Large, variably shaped yellow spot behind each eye; 2 broad yellow neck stripes extend behind eye.* Male has elongated claws on front feet.

Breeding: Habits poorly known. Females nest in early June on sandbars. A second clutch probably is deposited later in the season.

Habitat: Narrow-channeled, fast-moving sand- and clay-bottomed rivers with abundant brush snags and logjams.

Range: Pearl River system of Mississippi and Louisiana.

A powerful swimmer, it can move against a rapid current. It often shares basking sites with the Alabama Map Turtle. It feeds on aquatic insects and plants.

274, 277 False Map Turtle
(*Graptemys pseudogeographica*)
Subspecies: False, Ouachita, Sabine

Description: Males, 3½–5¾" (8.9–14.6 cm); females, 5–10¾" (12.7–27.3 cm). Carapace brown with light-yellow oval markings and dark blotches; distinct keel with blunt black spines (reduced in adults). Plastron cream to yellow; unpatterned in adults, intricately patterned in juveniles. *Short yellow bar or crescent-shaped mark behind eyes; neck stripes pass below mark and reach eye.* Male has elongated claws on front feet.

Breeding: Courting male approaches female from the rear, swims above her, turns to face her, and drums his claws against her snout. Nests May to July. Lays 1–3 clutches of 6–13 elliptical, leathery, soft-shelled eggs, about 1½" (38 mm) long. Hatchlings appear in early fall.

Habitat: Large rivers, lakes, ponds, and reservoirs, preferably with abundant aquatic vegetation.

Range: Mississippi, Missouri, and Ohio river drainages, south to Louisiana; Sabine River drainage of Texas and Louisiana.

Subspecies: False (*G. p. pseudogeographica*), yellow mark behind eye, 4–7 neck stripes reaching eye, no large spots on jaws; Missouri River Valley in North Dakota, South Dakota, Nebraska and Iowa, and upper Mississippi River. Ouachita (*G. p. ouachitensis*), recently recognized as a full species; rectangular yellow mark behind eye, 1–3 neck stripes reaching eye, large spot below eye, another spot on lower jaw; Nebraska, Iowa, Illinois, Indiana, Ohio, and West Virginia south to Ouachita River system of n. Louisiana. Sabine (*G. p. sabinensis*), oval or elongated yellow spot behind eye, 5–9 neck stripes reaching eye, transverse bars under the chin; Sabine River system of Louisiana and Texas.

Once common, its numbers have been reduced by water pollution. A gregarious basker, it feeds on aquatic plants, insects, crustaceans, and molluscs. Occasionally eaten by man.

284 Alabama Map Turtle
(*Graptemys pulchra*)

Description: Males, 3½–5″ (8.9–12.7 cm); females, 7–11½″ (17.8–29.2 cm). Carapace olive to dull green, with a black stripe down middle of vertebral scutes and *spinelike projections* on keel (except large females). Costal scutes bear reticulated pattern of black-bordered yellow-orange lines, while *marginals have light bars or C-shaped markings.* Plastron yellow with black-bordered scute seams; no spines on pectoral and abdominal scutes. Large yellow or greenish-yellow blotches behind eyes and patch atop head; blotches may be connected. Long stripes or bars from point of chin down throat. Adult females are much larger than males and develop greatly enlarged heads.

Breeding: Nesting begins in late April, peaks in June, and continues through August. Lays up to 7 clutches, averaging 6–8, 1½″ (38 mm) elliptical eggs in a 6″ (15 cm) deep, flask-shaped cavity dug on sandbars or sandy banks. Incubation takes 10–11 weeks. Males mature in 4 years; females in 14.

Habitat: Medium-sized creeks to large rivers with deep pools, sandbars, snags, and logjams and an abundance of molluscs.

Range: Rivers draining into the Gulf of Mexico from the Yellow River system of Alabama and extreme w. Florida to the Pearl River system of e. Louisiana.

Basking turtles are extremely wary and difficult to approach. At night, they have been observed sleeping near their basking sites, clinging to submerged branches several inches below the water's surface.

---

282 Texas Map Turtle
(Graptemys versa)

Description: Males, 2¾–3½" (7–8.9 cm); females, 4–5" (10.2–12.7 cm). Smallest map turtle. Carapace olive with netlike pattern of yellow lines; scutes convex; keel has low, blunt spines, often dark-tipped. Plastron yellow with thin black lines following scute seams. *Light yellow-orange stripe, often J-shaped, extends backward from each eye.* Yellow or orange, black-bordered longitudinal blotches on chin.

Breeding: Habits unknown.

Habitat: Colorado River system, Texas.

Range: Colorado River, Texas.

Although described in 1925, its natural history is poorly known. Apparently it is an opportunistic feeder, as captives eat an assortment of plant and animal matter.

298, 299, 300 Diamondback Terrapin
(Malaclemys terrapin)
Subspecies: Northern, Carolina, Florida East Coast, Mangrove, Ornate, Mississippi, Texas

Description: Males, 4–5½" (10.2–13.8 cm); females, 6–9⅜" (15.2–23.8 cm). Carapace keeled; light brown or gray to black; scutes bear *deep growth rings,* giving sculptured appearance. Plastron oblong; yellowish or greenish, with dark flecks or blotches; not hinged. Head and neck gray, peppered with black. Eyes black and prominent; jaws light-colored.

Breeding: Nests April to May in south, later in north. Lays 4–18 (average 9) pinkish-white, thin-shelled, leathery eggs, 1¼" (32 mm) long and blunt-ended. Nest cavities 4–8" (10–20 cm) deep are dug at sandy edges of marshes and dunes. Incubation takes 9 to 15 weeks. Females mature in about 7 years; males earlier.

Habitat: Salt-marsh estuaries, tidal flats, and lagoons behind barrier beaches.

Range: Cape Cod to Texas along Atlantic and Gulf coasts.

Subspecies:  Northern (*M. t. terrapin*), carapace wedge-shaped when viewed from above; Cape Cod to Cape Hatteras region of North Carolina. Carolina (*M. t. centrata*), similar to Northern but carapace more oval in appearance; Cape Hatteras to n. Florida. Florida East Coast (*M. t. tequesta*), carapace dark with central portion of scute somewhat lighter in appearance, scutes of upper shell lack conspicuous growth rings; Atlantic coast of Florida. Mangrove (*M. t. rhizophorarum*), neck striped, hind legs vertically striped; Florida Keys. Ornate (*M. t. macrospilota*), large orange or yellow blotches in center of large carapace scutes; Florida Bay to Florida panhandle. Mississippi (*M. t. pileata*), carapace dark brown or black, plastron yellow, skin of legs very dark; extreme w. Florida to w. Louisiana. Texas (*M. t. littoralis*), highest point of carapace toward rear, plastron nearly white, legs greenish-gray with heavy black spots; w. Louisiana to w. Texas.

Diamondback meat was highly esteemed as a delicacy at the turn of the century. As a consequence their numbers were greatly reduced. Then development of coastal marshes destroyed much of their habitat. Recent protective legislation has restored some populations. Adults are often seen

basking on mud flats. They feed on
marine snails, clams, and worms.

304, 306, 308, 309	**Eastern Box Turtle**

Eastern Box Turtle
(Terrapene carolina)
Subspecies: Eastern, Gulf Coast,
Three-toed, Florida

Description: 4–8½″ (10–21.6 cm). Terrestrial.
Movable plastron hinge allows lower shell
to close tightly against carapace.
Carapace high-domed and keeled; variable
in color and pattern. Plastron often as
long as carapace; tan to dark brown,
yellow, orange, or olive; patternless or
with some dark blotching. Males
usually have red eyes and depression in
rear portion of plastron; females have
yellowish-brown eyes.

Breeding: Nests May to July. Lays 3–8 elliptical,
thin-shelled eggs, averaging about 1⅜″
(35 mm), in a 3–4″ (76–102 mm) deep
flask-shaped cavity. Hatchlings
sometimes overwinter in the nest.
Females are capable of storing sperm
and can produce fertile eggs for several
years after a single mating. Mature in 5
to 7 years.

Habitat: Moist forested areas, but also wet
meadows, pastures, and floodplains.

Range: S. Maine south to Florida Keys and
west to Michigan, s. Illinois, Missouri,
and e. Kansas, Oklahoma, and Texas.
Isolated population in extreme se.
Wisconsin.

Subspecies: Eastern (*T. c. carolina*), carapace
brightly marked, 4 toes on hind feet; s.
Maine south to Georgia, west to
Michigan, Illinois, and Tennessee. Gulf
Coast (*T. c. major*), largest ssp., not
brightly marked, rear margin of
carapace noticeably flared, 4 toes on
hind feet; coastal plain from Florida
panhandle to Louisiana. Three-toed (*T.
c. triunguis*), carapace tan or olive with
obscure pattern, head and front legs
orange-spotted, usually 3 toes on hind

feet; Missouri south to Alabama and Texas. Florida (*T. c. bauri*), carapace brightly patterned with radiating lines, side of head with 2 stripes, usually 3 toes on hind feet; peninsular Florida and Keys.

Box Turtles are usually seen early in the day, or after rain; they often retire to swampy areas during the hot summer months. They are fond of slugs, earthworms, wild strawberries, and mushrooms poisonous to man—which habit has killed many a human who has eaten their flesh. New York Indians are responsible for eliminating this turtle from much of the area between Ohio and New England; they ate Box Turtle meat, used the shells for ceremonial rattles, and buried turtles with the dead. A few specimens are known to have lived more than 100 years, having served as "living records," with fathers then sons carving their names or other family records on the shell. If habitat conditions remain constant, a Box Turtle may spend its life in an area scarcely larger than a football field.

305, 307 Western Box Turtle
(*Terrapene ornata*)
Subspecies: Ornate, Desert

Description: 4–5¾" (10.2–14.6 cm). Carapace *high-domed*, keelless, with distinctive pattern of radiating yellowish lines on a brown or black background. Plastron has *distinct movable hinge;* is often as long as carapace; scutes continuously patterned, with radiating yellow lines. Male has red eyes, and hind portion of plastron is slightly concave; female's eyes are yellowish-brown.

Breeding: Nesting May to mid-July. Early nesters may lay a second clutch in July. Lays 2–8 somewhat brittle-shelled, ellipsoidal eggs, about 1⅜" (35 mm) in length, in shallow flask-shaped cavity dug in well-drained soil. Incubation takes 9 to 10 weeks. Maturity reached in 8 to 10 years.

Habitat: Primarily open prairies; also grazed pasturelands, open woodlands, and waterways in arid, sandy-soiled terrain.

Range: S. South Dakota, Iowa, and e. Illinois south to Louisiana and Texas, west to sw. Arizona. Separate population in nw. Indiana and adjacent Illinois.

Subspecies: Ornate (*T. o. ornata*), with radiating lines (5–8 on second costal scute) that sharply contrast with dark carapace; se. Wyoming and Indiana south to Louisiana and New Mexico. Desert (*T. o. luteola*), radiating lines less prominent (11–14 on second costal scute); s. Arizona to Trans Pecos region of Texas south into Mexico.

In the morning the Western Box Turtle basks briefly, then searches for food. By midday it seeks shady shelter. Where cattle share its habitat, it methodically searches dung piles for beetles. It also relishes grasshoppers, caterpillars, cicadas, mulberries, and carrion. This turtle is often seen crossing roads after a downpour; consequently many are killed by automobiles.

TORTOISE FAMILY
(Testudinidae)

Ten genera and 39 species worldwide.
One genus, *Gopherus*, gopher tortoises,
in North America.
Gopher tortoises are strictly terrestrial.
Their hind feet are stumpy and
elephantine, the front limbs bear thick
hard scales, and all lack webbing. The
shell is often dome-shaped, and the
thick carapace and plastron are united
by a well-developed bridge. Male
plastron is concave. Gopher tortoises
have flattened front limbs and are
accomplished burrowers. All are
herbivorous. Females may nest 2 or 3
times a season.

328 Desert Tortoise
(*Gopherus agassizii*)

Description: 9¼–14½" (23.5–36.8 cm).
Terrestrial, with *domed shell and round,
stumpy elephantine hind legs.* Front limbs
flattened for digging and heavily scaled;
all toes webless. Carapace oblong, horn-
brown; scute centers often yellowish.
Bridge well developed, single axillary
scute. Plastron yellowish, with brown
along scute margins; adult throat scute
project beyond carapace. Head small,
rounded in front, reddish-tan; iris
greenish-yellow. Front and hind feet
about equal in size. Male plastron
concave.

Breeding: Mates chiefly in spring; nests May to
July. Lays 2–3 clutches of 2–14 hard,
chalky, elliptical or spherical eggs, in
funnel-shaped nest, 6" (15.2 cm) deep
—sometimes located at mouth of
burrow. Hatching occurs mid-August
to October. Maturity reached in 15–20
years.

Habitat: Arid sandy or gravelly locales with
creosote, thorn scrub, and cacti;

also washes, canyon bottoms, and oases.

Range: Se. California and s. Nevada southeast into Mexico.

Desert Tortoises feed on grasses in early morning and late afternoon. During the heat of the day they retreat to a shallow burrow dug in the base of an arroyo wall. They have been known to dig horizontal tunnels up to 30′ (9.1 m) in length. In September they may congregate in a communal den to spend the winter, becoming active again in March. When two males meet, they bob their heads rapidly, rushing toward each other and striking their gular scutes together. One of the two may be overturned. An endangered species.

---

329 Berlandier's Tortoise
(*Gopherus berlandieri*)

Description: 4½–8¾″ (11.4–22.2 cm). Terrestrial, with *domed shell* and *round, stumpy elephantine hind legs*. Front limbs flattened for digging and heavily scaled; all toes webless. Carapace brown, *nearly as broad as long* (elongated in older males); scute centers may be yellowish. Bridge well developed, usually 2 axillary scutes. Plastron yellowish; adult gular scutes project beyond carapace. Head wedge-shaped, somewhat pointed in front, yellowish; iris of eye brown surrounded by greenish-yellow ring. Front feet slightly larger than rear feet. Male plastron concave.

Breeding: Nests April to September. Lays hard-shelled, elongate eggs in chambers 2½″ (6 cm) deep beneath overhanging bush. Up to 7 laid each season, but not at a single site. Matures in 3–5 years.

Habitat: Scrub woodlands with sandy soils; also chaparral and mesquite.

Range: S. Texas into Mexico.

Unlike other gopher tortoises, it does not dig an elaborate resting place. Instead, it modifies other animal burrows or simply scrapes away at the base of a clump of vegetation to form a gently sloping ramp. Such "pallets" offer some protection from the midday sun. Its diet includes prickly-pear cactus, grasses, and the droppings of other tortoises. Many thousands of these tortoises have been collected for the pet trade over the years, but few survive more than a few months.

330 Gopher Tortoise
(*Gopherus polyphemus*)

Description: 9¼–14½" (23.5–36.8 cm). Terrestrial, with *domed shell* and *round, stumpy elephantine hind legs.* Front limbs flattened for digging and heavily scaled; all toes webless. Carapace elongated, brown to tan; scutes often light-centered. Bridge well developed; usually 1 axillary scute. Plastron yellowish; adult gular scutes project beyond carapace. Head large and blunt in front, grayish-brown; iris of eye dark brown. Hind feet smaller than front feet. Male plastron concave.

Breeding: Mates in spring; nests late April to mid-July. Lays 2–7 brittle, spherical eggs in cavity 5" (13 cm) deep, occasionally at mouth of burrow. Hatching occurs August to September.

Habitat: Well-drained sandy areas between grasslands and forests.

Range: Coastal plain from s. South Carolina to Florida and west to extreme e. Louisiana.

A most adept digger, it makes unusually long burrows: a record one was 47½' (14.5 m) long, straight, unbranched, and ended in an enlarged chamber. The burrow serves as a shelter, where temperature and humidity remain relatively unchanged. Often other animals share the burrow— including small mammals, burrowing owls, snakes, gopher frogs, toads, and assorted invertebrates. On cool days, Gopher Tortoises bask at the burrow entrance before foraging for grasses and leaves.

SEA TURTLE FAMILY
(Chelonidae)

Four genera of 6 marine-dwelling species within the range of this guide: the loggerheads (*Caretta*), the green turtles (*Chelonia*), the hawksbills (*Eretmochelys*), and the ridleys (*Lepidochelys*).

Sea turtles are characterized by heart-shaped, scute-covered carapaces, paddlelike limbs with one or two claws, and, with the exception of the leatherback, are the largest living aquatic turtles. They dwell in tropical waters, ranging into temperate zones in the summer.

Sea turtles mate in shallow water off nesting beaches. The females come ashore two or more times a season to lay a hundred or more spherical, leathery-shelled eggs in a deep nest cavity dug with the hind feet. After filling the nests they lumber back to the sea. Hatchlings struggle to escape from the sandy chamber and then must face a horde of predators. Few reach the haven of water and still fewer survive their first year.

Unfortunately, sea turtles are vanishing. Some species are in great danger of extinction.

265 **Loggerhead**
(*Caretta caretta*)
Subspecies: Atlantic, Pacific

Description: 31–48″ (79–122 cm). Ocean-dwelling, with paddlelike limbs. Carapace elongated and heart-shaped, keeled (3 keels in young), *reddish-brown; 5 or more costal scutes, 1st touching nuchal;* 3 scutes on bridge. Plastron cream-yellow, with 2 ridges (lost with age). Two pairs of prefrontal scales between eyes. Male's tail extends well beyond shell.

Breeding: Nests along Atlantic and Gulf coasts

from New Jersey to Texas. At night, from May to August, during periods of high tides, females come ashore on wide, sloping beaches and lay about 125 spherical, 1⅝" (41 mm) eggs. Hatching takes 7–9 weeks.

Habitat: Coastal bays, lagoons, estuaries, open seas.

Range: Warm waters of Atlantic and Pacific. In summer, ranges north to New England (occasionally Newfoundland), and to s. California.

Subspecies: Atlantic (*C. c. caretta*), carapace with average of 12 marginal scutes on each side; warm waters of Atlantic Ocean, reaches New England in summer months. Pacific (*C. c. gigas*), averages 13 marginal scutes; Pacific, Indian, and e. Atlantic oceans, occasionally reaches s. California.

There are reports of Loggerheads taken in the past weighing 1,000 pounds (455 kg) or more. Such giants are probably gone, now that their numbers have been severely reduced. Development of beaches and coastal islands has destroyed many nesting sites. Many turtles drown in shrimp nets. Hatchlings, normally attracted to light reflected off the ocean, are sometimes confused by artificial highway lights and crawl toward them. Many are accidentally killed. They are omnivorous, their diet including sponges, mulluscs, crustaceans, sea urchins, and marine plants.

267 **Green Turtle**
(*Chelonia mydas*)
Subspecies: Atlantic, Pacific

Description: 28–60¼" (71–153 cm). Ocean-dwelling, with paddlelike limbs. Carapace broad and oval-shaped, unkeeled; olive to dark brown and may have mottled or radiating pattern; 4

costal scutes, 1st not touching nuchal; 4 scutes on bridge. Plastron white or yellowish. *Pair of prefrontal scales between eyes.* Young have vertebral keel and pair of keels on plastron; flippers are edged in white. Male's tail is tipped with a flattened nail and extends well beyond shell.

Breeding: Nests in small numbers along se. coast of Florida. Females nest every 2–4 years, 1–8 times a season. At night they crawl onto the beach, dig an urn-shaped nest hole, and lay about 100 spherical eggs, the size of golf balls, which hatch in about 2 months. Sexual maturity takes 20–30 years.

Habitat: Shallow waters with abundant aquatic plants, and open seas during migrations.

Range: Warm waters of Atlantic and Pacific.

Subspecies: Atlantic (*C. m. mydas*), carapace brown and not noticeably indented above hind limbs; warm waters of the Atlantic, occasionally reaching New England. Pacific (*C. m. agassizi*), carapace darker and indented above hind limbs; warm waters of Indo-Pacific oceans, reaches s. California.

Hatchlings spend a "lost" year at sea; they drift with large floating mats of seaweed, feeding on small invertebrates and plants carried by the currents to no one knows where. A year later they reappear in shallow waters, where they graze on turtle grass and other marine plants until reaching maturity. With pinpoint accuracy they then return to the beaches where they hatched, having traveled perhaps 2,000 miles (3,200 km) from their pastures. In times past the Green Turtle provided fresh meat for seafarers. Today, its flesh and eggs are an important protein source for impoverished peoples in the tropics. Because its feeding and nesting grounds are well known and unchanging, it is extremely vulnerable to human predation. Its products—gourmet meat, flipper leather, cosmetic and cooking oils, and "turtle soup" calipee —are in demand worldwide. As a consequence the species is rapidly vanishing.

266 Hawksbill
(*Eretmochelys imbricata*)
Subspecies: Atlantic, Pacific

Description: 30–36" (76–91 cm). Ocean-dwelling, with paddlelike limbs. Carapace shield-shaped, keeled, greenish-brown with mottled or radiating pattern; *scutes overlap* (except in hatchlings and old animals); *4 costal scutes,* 1st not touching nuchal; 4 scutes on bridge. Plastron yellow; has 2 ridges in young. Snout resembles hawk's beak. *Two pairs of prefrontal scales between eyes.* Male's tail extends well beyond shell's margin; male's plastron slightly concave.

Breeding: Mates in shallow water off nesting beaches. Lays 50–200+ spherical eggs, about 1½" (38 mm) in diameter, in a chamber 2' (61 cm) deep. Hatchlings emerge in 8–11 weeks. Rarely nests on Florida beaches.

Habitat: Shallow coastal waters with rocky bottoms, coral reefs, mangrove-bordered bays and estuaries.

Range: Primarily warm waters of Atlantic and Pacific oceans. Stragglers reach New England and occasionally s. California.

Subspecies: Atlantic (*E. i. imbricata*), with nearly straight-sided carapace that tapers posteriorly, keel not continuous except on last 4 vertebral scutes; warm waters of the Atlantic, occasionally reaches Massachusetts. Pacific (*E. i. bissa*), carapace more heart-shaped, keel continuous; tropical waters of Indian and Pacific oceans, stragglers rarely reach s. California.

An endangered species. In demand for its meat and eggs, eaten locally, its calipee, used in turtle soup, and to make "tortoiseshell" jewelry. The Hawksbill feeds on many invertebrates, including toxic sponges, so its flesh may be poisonous to humans. It bites without hesitation when captured.

262 Atlantic Ridley
(*Lepidochelys kempi*)

Description: 23½–29½" (60–74.9 cm). Ocean-dwelling, with paddlelike limbs. Carapace *wide* and heart-shaped, keeled, *gray; 5 costal scutes, 1st touching nuchal; 4 scutes with pores on bridge.* Plastron white. Two pairs of prefrontal scales between eyes. Young have 3 ridges on carapace, 4 on plastron. Males have tails extending well beyond margin of shell.

Breeding: Nests April to July, during the day, along short stretch of the Tamaulipas beach in Gulf of Mexico. Lays 2–3 clutches of about 110 spherical, 1½" (38 mm) eggs.

Habitat: Prefers shallow coastal waters.

Range: Gulf of Mexico, ranging north in summer months to New England and occasionally Nova Scotia.

Highly endangered. In the past, an estimated 40,000 turtles would congregate off the nesting beach, while 10,000 females nested ashore. Today the species is almost extinct as a consequence of egg robbing, the slaughter of nesting females for food, and drownings in shrimp boat nets. Recently the Mexican and U.S. governments have begun last-ditch efforts to save the species.

264 Olive Ridley
(*Lepidochelys olivacea*)

Description: 24½–28" (62–71 cm). Ocean-
dwelling, with paddlelike limbs.
Carapace wide and heart-shaped, keeled,
olive; 6–8 costal scutes, 1st touching
nuchal; 4 scutes with pores on bridge.
Plastron greenish-white. Two pairs of
prefrontal scales between eyes. Young
have 3 keels on carapace, 4 on plastron.
Males have concave plastrons and tails
extending well beyond margin of shell.

Breeding: Nests outside range of this guide.

Habitat: Bays, lagoons, and shallow water
between reefs and shoreline.

Range: Warm waters of Atlantic and Pacific
oceans; occasionally reaches s.
California.

Highly carnivorous, it feeds on fish,
crustaceans, molluscs, and sea urchins.
In turn, hatchlings are eaten by crabs,
birds, and fish. Also, man raids the
Ridleys' nests for eggs and kills adults
for meat and flipper skin.

LEATHERBACK TURTLE FAMILY
(Dermochelyidae)

One living species, *Dermochelys coriacea*, the largest living turtle. Pelagic. Although chiefly seen in tropical waters, it ranges into temperate waters during the summer.
The Leatherback is the most specialized aquatic turtle. Instead of horny scutes it is covered with smooth leathery skin. The carapace is made up of many small bony platelets embedded in the skin. Lacking a rigid shell, its ribs and vertebrae, unlike those of other turtles, are not attached to the carapace. Males have concave plastrons and tails longer than the hind limbs.

263 Leatherback
(*Dermochelys coriacea*)
Subspecies: Atlantic, Pacific

Description: 50–84″ (127–213 cm). *Largest living turtle.* Adults weigh 600–1,600 lbs. (273–727 kg). Ocean-dwelling, with paddlelike, clawless limbs. Carapace elongated and triangular, slate to blue-black, *covered with smooth skin* (not scutes), with 7 *prominent keels.* Plastron white, with 5 ridges. Male has concave plastron and tail longer than hind limbs.

Breeding: Occasionally nests on the east and west coasts of Florida. Nests April to November between 9:00 P.M. and midnight. Lays 50–170 spherical eggs, 2–2½″ (51–63 mm) in diameter, in cavity 3′ (1 m) deep. Lays several clutches a season about 10 days apart. Eggs hatch in 8–10 weeks.

Habitat: Open seas and bays and estuaries.

Range: Atlantic and Pacific oceans; chiefly in tropical waters but can be seen as far north as Newfoundland and British Columbia in summer.

Subspecies: Atlantic (*C. c. coriacea*), warm parts

of Atlantic Ocean north to
Newfoundland. Pacific (*D. c. schlegelii*),
tropical waters of Pacific and Indian
oceans, north to British Columbia.

A powerful swimmer, it wanders great
distances at sea. Long backward-
projecting spines line its mouth and
esophagus and help the Leatherback to
swallow jellyfish, its main food. When
captured, it flails at the assailant with
its flippers, and emits groans and
belches. Not generally eaten, it is taken
for oils of cosmetic value. Its eggs are
prized. Critically endangered.

SOFTSHELL TURTLE FAMILY
(Trionychidae)

Seven genera of 23 species worldwide; 3 species of *Trionyx* in North America. Freshwater.

Softshells are easily identified by a nearly circular, pancakelike carapace covered with soft leathery skin instead of horny scutes. Feet are paddlelike, fully webbed, have 3 claws. The snout is tubular, the beak sharp and enclosed in fleshy lips.

Strong swimmers, softshells cruise submerged, breathing through snorkel-like snouts. They like to bask near the shore, but are easily disturbed, and display great agility and speed in retreating to the safety of water. One to three clutches of hard-shelled, spherical eggs are laid each season. Female softshells grow significantly larger than males and as adults their original carapace patterns become obscured by blotches or mottling. Males tend to retain the juvenile pattern; they have long thick tails with anal opening near the tail tip.

Extreme care should be taken in handling softshells. The long neck and sharp jaws surprise many a careless collector.

272, 273 Florida Softshell
(*Trionyx ferox*)

Description: Males, 5⅞–12⅞″ (15.1–32.7 cm); females, 10⅞–19⅝″ (27.7–49.8 cm). Shells covered with soft leathery skin, not horny scutes. Carapace dark brown or brownish-gray, may have darker blotches; *ridge along margin and numerous blunt tubercles on leading edge.* Dark-bordered yellow or red band extends from eye to base of lower jaw. Lateral ridge on nostrils extends from nasal septum.

Breeding: Nests mid-March to July, in the morning, on sandy ground exposed to sunlight. Lays 4–22 thin-shelled, spherical eggs, about 1⅛″ (29 mm) in diameter, in cavity 5″ (13 cm) deep. Incubation takes about 9 weeks.

Habitat: Sandy or muddy-bottomed lakes, ponds, and canals, and big springs with overhanging foliage.

Range: S. South Carolina and Georgia through Florida (except the Keys), west along coastal plain to Mobile Bay, Alabama.

It is frequently seen floating on the water's surface or basking along banks. Below water, it rests buried in mud or sand, with only its head poking out. Its diet includes crayfish, snails, frogs, fish, and occasionally young waterfowl.

268, 269 **Smooth Softshell**
(*Trionyx muticus*)
Subspecies: Midland, Gulf Coast

Description: Males, 4⅜–7″ (11.1–17.8 cm); females, 6⅝–14″ (17–35.6 cm). Shell covered with soft leathery skin, not horny scutes. *Carapace smooth, without the spines or bumps* of other American softshells; olive to orange-brown, with pattern of darker dots and dashes. Large females have mottled or blotched pattern on carapace. Nostrils round.

Breeding: Female matures in 7 years. Nests on sandbars, May to July. Lays 1–3 clutches of 4–33 hard-shelled, spherical eggs, each about ⅞″ (22 mm) in diameter, in a cavity 6–9″ (15–23 cm) deep. Number of eggs in clutch is proportional to female's size. Turtle's hind feet dig nest and guide and arrange eggs in cavity. Incubation period 2–2½ months.

Habitat: Prefers rivers and large streams with moderate to fast currents and sandy or muddy bottoms.

Range: Drainage systems of the Ohio,

Mississippi, Missouri, Arkansas, and Alabama rivers.

Subspecies: Midland (*T. m. muticus*), pale stripes on snout in front of eyes, pale stripe behind eye with narrow dark borders; c. United States from w. Pennsylvania to w. Wisconsin and sc. North Dakota, south to n. Alabama and e. Texas. Gulf Coast (*T. m. calvatus*), lacks light stripes on snout, pale stripe behind eye with wide dark borders; coastal plain streams from the Escambia River system of s. Alabama and Florida panhandle to the Pearl River drainage of Mississippi and Louisiana.

Highly aquatic; rarely seen away from water. It basks on sandbars within 10 feet (3 m) of the waterline and rapidly retreats to the water when alarmed. It feeds primarily on crayfish and other small invertebrates, also frogs and fish.

270, 271 Spiny Softshell
(*Trionyx spiniferus*)
Subspecies: Eastern, Western, Gulf Coast, Pallid, Guadalupe, Texas

Description: Males, 5–9¼″ (12.7–23.5 cm); females, 6½–18″ (16.5–45.7 cm). Shell covered with soft leathery skin, not horny scutes. Carapace olive to tan, with black-bordered "eye spots" or dark blotches and dark line around shell's rim; *spiny tubercles on leading edge;* 2 dark-bordered light stripes on each side of head. Nostrils have a lateral ridge extending from nasal septum.

Breeding: Nests May to August. Digs flask-shaped cavity in bank of sand or gravel exposed to full sunlight and lays 4–32 spherical eggs, 1⅛″ (29 mm) in diameter. May nest more than once a season. Hatchlings emerge late August to October or following spring.

Habitat: Likes small marshy creeks and farm

ponds as well as large, fast-flowing rivers and lakes.

Range: Throughout c. United States as far west as the Continental Divide. Separate populations in Montana, s. Quebec, Delaware, and the Gila–Colorado River system of New Mexico and Utah.

Subspecies: Eastern (*T. s. spiniferus*), large carapace with black-bordered eye spots and one dark rim line; w. New York to Wisconsin, south to Tennessee River, with separate population in Lake Champlain and lower part of Ottawa River in Canada; introduced into s. New Jersey. Western (*T. s. hartwegi*), carapace with small dots, eye spots, and one dark rim line; Minnesota to Arkansas, west to se. Wyoming, e. Colorado and ne. New Mexico, separate population in Missouri River drainage in Montana. Gulf Coast (*T. s. asper*), 2 or more dark lines around rear rim of carapace, head stripes fused on side of head; sc. North Carolina to Florida panhandle, west to Mississippi and e. Louisiana. Pallid (*T. s. pallidus*), pale with white tubercles on rear half of carapace; w. Louisiana, s. Oklahoma, and most of n. and e. Texas. Guadalupe (*T. s. guadalupensis*), small white tubercles surrounded by narrow black rings on most parts of carapace; sc. Texas in Nueces and Guadalupe–San Antonio drainage systems. Texas (*T. s. emoryi*), white tubercles on back third of carapace, pale marginal rim becomes conspicuously widened along rear edge; Rio Grande drainage in Texas and New Mexico, and the Gila–Colorado River system in Arizona, New Mexico and extreme sw. Utah. Characteristics intergrade where ranges of subspecies overlap.

Difficult to approach and fast-moving on land and in water. It is fond of basking on banks, logs, and floating debris. Carnivorous. Captive longevity exceeds 25 years.

Lizards
(order Squamata; suborder Lacertilia)

There are approximately 3,000 species
of lizards worldwide; most inhabit
warm dry regions and the tropics. They
belong to 17 families, 8 of which are
native to North America; 115 species,
including established exotics, may be
encountered north of Mexico.
Known from the Triassic Period to the
present, lizards today comprise the
largest living group of reptiles. They
come in a bewildering array of sizes,
shapes, and colors: from tiny gecko
species less than 3″ (7.6 cm) long to the
giant dragon lizard of Komodo nearly
10′ (3 m) long. Typical lizards
superficially resemble salamanders, but
their dry scaly skin, clawed feet, and
external ear openings quickly separate
them from their distant moist-skinned
ancestors. Legless lizards may be
confused with snakes, but unlike
snakes, they possess movable eyelids.
The pattern and color of lizards vary
greatly: males and females of the same
species often show color differences;
juveniles are frequently distinct from
adults. Subspecies, too, may differ
strikingly.
Lizards have varied life styles. Although
generally diurnal, the majority of
geckos, the night lizards, and the Gila
Monster are nocturnal. Only two
species are venomous—the Gila
Monster of southwestern United States
and Mexico and the Beaded Lizard of
Mexico. Courtship is brief, and
fertilization is internal. Most lizards are
egg-layers, but occasionally young are
born alive.

GECKO FAMILY
(Gekkonidae)

Eighty-nine genera with 750 species scattered throughout the tropics, including many oceanic islands. Three genera—*Coleonyx, Phyllodactylus,* and *Sphaerodactylus*—with 5 species are native to our range. Another 2 genera with 5 species have been introduced.

Geckos typically have flattened bodies and short limbs. In addition to claws, many species have expanded toe pads. On the bottom of each toe pad are scales covered with a myriad of microscopic hairlike bristles. Minute suction cups on the tips of the bristles permit geckos to walk up walls and across ceilings. Most species lack movable eyelids; instead the perpetually open eyes are protected by a transparent scale, the spectacle. The American genus *Coleonyx* is an exception; it has eyelids. Among diurnal species the pupil of the eye is round; in nocturnal species it is vertically elliptical.

Many geckos appear fragile; their thin soft skin tears easily and the tail breaks so readily it seems cast off even before it is grasped. Among some populations many individuals will be found to have the tail in some stage of regeneration. Geckos are the most vocal of lizards. Voices vary from the sound that prompted the family name—the raucous "geh-oh" of the giant Asian Tokay Gecko—to the cricketlike chirps small species give when defending a feeding site.

Most geckos lay two eggs at a time; *Sphaerodactylus* and *Gonatodes* lay only single eggs.

393, 394 Texas Banded Gecko
(*Coleonyx brevis*)

Description: 4–4¾" (10–12 cm). Sleepy-looking
lizard with *protruding white-rimmed
eyelids*. Skin supple; back scales
uniformly granular. Pinkish with dark
brown crossbands, most evident in
juveniles, breaking up with age into
nearly uniform mottling. Preanal pores
in male separate at midline. Toes
slender, no pads. Females larger.

Voice: Squeaks when alarmed.

Breeding: Clutches of 2 eggs, laid April to June.

Habitat: Rock outcrops and canyon beds in
desert areas. Found beneath shelving
rocks, vegetative debris, and discarded
boards.

Range: S. New Mexico through s. Texas
(almost to coast) and ne. Mexico.

Nocturnal. Most easily encountered
crossing highways during nightly
movements. This gecko feeds on insects
and small spiders which it stalks with a
feline twitching of the tail. It eats its
shed skin: loose skin is grasped in the
jaws, pulled free of the body, then
chewed and swallowed.

Big Bend Gecko
(*Coleonyx reticulatus*)

Description: 5½–6⅝" (13.9–16.8 cm). *Protruding
eyelids*. Skin supple; back scales small
and granular, interspersed with
lengthwise rows of enlarged tubercles.
Light tan with dark brown crossbands,
most evident in juveniles. Pattern
breaks up with age into profusion of
dark spots and blotches. Toes slender,
no pads.

Voice: Squeaks.

Breeding: Poorly known.

Habitat: Rocky outcrops in desert areas,
particularly loose rocks that provide
shelter from the heat of day.

Range: Big Bend region of Texas (Brewster and Presidio counties) into Mexico.

Nocturnal. Heavy rains seem to bring this gecko out of hiding. The first known specimen was captured in a mouse trap in 1956; none was collected again until 1971.

392, 395 **Banded Gecko**
(*Coleonyx variegatus*)
Subspecies: Desert, San Diego, Utah, Tucson

Description: 4½–6″ (11.4–15 cm). Medium-sized lizard with *protruding eyelids*. Light tan with dark brown to black crossbands, most evident in juveniles, breaking up with age into blotches, spots, and variegations. Skin supple; dorsal scales uniformly granular. Preanal pores continuous across midline of males. Toes slender; no pads.

Voice: Chirps when caught.

Breeding: Lays 1–3 clutches of 2 eggs, May to September. Hatchlings appear in 45 days, July to November.

Habitat: Rocky tracts, canyon walls, and sand dunes in deserts and semi-arid areas. Avoids the heat of day by hiding in rock crevices or under logs, fallen limbs, or rubbish.

Range: S. California, Nevada, Utah, and Arizona, south into Baja California and Mexico.

Subspecies: Eight; 4 in our range.
Desert (*C. v. variegatus*), s. California (except coast), sw. Nevada, and w. Arizona to Gulf of California.
San Diego (*C. v. abbotti*), Pacific slopes of s. California into n. half of Baja peninsula.
Utah (*C. v. utahensis*), sw. Utah and adjacent corners of Nevada and Arizona.
Tucson (*C. v. bogerti*), se. Arizona and sw. New Mexico.

Nocturnal. Often encountered at night, silhouetted by auto headlights against the black asphalt of desert roads. Feeds on insects and spiders. When stalking prey, a Banded Gecko waves its tail like a prowling cat. A constriction at the tail's base marks the place where it breaks away when grabbed.

402 Yellow-headed Gecko
(*Gonatodes albogularis*)

Description: 2¾–3½" (6.9–8.9 cm). Only North American gecko lacking moveable eyelids and expanded toe pads. Brown to almost black; females mottled by gray and cream, usually have light collar stripe. *Males darker, with yellow to orange head.* Tail white-tipped. Scales granular, do not overlap. Toes have sharp claws.

Breeding: Single eggs are laid throughout year.

Habitat: Tree trunks within 6' (2 m) of ground; piles of brush and rock. Also abundant around houses, in stacks of lumber and trash.

Range: Introduced into mainland Florida and the Florida Keys; native to Central and South America.

Primarily diurnal. This lizard can be seen feeding on crawling insects and spiders near its hiding place. It changes color in response to temperature and light conditions. Basking Yellow-heads are dark brown to black. At night or in the shade, they fade to gray (females) or blue-green (males).

401 Indo-Pacific Gecko
(*Hemidactylus garnoti*)

Description: 4–5½" (10.2–14 cm). No eyelids. *Toes webbed at base;* clawed toe tips free of expanded pads. Light brown to gray, occasionally with dark or light spotting. Back scales small, granular. Tail flattened with lateral saw-tooth fringe; tail underside orange to pink.

Breeding: Unisexual; all females. No mating. Clutches of 2 eggs in communal nest throughout summer; hatching occurs late summer to midwinter.

Habitat: Tree crevices or loose bark. Also found on buildings.

Range: Introduced into peninsular Florida; native to s. Asia.

Nocturnal. No males are known for the species. Like the Mediterranean Gecko, it commonly shares human habitations. It frequently can be seen at night on walls and ceilings, darting after insects attracted to lights and open windows.

397 Mediterranean Gecko
(*Hemidactylus turcicus*)

Description: 4–5" (10.2–12.7 cm). No eyelids. *Toes unwebbed;* clawed toe tips free of expanded pads. Translucent pinkish to white, with some darker blotching. Scales small, granular; rows of white keeled knobby tubercles down back. Round tail ringed with keeled tubercles and banded, especially in juveniles.

Voice: Males squeak while fighting.

Breeding: Mates March to July in U.S. Clutches of 1–2 eggs are laid April to August.

Habitat: Under palm leaves and in crevices of tree bark and rocky outcrops. Most common in occupied buildings.

Range: Introduced into peninsular Florida, Louisiana, and Texas. Native to Mediterranean area, Middle East, and India.

The most conspicuous gecko in North America. From twilight to dawn it can be seen darting along walls and ceilings

to feed on insects attracted by lights.
This lizard is highly territorial; males
vigorously defend a favorite foraging
area. Females are easily identified in
spring and summer by white eggs that
can be seen through the translucent
skin of the abdomen.

391 Leaf-toed Gecko
(*Phyllodactylus xanti*)
Subspecies: California

Description: 4–5″ (10–12.7 cm). No eyelids, *2 expanded leaf-shaped pads on toe tips.* Translucent pink to gray, occasionally with dark brown blotches. Back scales small and granular, with lengthwise rows of wartlike tubercles.

Voice: Squeaks when alarmed.

Breeding: 1 or 2 eggs are laid May to July; hatchlings appear June to August.

Habitat: Cracks and crevices in granite outcrops in desert. Occasionally found beneath tree bark or dead prickly pear pads.

Range: S. California to tip of Baja California and islands in Gulf of California.

Subspecies: Nine; 1 in our range. California (*P. x. nocticolus*).

Nocturnal. Emerges shortly after dark to search for insects and spiders, which are captured after a short stalk.

Ocellated Gecko
(*Sphaerodactylus argus*)
Subspecies: Antillean

Description: 1¾–2¾″ (4.4–6.9 cm). Miniscule brown gecko. Many *tiny eyelike white spots on head and neck may fuse to become light lines on head,* neck, or back. Underside of tail orange or red. Snout pointed; small spine over each large lidless eye. Back scales small,

overlapping, keeled. Expanded toe
pads.

Breeding: 1 egg at a time is laid in summer.

Habitat: Leaf litter of tropical hammocks, coastal
vegetation; rubbish around human
dwellings.

Range: Introduced into Florida Keys (Key
West); native to Jamaica.

Subspecies: Two; 1 in our range.
Antillean (*S. a. argus*).

Primarily diurnal. The species is
definitely established in Key West but
seems to be confined to only a few
locales, so that it may easily be missed.

396, 398 Ashy Gecko
(*Sphaerodactylus elegans*)

Description: 2½–2⅞" (6.4–7.3 cm). A tiny gecko
with *fine salt-and-pepper speckling.* Head
and tail seem white to yellow with red
to brown speckling, whereas trunk
appears the reverse, dark with light
speckling. Juveniles creamy with dark
red crossbands and a vivid red tail.
Snout pointed; small spine over each
lidless eye. Back scales small and
granular, do not overlap. Expanded toe
pads.

Breeding: Single eggs are laid in August, often in
communal nests.

Habitat: Loose leaf litter and rock piles around
houses and in tropical hammocks.

Range: Introduced into Florida Keys; native to
Hispaniola and Cuba.

Primarily diurnal, this gecko is
occasionally discovered at night feeding
on insects attracted to a patio light or
lit window. Early traders carried the
species to the Keys, probably in
shipments of produce or lumber.
Juveniles are so strikingly different in
color that once they were considered a
distinct species.

399, 400 Reef Gecko

(*Sphaerodactylus notatus*)
Subspecies: Florida

Description: 1¾–2½" (4.4–6.3 cm). Smallest North American lizard. Snout pointed; small spine over each large lidless eye. Drab brown with dark speckling; tail underside pale orange. Female and young show 3 light-centered dark stripes on head; *neck has dark blotch surrounding 2 white eyelike spots.* Markings fade with age in male until only scattered dark speckles remain. Juvenile tail has white tip. Back scales large, overlapping, and strongly keeled. Expanded toe pads.

Breeding: Eggs are laid singly, March to December, in communal nests in rotten logs or under debris. Eggs are large; can pass through female vent because of flexible shell. Exposure to air causes shell to resume its full oval shape and harden.

Habitat: Leafy, coarse debris on shaded floor of hammocks, coconut groves, ornamental gardens.

Range: Florida Keys; se. coastal ridge of mainland Florida.

Subspecies: Five; 1 in our range. Florida (*S. n. notatus*).

Primarily diurnal, but so secretive it seems less common than it actually is. Scuffling through ankle-deep leaf litter will bring it to view scurrying to other shelter.

IGUANID FAMILY
(Iguanidae)

Sixty genera of approximately 628
species in the Americas, Madagascar,
and Fiji. Fourteen genera with 44
species are native to our range and
another 4 genera with 8 species have
been introduced.
Iguanids range from 4–72″ (10 cm–
2 m) in length. A typical iguanid
is of moderate size, has 5 clawed toes
on each of 4 legs, and a long tail; its
teeth are attached to a ledge on the
inside of the jaw. Most species are
either arboreal or terrestrial; they feed
on insects and other invertebrates, but
some, like *Dipsosaurus* and *Sauromalus,*
eat leaves, fruit, and blossoms. Except
for a very few species that live in cool
mountain habitats and give birth to
living young, iguanids are egg-layers.
Clutches of many eggs or offspring are
the rule, but *Anolis* lays only one egg
every couple of weeks.
Iguanids are possibly the most visually-
oriented of all lizards. They
communicate at a distance by a show of
color and behavioral signals. Mates are
courted, territories defended, and
interlopers driven off by elaborate and
precisely timed combinations of head-
bobbing, body push-ups, and open-
mouth displays that are unique to each
species. Some further enhance the effect
by curling the tail, inflating the chest
and throat, or extending the throatfan,
all of which expose a bright patch of
color to the view of another lizard.
Finally, many species exhibit intense
color during mating season.

383, 385 **Green Anole**
(*Anolis carolinensis*)

Description: 5–8″ (12.7–20.3 cm). A slender lizard
with *extensible pink throatfan,* large toe

pads. Snout long, wedge-shaped. No back crest. Usually green, but in seconds can change to brown or intermediate colors. Tail round.

Breeding: Mates March to September. Single eggs are laid every 14 days, April to September, in leaf litter, trash, rock piles, moist debris. Incubation takes 5–7 weeks.

Habitat: Arboreal. Encountered on vertical surfaces like fence posts and walls; but favors tree boles, shrubs, vines, tall grasses, palm fronds.

Range: S. Virginia to the Florida Keys, west to c. Texas and Oklahoma.

Diurnal, but easily collected by night with the aid of a light; moisture on the skin makes these anoles shine as though covered with reflecting yellow paint. Adults prefer shaded perches. Juveniles prefer sunnier locations closer to ground. Basking anoles are typically brown; fighting males turn green with a black patch behind the eyes. They slowly stalk their prey: flies, beetles, moths, spiders, even small crabs.

390 Crested Anole
(Anolis cristatellus)

Description: 4¾–8″ (12.0–20.5 cm). *Extensible throatfan, yellow to orange.* Enlarged toe pads. Snout moderate. Gray, gray-green, or green, with dark line between eyes and U-shaped dark line around back of head. Females and juveniles may have light stripe down back or dark crossbands. Back has a fold of skin rather than a crest, but in males the compressed tail has a high crest visibly supported by bony rays.

Breeding: 2 or 3 eggs are laid in leaf litter, under logs, or beneath loose bark.

Habitat: Sunny tree trunks, fence posts, walls.

Range: Introduced into Miami, Florida; native to Caribbean islands.

Diurnal. This lizard usually rests head down on a tree or post. It climbs higher than 10 feet (3 m) only reluctantly. It is a sun-loving species, seldom found in even partial shade. Diet includes insects, spiders, fruit, and other lizards.

Large-headed Anole
(*Anolis cybotes*)

Description: 7–8″ (18–20.3 cm). *Chunky lizard with pale yellow extensible throatfan;* enlarged toe pads, short snout. Gray to brown, with light stripe down center of back in females and juveniles. Mature males nearly uniform brown, occasionally show dark markings or green lateral stripes. Back crest prominent on neck and male tail; weak elsewhere.

Breeding: Mates March to September. Hatchlings emerge April to October.

Habitat: Arboreal. Trees, vines, fences, and piles of garden trash.

Range: Introduced into ne. Dade County, Florida, from Hispaniola.

Diurnal. A highly territorial species. Males can be seen displaying on streetside banyan (*Ficus*) trees and ornamental shrubs, and on the ground in residential areas.

Bark Anole
(*Anolis distichus*)
Subspecies: Florida, Green

Description: 3¼–5″ (9–12.7 cm). *A slender lizard with yellow or pale orange extensible throatfan;* enlarged toe pads, short snout. Gray, brown, or pale green; dark line between eyes, sometimes dark chevrons on back. Tail round or slightly compressed. No back crest.

Breeding: Single eggs are laid every 14 days,

March to October, in leaf litter near
base of trees; hatch in 5–8 weeks.

Habitat: Arboreal. Vertical surfaces in partial or
deep shade: tree trunks, limbs, vines,
less frequently on walls.

Range: Introduced into s. Florida from
Bahamas and Hispaniola.

Subspecies: Eighteen; 2 in our range.
Florida (*A. d. floridanus*), gray or
brown, never green, throatfan pale
yellow; se. Florida (Dade and Monroe
counties).
Green (*A. d. dominicensis*), gray or pale
green, throatfan pale orange;
introduced into Miami, Florida.

Diurnal. Adult males defend breeding
territories by extending the throatfan
and bobbing the head up and down.
During a vigorous contest, a male will
do push-ups with its entire body while
clamping its teeth on its protruding
tongue. This display usually scares off
an intruder. Bodily contact or biting
seldom take place. This species prefers
to eat crawling insects.

389 Knight Anole
(*Anolis equestris*)

Description: 13–19⅜" (33.0–49.2 cm). *Large lizard
with wrinkled bony head, pale pink
extensible throatfan,* and enlarged toe
pads. Snout long and wedge-shaped.
Bright green; can change to brown.
Yellow or white stripe under eye and
over shoulder. Tail slightly compressed.
Low crest, most pronounced on neck
and shoulders.

Breeding: Summer.

Habitat: Arboreal. Under shady canopies of large
trees.

Range: Introduced into Dade and Broward
counties, Florida; native to Cuba.

Diurnal. This lizard becomes fiercely
defensive when a snake—or anything

suggestive of a snake, such as sticks or a garden hose—gets too close. It turns broadside to the threat, extends the throatfan, raises the back crest, and gapes menacingly. Knight Anoles are relatively slow and can be easily caught by hand, but their strong jaws and sharp teeth should give collectors pause. In Florida, the species does not seem to survive cold winters in great number.

384 Brown Anole
(*Anolis sagrei*)
Subspecies: Cuban, Bahaman

Description: 5–8¾″ (13–21.3 cm). Extensible throatfan yellow to red-orange, *with white line down center*. Enlarged toe pads; short snout. Tan to dark brown; with dark-bordered, interconnected light diamonds or stripe down back; pattern fades in mature males to a uniform tan. Tail compressed, with prominent crest in male. No crest on back.

Breeding: Mates throughout spring and summer. Single eggs, laid from June to September, hatch in 30 days.

Habitat: Trees, shrubs, fences, walls, lumber stacks, rock piles. Usually within 6′ (2 m) of the ground.

Range: Introduced into peninsular Florida; native to Jamaica, Cuba, the Bahamas.

Subspecies: Five; 2 in our range.

Cuban (*A. s. sagrei*), bright red-orange throatfan; s. Florida.
Bahaman (*A. s. ordinatus*), throatfan mustard or yellow with red splotches; Lake Worth, Miami, and Chokoloskee Island, Florida.

Diurnal. Although frequently found on trees and shrubs, the Brown Anole is a terrestrial species. It never ventures far from the ground and rests head down so that it can flee earthward when threatened. Males vigilantly protect territory, driving intruders away by a ritual of intense headbobs or push-ups, followed by displays of the colorful throatfan. Ants, beetles, grasshoppers, spiders, and other prey are caught by swift dashes.

362 Zebra-tailed Lizard
(*Callisaurus draconoides*)
Subspecies: Nevada, Mojave, Arizona

Description: 6–9⅛" (15.2–23.2 cm). A large lizard with granular scales, 2 folds across throat, *external ears*. Gray, usually with paired dusky spots running down back, becoming crossbands on tail. *Black crossbars on white underside of flattened tail.* Male has pair of black bars on side, extending into large blue blotches on belly. Female lacks blue blotches; black bars faint to absent.

Breeding: Clutch size 2–8, average 4, laid June to August. Multiple clutches common in southern part of range. Hatchlings appear July to November.

Habitat: Areas of hard-packed soil with little vegetation; occasionally among small rocks.

Range: C. Nevada and extreme sw. Utah, south through Arizona and se. California into Mexico.

Subspecies: About 12; 3 in our range.
Nevada (*C. d. myurus*), nc. Nevada.
Mojave (*C. d. rhodostictus*), s. Nevada and extreme sw. Utah to se. California and w. Arizona.
Arizona (*C. d. ventralis*), sc. Arizona and n. Mexico.

Diurnal. These lizards are swift runners, curling their tails over their backs to expose the "zebra" stripes. When disturbed, they curl tail and wag it. They eat anything they can catch: insects, spiders, smaller lizards. Diet also occasionally includes flowers.

361 Greater Earless Lizard
(*Cophosaurus texanus*)
Subspecies: Texas, Southwestern

Description: 3¼–7¼" (8–18.4 cm). Relatively large, with *smooth granular back scales, 2 folds across throat, no external ears.* Light flecks on gray to brown, depending on color of habitat. *Wide black crossbands under flattened tail.* Male has a pair of curved black bars within blue blotch on side just in front of hind leg. Female lacks blue blotch, usually has no black bars; has orange throat and pinkish sides when gravid.

Breeding: Some 5 eggs are laid monthly from March to August, for season total of 25.

Habitat: Stretches of broken rock, limestone cliffs, dry sandy streambeds, rocky washes.

Range: C. Arizona through c. and s. Texas and into Mexico.

Subspecies: Three; 2 in our range.
Texas (*C. t. texanus*), e. New Mexico and nc. Texas into Mexico. Southwestern (*C. t. scitulus*), se. Arizona to sc. New Mexico and extreme w. Texas into Mexico.

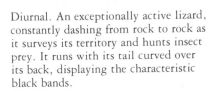

Diurnal. An exceptionally active lizard, constantly dashing from rock to rock as it surveys its territory and hunts insect prey. It runs with its tail curved over its back, displaying the characteristic black bands.

355, 356 Collared Lizard
(*Crotaphytus collaris*)
Subspecies: Eastern, Western, Yellow-headed, Chihuahuan, Sonoran

Description: 8–14" (20–35.6 cm). Large head; *conspicuous black-and-white collar across back of neck. Inside of mouth dark.* Tail not flattened side to side. Yellow-brown to green with bluish highlights and usually light spots and dark bands.

Mature male has blue-green or orange throat without black center seen in Desert Collared Lizard. Gravid female has red-orange spots and bars on sides. Young show alternating dark and light crossbanding.

Breeding: Mates April to June; lays 1–12 eggs in midsummer. Hatchlings are about 3½" (8.9 cm) long.

Habitat: Hardwood forests to arid areas with large rocks for basking. More frequent in hilly regions, especially among limestone ledges that provide crevices for good cover.

Range: E. Utah and Colorado to extreme sw. Illinois, south through c. Texas, into Mexico and west into c. Arizona.

Subspecies: Six; 5 in our range.
Eastern (*C. c. collaris*), s. Missouri through n. Arkansas, west to n. and c. Texas.
Western (*C. c. baileyi*), c. Arizona and wc. New Mexico.
Yellow-headed (*C. c. auriceps*), upper Colorado and Green River basins of e. Utah, w. Colorado, and n. New Mexico and Arizona east of Colorado River.
Chihuahuan (*C. c. fuscus*), se. Arizona, sw. New Mexico and extreme w. Texas, and south into Mexico.
Sonoran (*C. c. nebrius*), s. Arizona into Mexico.

Diurnal. A wary, feisty lizard that will bite readily and hard, given the chance. It feeds on insects and other lizards. When fleeing would-be captors, it lifts body and tail and dashes along on its hind legs, giving it the appearance of a fierce little dinosaur.

---

359 Desert Collared Lizard
(*Crotaphytus insularis*)
Subspecies: Mojave, Baja

Description: 6–13" (15.2–33.0 cm). Large head; *conspicuous black-and-white collar across*

back of neck. Inside of mouth light-colored.
Tail flattened from side to side.
Tan to olive, with pale yellow
crossbands. Mature males have blue-
gray throats with a black center; also
large dark blotches in the groin. Gravid
females show red-orange spots on sides.
Young usually have vivid crossbanding,
lack dark throat coloring and groin
blotches.

Breeding: Mates May to June; lays 3–8 eggs in
midsummer.

Habitat: Rocky terrain; canyons, gullies, similar
protected areas where vegetation is
sparse.

Range: Se. Oregon, adjacent Idaho, and w.
Utah, south into Arizona and se.
California.

Subspecies: Three; 2 in our range.
Mojave (*C. i. bicinctores*), California,
Oregon, Idaho, Nevada, and w. Utah
and Arizona.
Baja (*C. i. vestigium*), Riverside County,
California, south through Baja
peninsula.

Diurnal. These lizards are frequently
found basking on large boulders. They
are wary and hard to catch during
midday when temperatures are warm.
They feed on insects and make hardy
pets, but will eat other lizards.

360 Reticulate Collared Lizard
(*Crotaphytus reticulatus*)

Description: 8–16¾" (20–42.5 cm). Large head,
slender neck. Tail slightly flattened side
to side. Gray to reddish-brown, with an
*open netlike pattern of narrow light-colored
lines.* Large black spots arranged in rows
across back. Male has black collar
markings, yellow-orange tint on chest
during breeding season. Female has
faint black collar or none. Gravid
female has red bars on sides.

Breeding: Mates April and May; 8–11 eggs, laid

in midsummer, hatch 60–90 days
later. Hatchling size about 3½–4″
(8.9–10.2 cm).

Habitat: Semidesert brushland, escarpments,
isolated rock piles, pack rat burrows.

Range: Rio Grande valley of s. Texas and
adjacent Mexico.

Diurnal. These are shy lizards that dash
for cover at the slightest hint of danger.
When retreat is impossible, they defend
themselves by arching the back and
opening the mouth. They do not
hesitate to bite. Voracious predators,
they feed chiefly on grasshoppers but
also relish insects, spiders, reptiles, and
small mammals.

333 Spiny-tailed Iguana
(*Ctenosaura pectinata*)

Description: 12–48″ (30.5–121.9 cm). One of the
largest lizards in our range. Gray
to brown to yellow, with ill-defined
broad dark crossbands. Raised crest of
scales down back, most prominent in
males. Small smooth scales on trunk;
large keeled spines ringing tail. *3 rows
of small scales between rows of enlarged
spines around tail base.* Limbs dark.
Hatchlings green.

Breeding: Mates early spring. Clutches of up to
50 eggs are laid in burrows, April to
May.

Habitat: Open rocky terrain with holes and
crevices deep enough to provide cover.

Range: Introduced into s. Texas at Brownsville
and s. Florida at Miami; native to
Mexico.

Diurnal. This big lizard usually takes
refuge among rocks when approached.
But where trees are available, it may
leave the ground and climb out of
reach.

345 Desert Iguana
(*Dipsosaurus dorsalis*)

Description: 10–16″ (25.4–40.6 cm). Large round-bodied lizard with long tail and low crest of slightly enlarged keeled scales down back. Head relatively small; snout short. *Brown around head, giving way to reddish-brown netlike pattern and gray or white spotting on neck and trunk.* Reticulations may break into dark lengthwise lines on sides. Tail gray or white with encircling rows of dark spots. Breeding adults show pink on sides of belly.

Breeding: Mates April to May. Clutch of 3–8 eggs is laid June to August. Hatchlings appear August to September.

Habitat: Arid and semiarid regions of sand, scattered rocks, and creosote bush.

Range: S. California, Nevada, and w. Arizona south into Mexico.

Subspecies: Three; 1 in our range, *D. d. dorsalis.*

Diurnal. These are wary lizards that flee to the nearest rodent burrow or bush at the slightest hint of danger. They are tolerant of high temperatures and are active even at 115°F (46°C). When surface temperatures do get too hot for them, Desert Iguanas climb into bushes to reach cooler air layers. Food consists primarily of vegetable matter; insects and carrion are also eaten.

347 Blunt-nosed Leopard Lizard
(*Gambelia silus*)

Description: 8–9¼″ (20.3–23.5 cm). Head short, nose blunt. Tail round, not flattened. Narrow white to yellow crossbars separated by *wide gray or brown crossbars containing numerous dark-edged brown spots.* Throat blotched with gray.

Breeding: Mates April to June. 2–5 eggs are laid in June or July, hatch in August or September.

Habitat: Sandy areas, alkali flats, canyon floors, foothills, with sparse, open vegetation.

Range: San Joaquin Valley and surrounding foothills, California.

Diurnal. This lizard usually "freezes" when danger threatens, only to dash for cover if closely approached. Grasshoppers are among its favorite foods. Because much of its habitat has been converted to farms and communities, the Blunt-nosed Leopard Lizard is threatened with extinction. An endangered species.

357 Leopard Lizard
(*Gambelia wislizenii*)
Subspecies: Long-nosed, Cope's, Lahontan Basin, Pale

Description: 8½–15⅛″ (22–38.4 cm). Large and slender. *Gray or brown with many dark spots on body and tail.* Color darkens in cool temperatures. White crossbars usually evident on back, sides, and tail. Longitudinal gray streak down throat. Gravid females have red-orange spots and bars on sides. Tail round, not flattened.

Breeding: In spring; 4–7 eggs are laid May to July. Occasionally second clutches are laid in August.

Habitat: Semiarid regions where soil is sandy or gravelly, and vegetation sparse or in clumps.

Range: S. Oregon and Idaho to sw. Wyoming, south to w. Texas and into Mexico, west through s. California and Baja California.

Subspecies: Five; 4 in our range.
Long-nosed (*G. w. wislizenii*), large dark spots encircled by white dots; extreme s. Nevada and sw. California through Arizona and New Mexico to Big Bend region of Texas and adjacent Mexico.
Cope's (*G. w. copei*), dark spots fragmented, each part with

encircling white dots; extreme s.
California and Baja peninsula.
Lahontan Basin (*G. w. maculosus*),
large squarish dark spots without
encircling white dots; se. Oregon and
s. Idaho, w. Nevada, and ne. California.
Pale (*G. w. punctatus*), small dark spots
without encircling white dots; upper
Colorado River basin of se. Utah and
w. Colorado, south into n. Arizona and
extreme nw. New Mexico.

Diurnal. An agile lizard that darts from
bush to bush in search of insects or lies
hidden in the shade awaiting unwary
prey. It frequently eats smaller lizards.

348 Spot-tailed Earless Lizard
(*Holbrookia lacerata*)
Subspecies: Northern, Southern

Description: 4½–6″ (11–15.2 cm). *Smooth granular
back scales,* 2 folds across throat, *no
external ears.* Gray with paired dark
blotches extending from head to tail.
Up to 6 dark spots or streaks on side of
belly. *Round dark spots under tail.* Egg-
bearing female has yellow-green neck
and body.

Breeding: 4–12 eggs are laid May to June, then
again July to August. Incubation lasts
4–5 weeks; hatchlings measure 1½″
(3.8 cm).

Habitat: Arid areas with sparse vegetation;
seasonally dry prairie brushland.

Range: C. and s. Texas and adjacent Mexico.

Subspecies: Northern (*H. l. lacerata*), Edwards
Plateau region of c. Texas.
Southern (*H. l. subcaudalis*), Texas
south of the Balcones Escarpment and
into n. Mexico.

Diurnal. This wary lizard scrambles for
cover at the first sign of a human.
However, in open areas where cover is
scarce, it tires after a 50–100 yard
(50–100 m) chase and can be caught by

hand. Some elude capture by
"swimming" out of sight into loose
sand.

346, 366, 370 **Lesser Earless Lizard**
(*Holbrookia maculata*)
Subspecies: Northern, Eastern,
Speckled, Western, Bleached

Description: 4–5⅛" (10–13 cm). Small; with
smooth granular back scales, 2 folds
across throat, *no external ears*. Gray to
brownish, depending on earth color.
Usually has lengthwise rows of dark
blotches separated by pale stripe down
center of back and dorsolateral light
stripes. No spots under tail. Male has
pair of blue-bordered black marks on
each side of belly behind foreleg.
Female has orange throat during
breeding season.

Breeding: Mates during spring and summer. Lays
an average of 5–7 eggs, April to
September; hatchlings appear May to
October.

Habitat: Sandy soil areas in grassy prairie,
cultivated fields, dry streambeds, desert
grasslands.

Range: S. South Dakota through the Great
Plains to c. Texas, west through most
of New Mexico and Arizona into
Mexico.

Subspecies: Northern (*H. m. maculata*), s. South
Dakota to Texas panhandle, west into
e. New Mexico, Colorado, and extreme
se. Wyoming.
Eastern (*H. m. perspicua*), ec. Kansas
south through to c. Texas.
Speckled (*H. m. approximans*), c.
Arizona through New Mexico to
extreme w. Texas, south into Mexico.
Western (*H. m. thermophila*), s. Arizona
below 5,000' (1,500 m) and south
into Mexico.
Bleached (*H. m. ruthveni*), restricted to
White Sands region (Otero County),
New Mexico.

Diurnal. Loss of the external ear may be an adaptation to this lizard's habit of burrowing headfirst into sand. It subsists on insects and small spiders.

365 Keeled Earless Lizard
(*Holbrookia propinqua*)
Subspecies: Texas

Description: 4½–5⁹⁄₁₆″ (11–14.1 cm). A rather small lizard with *keeled granular back scales,* 2 folds across throat, *no external ear.* Gray to brown. *No spots under relatively long tail.* Male has unbordered black marks on side of belly behind foreleg. Female usually paler, lacks paired side marks.

Breeding: Midsummer. Hatchlings about 1½″ (3.8 cm).

Habitat: Sand dunes and barrier beaches.

Range: S. Texas and into Mexico.

Subspecies: Two; 1 in our range. Texas (*H. p. propinqua*).

Diurnal. This species is abundant among the sand dunes of Padre Island, where it forages for insects in the beach vegetation.

388 Common Iguana
(*Iguana iguana*)

Description: 40–79″ (101–200 cm). Large and blunt-snouted. Green to blue-gray with crest of comblike spines down back and tail, most prominent on neck. Dark bands across shoulders and tail. Color darkens with age, becoming nearly black. *Enormous black-edged scale on lower jaw below ear.* Jagged fringe on edge of dewlap. Male has larger head with splashes of orange or yellow, row of glandular pores under hind legs. Young bright green; low crest down back.

Breeding: Mates in fall. Clutches of 28–40 eggs

laid in holes excavated in earth, January to April; hatch 90 days later.

Habitat: Arboreal. Most frequently seen in large trees with dense canopies, especially in humid areas. Seems to prefer trees near or overhanging water.

Range: Introduced into Miami and Fort Lauderdale, Florida; native to Central and n. South America.

Diurnal. Adult Iguanas are agile climbers, but juveniles less than 12" (30.5 cm) long seem more terrestrial. Large Iguanas will readily take to water to avoid capture, diving and remaining on the bottom until the danger has passed. Although herbivores, Iguanas will not hesitate to bite if caught. The lashing tail is a formidable weapon, as are the sharp claws.

---

364 Curly-tailed Lizard
(Leiocephalus carinatus)
Subspecies: Northern

Description: 7–10½" (18.0–26.6 cm). Large keeled and *pointed scales on back form raised crest* on vaguely banded tail. Gray to brown; darker spotting or bands on head and sides. Throat speckled.

Breeding: Mates March to April. Eggs laid in midsummer.

Habitat: Open woods, coconut strands, beach dunes. Also, lawns and gardens.

Range: Introduced into s. Florida; native to Cuba, Bahamas.

Subspecies: Twelve; 1 in our range.
Northern (*L. c. armouri*).

Diurnal. The Curly-tail is usually found sunning itself atop a rock or log. Although it will climb, it is primarily terrestrial. Males curl the tail over the back like a loose watch spring in displays that attract females and drive away other males. The species feeds on any small prey it can catch.

358 Banded Rock Lizard
(*Petrosaurus mearnsi*)

Description: 8½–11½" (21.5–29.0 cm). A large flat-bodied lizard with a *single black collar and prominently banded tail*. Gray to olive, with wavy dark crossbands and many small white or bluish spots. Scales small and granular on back; large, keeled, and pointed on tail. Gravid females show orange on throat and over eyes.

Breeding: A clutch of 2–6 eggs is laid June to July.

Habitat: Rocks and large boulders in arid canyons.

Range: S. California into n. Baja California.

Diurnal; most active in the early morning. This lizard is adept at dashing about right side up or upside down among the rocks. If not hurried, it waddles with body close to the rock surface, limbs extended well out from sides and hindquarters swinging.

340 Texas Horned Lizard
(*Phrynosoma cornutum*)

Description: 2½–7⅛" (6.3–18.1 cm). Flat-bodied lizard with large crown of spines on head; *2 center spines longest.* 2 rows of pointed scales fringe each side. *Belly scales keeled.* Red to yellow to gray; dark spots have light rear margins. *Dark lines radiate from eye.*

Breeding: Mates April to May. Clutch of 14–37 eggs is laid in burrow dug by female May to July. Young hatch in some 6 weeks, measure about 1¼" (3.1 cm).

Habitat: From sea level to 6,000' (1,800 m) in dry areas, mostly open country with loose soil supporting grass, mesquite, cactus.

Range: Kansas to Texas and west to se. Arizona. Isolated population in Louisiana; introduced in n. Florida.

Diurnal. This lizard is the common "horned toad" of the pet trade. But since it feeds almost exclusively on live large ants—generally unavailable to the pet owner—most pet horned lizards slowly starve to death over a period of months.

342 Coast Horned Lizard
(*Phrynosoma coronatum*)
Subspecies: San Diego, California

Description: 2½–6⅜" (6.4–16.1 cm). Flat-bodied; large crown of spines on head; *2 center spines longest. Large dark spot on sides of neck.* 2 rows of pointed scales fringe trunk. Reddish, brown, yellow, or gray; dark blotches on back. Belly scales smooth.

Breeding: Clutches of 6–21 eggs are laid May to June, hatch July to September.

Habitat: Open areas of sandy soil and low vegetation; frequently found near ant colonies.

Range: Most of w. California into Baja California.

Subspecies: Several; 2 in our range.

San Diego (*P. c. blainvillei*), head scales smooth and convex, larger toward center; along coast of s. California, from Los Angeles to Baja.
California (*P. c. frontale*), head scales rough and equal in size; along the coast north of San Francisco Bay to Los Angeles, also inland in the Sacramento Valley.

Diurnal. This lizard inflates with air when frightened, making it a difficult prey to swallow. It threatens would-be captors with open mouth and hissing noises. It will bite in defense and, if that does not suffice, will spray an intruder with blood from the corners of its eyes.

337, 338 Short-horned Lizard
(*Phrynosoma douglassi*)
Subspecies: Pygmy, Eastern, Mountain,
Desert, Salt Lake

Description: 2½–5⅞" (6.3–14.9 cm). Flat-bodied;
head crowned by stubby spines interrupted at
rear by deep notch in skull. 1 row of
pointed scales fringes trunk. Belly
scales smooth. Gray, yellowish, or
reddish-brown; 2 rows of large dark
spots down back.

Breeding: Litters of 6–31 are born alive, July to
August.

Habitat: Varies, from open rocky or sandy plains
to forested areas; from sea level to above
9,000' (2,700 m).

Range: S. British Columbia to n. California, s.
Idaho, and most of Utah; s.
Saskatchewan, southeast to Kansas, and
south into Mexico. Separate populations
in w. Texas.

Subspecies: Numerous; 5 in our range.

Pygmy (*P. d. douglassi*), brownish to
bluish-gray, with very small head
spines pointed vertically; s. British
Columbia through Washington, e.
Oregon and s. Idaho to n. California
and n. Nevada.
Eastern (*P. d. brevirostre*), ill-defined
light border around dark spots, spines
horizontal; c. Montana and adjacent
Canada south to w. Dakotas, Nebraska,
nw. Kansas, and e. Colorado.
Mountain (*P. d. hernandesi*), reddish,
spines prominent and horizontal; s.
Utah and w. Colorado south through
Arizona, nw. New Mexico into Mexico.
Desert (*P. d. ornatissimum*), dark spots
partly light-bordered, spines horizontal;
extreme sc. Wyoming through w.
Colorado to New Mexico and e.
Arizona, with disjunct populations in
se. New Mexico and w. Texas.
Salt Lake (*P. d. ornatum*), nearly
uniform gray, spines horizontal; c. and
n. Utah, ne. Nevada, and extreme se.
Idaho.

Diurnal. This species is most active
during the midday warmth. At night it
burrows into the soil. It feeds primarily
on ants, but occasionally eats other
insects, snails, sow bugs, even small
snakes.

334 Flat-tailed Horned Lizard
(*Phrynosoma m'calli*)

Description: 3–4¾" (7.6–12.0 cm). Flat-bodied;
long slender spines crown head, 2 rows of
pointed scales fringe trunk. Midbelly
scales smooth. Rusty brown to gray;
dark stripe down spine. *Tail long, flat.*
Breeding: Mates April to May. Clutches of 7–10
eggs are laid May to June.
Habitat: Dunes and other regions of fine
windblown sand with little vegetation.
Range: Se. California and adjacent Arizona and
Mexico.

Diurnal. In midday this lizard will
burrow into sand to avoid the heat.
Burrowing is usually accomplished by
quick side-to-side shuffling. It also
burrows to escape the cool of night.

336 Round-tailed Horned Lizard
(*Phrynosoma modestum*)

Description: 3–4⅛" (8.0–10.5 cm). Flat-bodied;
*short crown of spines on head. No pointed
scales fringing sides.* Light brown to
gray, depending on surrounding soil;
usually a dark blotch on neck, another
in groin. Tail broad at base, becoming
round and slender. Belly scales
smooth.
Breeding: Mates in May; clutch of 9 eggs laid
June to July, hatches July to August.
Habitat: Sandy, gravelly washes and other semiarid regions of scrub vegetation.
Range: Se. Arizona through s. New Mexico to
w. Texas, and south into Mexico.

Diurnal. Camouflage coloration is the chief protection of the Round-tailed. When danger threatens, it flattens itself against the sand and remains motionless, practically disappearing from sight. Like many other horned lizards, it feeds primarily on ants.

339, 341 Desert Horned Lizard
(*Phrynosoma platyrhinos*)
Subspecies: Northern, Southern

Description: 3–5⅜" (7.6–13.6 cm). Flat-bodied; *relatively short spines crown head;* 1 row of pointed scales fringes trunk. Belly scales smooth. Red, tan, or dark gray, with wavy crossbands on sides of head, dark blotches on sides of neck.

Breeding: Mates April to May; 6–10 eggs are laid June to July.

Habitat: Areas of sandy, gravelly soil, windblown sand; flat arid stretches where rocks or scrub vegetation are present.

Range: Se. Oregon and s. Idaho south through e. California and w. Arizona into Mexico.

Subspecies: Three; 2 in our range.

Northern (*P. p. platyrhinos*), speckled belly, rounded tail; Oregon and s. Idaho south through w. Utah and most of Nevada.
Southern (*P. p. calidiarum*), speckled belly, slightly flattened tail; extreme s. Nevada and Utah to w. Arizona and se. California into n. Baja California.

Diurnal. If discovered in the open this lizard usually sits quietly, depending on camouflage for safety. When near vegetation, it will dash for cover under the nearest bush. If provoked, it hisses, threatens to bite.

335 **Regal Horned Lizard**
(*Phrynosoma solare*)

Description: 3½–6½" (8.8–16.7 cm). Flat-bodied;
head crowned by large, close-set spines.
Single row of pointed scales fringes
trunk. Scales smooth at midbelly,
weakly keeled on chest.

Breeding: Mates in summer. Clutches of 17–28
eggs are laid July to August, hatch
September to October.

Habitat: Rocky and gravelly areas supporting
scrub vegetation or succulents;
sometimes sandy regions.

Range: S. Arizona into Mexico.

Diurnal. This species is most active
during early morning, with a second,
smaller show of activity just before
sunset. When caught, it may become
rigid, with lungs deflated and legs
extended. If set down in this condition,
it will flop on to its back. Some will
squirt blood from the corner of
the eye.

331 **Chuckwalla**
(*Sauromalus obesus*)
Subspecies: Western, Glen Canyon,
Arizona

Description: 11–16½" (27.9–41.9 cm). *A large
potbellied lizard with loose folds of skin
around neck and shoulders.* Tail thick at
the base and blunt at the tip. No
enlarged rostral scale on nose. Male
black on head, forelegs, and forward
portion of trunk; red, gray, or yellow
toward tail. Female and young tend to
be crossbanded with gray and
yellow.

Breeding: Mating is believed to occur May to
June. Clutches of 5–10 eggs are laid
June to August but females may lay
only every second year.

Habitat: Open flats and rocky areas, especially
where large boulders are present.

Range: Se. California, s. Nevada, Utah, w. Arizona and adjacent Mexico.

Subspecies: Four; 3 in our range.

Western (*S. o. obesus*), more than 50 scales around mid-foreleg, single row of femoral pores; California, Nevada, Utah, and w. Arizona.

Glen Canyon (*S. o. multiforaminatus*), more than 50 scales around mid-foreleg, 2 rows of femoral pores; Colorado River gorge between Garfield County, Utah, and Page County, Arizona.

Arizona (*S. o. tumidus*), fewer than 50 scales around mid-foreleg, male red toward tail; s. Arizona and into Mexico.

Diurnal. On emerging in the morning, this lizard basks until its preferred body temperature of 100°F (38°C) is reached, whereupon it begins searching for food. Strictly herbivorous, it browses on leaves, buds, flowers, fruit. A frightened Chuckwalla retreats into a rocky crevice and wedges itself in sideways by inflating its body. It can sometimes be coaxed into backing out of the crack by repeatedly tapping its snout with a stick.

349, 372 Clark's Spiny Lizard
(*Sceloporus clarki*)
Subspecies: Sonoran, Plateau

Description: 7½–12¹³⁄₁₆″ (19.0–32.5 cm). *Greenish sheen on back; narrow dark crossbands on forelegs and feet.* Less distinct crossbands on back, hind legs, tail. Black mark on shoulder may extend to neck to form incomplete collar. Male has black and blue throat, blue belly patches; female usually lacks belly patches.

Breeding: Eggs are laid in clutches of 4–22, May to November. More than 1 clutch may be laid during season.

Habitat: Semiarid regions. Found in trees as well

as on ground among rock piles and outcroppings.

Range: C. and s. Arizona to sw. New Mexico and into Mexico.

Subspecies: Three; 2 in our range.
Sonoran (*S. c. clarki*), adults lose dark crossbands on body; se. Arizona and sw. New Mexico except Colorado Plateau, south into Mexico.
Plateau (*S. c. villaris*), adults retain dark crossbands; c. Arizona on edge of Colorado Plateau.

Diurnal. More at home in trees than on the ground, this wary lizard is seen as high as 30′ (10 m) from the ground. It takes refuge in rat nests built among the tree limbs. When approached, it flees to the opposite side of the tree. It feeds not only on insects but leaves and blossoms as well.

352 Blue Spiny Lizard
(*Sceloporus cyanogenys*)

Description: 5–14¼″ (13–36.2 cm). Largest of the spiny lizards. *White-bordered black collar on neck. Light spots on head and back.* Bright blue over shoulders. Indistinct dark bands on tail. Males have *metallic blue-green sheen on back,* blue chins and throats, blue patches on sides of belly. Female lacks metallic blue sheen and (usually) belly patches; has gray throat.

Breeding: Young are born alive in litters of 6–18, February to June.

Habitat: Stone piles, rock outcroppings, rock walls, dry earth banks. Also found in construction rubble.

Range: S. Texas along the Rio Grande and into Mexico.

Diurnal. In rocky areas, these big lizards shelter in holes beneath and between the rocks. Otherwise they secret themselves in deep ground crevices. Although their diet includes a

wide variety of invertebrates, flying
insects seem preferred.

377 Sagebrush Lizard
(*Sceloporus graciosus*)
Subspecies: Northern, Dunes, Southern

Description: 5–6³⁄₁₆″ (12.7–15.75 cm). A spiny
lizard. *Granular scales do not overlap on
rear of thigh.* Grayish-green to brown;
some darker spots and crossbars. *Faint
light dorsolateral stripes.* Sides reddish-
orange behind forelegs. Males usually
have light-blue mottling (not patches)
on throat and darker blue belly patches.
Females have pinkish orange on sides
and neck.

Breeding: Single clutch of 2–7 eggs is laid June
to July, hatches July to August.

Habitat: Primarily areas of sagebrush and
gravelly soils or fine-sand dunes. Never
far from shelter such as stony piles,
crevices, animal burrows.

Range: S. Montana to nw. New Mexico and
west to Washington, Oregon,
California, and Baja California.

Subspecies: Northern (*S. g. graciosus*), with blue
patches that do not meet across belly
and do not meet blue of throat;
Washington to Montana, south to New
Mexico, west to e. California, north to
coast in n. California and Oregon.
Dunes (*S. g. arenicolus*), with blue
patches that do not meet across belly,
throat without blue; found in sand
dunes of w. Texas and se. New Mexico.
Southern (*S. g. vandenburghianus*), with
blue patches that usually meet across
belly and touch blue of throat; s.
California into n. Baja peninsula.

Diurnal. Primarily terrestrial, these
lizards occasionally climb trees or
bushes in pursuit of insect prey.

371 Mesquite Lizard
(*Sceloporus grammicus*)
Subspecies: Northern

Description: 4–6⅞" (10–17.5 cm). A spiny lizard. Granular scales on rear of thigh do not overlap. Gray to olive, with 4–6 dark crossbars on back, more distinct in females. *No dorsolateral light stripes.* Males show small dark triangular patch in front of shoulder, blue belly patches, some black and blue on throat.

Breeding: Mates in fall. Young born alive, January to March. Litter size 4–12, average 6.

Habitat: Arid and semiarid terrain supporting mesquite and other low-growing trees and shrubs.

Range: Extreme s. Texas and into Mexico.

Subspecies: Three; 1 in our range. Northern (*S. g. disparilis*), scales on side of neck abruptly smaller than those on back of neck.

Diurnal. This lizard, which spends most of its time high in trees, is perfectly colored to blend with the mottled bark of mesquite. Shy and unobtrusive, it is not easy to find, especially since it flees to the backside of trunk or limb long before it can be seen.

353 Yarrow's Spiny Lizard
(*Sceloporus jarrovi*)

Description: 5–8¾" (12.7–22.2 cm). Striking, with *black-edged light-blue or pinkish scales on back.* Black collar with white rear edge. Male has blue throat, blue sides of belly; female is more subdued in pattern and color, may lack black collar.

Breeding: 3–13 young are born alive in summer, usually in June.

Habitat: Found among rocky outcrops in woodlands, at elevations of 5,000–

10,000′ (1,500–3,000 m).

Range: Se. Arizona and sw. New Mexico.

Subspecies: One in our range, *S.j. jarrovi.*

Diurnal. The Yarrow is not nearly so wary as most other lizards. It can be observed for considerable lengths of time sunning itself on rocks and tree stumps that also serve as shelter.

350 Desert Spiny Lizard
(*Sceloporus magister*)
Subspecies: Desert, Twin-spotted, Barred, Yellow-backed, Orange-headed

Description: 7–12″ (17.8–30.5 cm). A large rough-scaled lizard. Yellow to brown, with some crossbanding and a *black triangular mark with light rear edge on each shoulder.* Blue throats and blue patches on sides of belly; absent in females and young, who show more prominent crossbanding.

Breeding: Mates in spring and early summer; 7–19 eggs are laid May to July, incubation lasts 8–11 weeks. More than 1 clutch may be laid in a season.

Habitat: Arid and semiarid areas at low elevation where vegetation and rocks provide adequate cover.

Range: S. Nevada south into Baja California and southeast through Arizona, New Mexico, and w. Texas. Isolated population in c. California.

Subspecies: Nine; 5 in our range.

Desert (*S. m. magister*), male has dark purple to black band down center of back, bordered by narrow light stripes; sw. Arizona into Mexico.
Twin-spotted (*S. m. bimaculosus*), male has 2 rows of prominent dark spots down back and usually a dark stripe behind eye; se. Arizona through c. New Mexico to sw. Texas, south into Mexico.
Barred (*S. m. transversus*), male usually has several distinct crossbands on back

and dark patches near forelegs; small area of ec. California and adjoining wc. Nevada.

Yellow-backed (*S. m. uniformis*), male uniform yellowish-tan, sometimes with vague darker blotches; c. and s. Nevada, extreme sw. Utah to c. Arizona, se. California and ne. Baja California, isolated population west of San Joaquin Valley, California.

Orange-headed (*S. m. cephaloflavus*), head yellowish-orange, male has several chevron-shaped bars along back; se. Utah, extreme sw. Colorado, nw. New Mexico, and ne. Arizona.

Diurnal. These are wary lizards that dart into rocky crevices, rodent holes, or vegetative cover when startled. They readily climb trees or walls in search of insect prey. Occasionally they eat flowers and leaves.

368 Canyon Lizard
(*Sceloporus merriami*)
Subspecies: Merriam's, Big Bend, Presidio

Description: 4½–6¼" (11–15.9 cm). A spiny lizard with *scales on sides of neck abruptly smaller than those on top of neck.* Back scales keeled, very small; side scales granular, do not overlap. Gray to bluish, with 2 lines of about 10 small dark spots running down middle of back. Dark vertical line on neck in front of shoulder.

Breeding: Eggs are laid in spring and early summer; hatch in mid-summer. More than 1 clutch may be laid during season.

Habitat: Canyon walls, rocky slopes, with well-drained soils and near absence of vegetation.

Range: Sw. Texas and adjacent Mexico.

Subspecies: Four; 3 in our range.
Merriam's (*S. m. merriami*), no dark

bands under tail or dark bars on throat;
Valverde, Crockett, Terrell, and e.
Brewster counties, Texas, and extreme
nw. Mexico.
Big Bend (*S. m. annulatus*), broad dark
bands under tail and dark bars on
throat, squarish dark marks on back; w.
Brewster County, Texas.
Presidio (*S. m. longipunctatus*), dark bars
on throat, incomplete dark bars under
tail, and paired marks on back like
arched eyebrows; Presidio Co., Texas.

Diurnal. Despite the sparsity of cover
where these lizards are found, they are
not particularly wary. Often
they can be approached closely and
caught by hand.

379 Western Fence Lizard
(*Sceloporus occidentalis*)
Subspecies: Northwestern, Island, San
Joaquin, Coast Range, Great Basin,
Sierra

Description: 6–9¼" (15.2–23.5 cm). Spiny; *scales
on back of thigh abruptly smaller*. Scales
on back same size as those on sides and
belly. Olive, brownish, or black, with
pattern of paired blotches or wavy
crossbars down back and occasionally
some striping. Undersurfaces of legs
yellowish-orange. Blue patches on sides
of belly; adult male has blue patch on
throat.

Breeding: Mates early spring. Single clutch of 3–
14 eggs, laid May to July, hatches July
to September.

Habitat: Rocky and mixed forest areas from sea
level to above 9,000' (2,700 m).
Adapts to wide variety of conditions
but not to desert. Frequents stone
fences, fence posts, old buildings.

Range: C. Idaho south through Nevada and
west to the Pacific coast.

Subspecies: Northwestern (*S. o. occidentalis*), 2 blue
throat patches often with light-blue

connecting band, belly light; c. Washington to c. California.

Island (*S. o. becki*), black throat patch; only on Channel Islands off coast of s. California.

San Joaquin (*S. o. biseriatus*), 1 blue throat patch, belly gray or black with blue patches; only in lower San Joaquin Valley, California.

Coast Range (*S. o. bocourti*), throat patches small in males, absent in females; from San Mateo to Santa Barbara County, California.

Great Basin (*S. o. longipes*), 1 large blue throat patch in males, belly gray to black with blue patches; c. Oregon and extreme se. Washington to c. Idaho south through e. California, Nevada, and w. Utah, also c. California south along coast into n. Baja California.

Sierra (*S. o. taylori*), entire belly and throat blue in adult males; in Sierra Nevada of California, usually above 7,000′ (2,100 m).

Diurnal. Easily encountered; may be observed even in midwinter on mild days. Commonly called the blue-belly, it is often seen displaying to attract females or drive off male intruders; it bobs its head and flattens its sides, showing off the blue patches.

381 Texas Spiny Lizard
(*Sceloporus olivaceus*)

Description: 7½–11″ (19.0–27.9 cm). A large gray to rusty brown lizard with rough scales. *Most males have pale dorsolateral stripes*, and a small blue patch on each side of belly. Females have dark wavy lines across back.

Breeding: Mating and egg-laying occur during most of spring and summer. 1–4 clutches are laid each year, average 11 eggs for yearlings, and about 25 eggs for four-year-olds.

Habitat: Primarily arboreal; in mesquite, live oak, and other trees; also on man-made structures that provide shelter.

Range: Extreme sc. Oklahoma through the prairie of c. Texas and into Mexico.

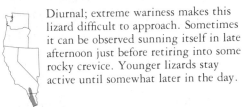

Diurnal. These lizards are skillful climbers, usually fleeing upward to avoid capture.

351 Granite Spiny Lizard
(*Sceloporus orcutti*)

Description: 7½–10⅝" (19–26.9 cm). A big lizard with large rounded *scales notched at rear on both sides of keel*. Dark copper to black; color may mask dark triangle on each side of neck and crossbands on trunk and tail. Belly and throat blue. *Male may have purple band down middle of back* and blue spots on scales.

Breeding: Mates in April. Single clutch of approximately 11 eggs is laid May to July, hatches July to September.

Habitat: Areas of large boulders and granite cliffs supporting mixed vegetation; pine and oak in more moist locations, palm and mesquite in drier ones.

Range: S. California and Baja peninsula.

Subspecies: Two, 1 in our range, *S. o. orcutti*.

Diurnal; extreme wariness makes this lizard difficult to approach. Sometimes it can be observed sunning itself in late afternoon just before retiring into some rocky crevice. Younger lizards stay active until somewhat later in the day.

354 Crevice Spiny Lizard
(*Sceloporus poinsetti*)

Description: 8½–11¼" (21.6–28.6 cm). Large rough-scaled lizard with prominent *white-edged black collar* and clearly marked *dark crossbands on tail*. Drab olive to reddish. Blue patches on throat and belly are absent in female. Young and females show more distinct crossbanding.

Breeding: 7–11 young are born alive, June to July.

Habitat: Limestone and other exposed rocky outcrops in arid and semiarid areas.

Range: S. New Mexico to c. Texas and south into Mexico.

Subspecies: Three; 1 in our range, *S. p. poinsetti*.

Diurnal. These are elusive lizards, quick to hide among stones and in crevices. They eat insects and occasionally consume blossoms or leaves that blow by.

367, 378 Bunch Grass Lizard
(*Sceloporus scalaris*)

Description: 2–3¾" (5.1–9.6 cm). A small-bodied, rather flat lizard with pointed, keeled, overlapping scales on back. *Scales on sides arranged in rows paralleling back scales.* Throat fold incomplete. Brown to yellowish-brown, usually with crescent-shaped blotches down the back and 2 pale orange stripes along sides. Dark blotch at front of shoulder. Male usually shows blue patch on belly; in female blue is reduced or absent.

Breeding: Clutch of 9–12 eggs is laid in June to July. Incubation lasts 7 weeks.

Habitat: High elevations: 4,300–10,000′ (1,300–3,000 m). Normally frequents bunch grass of sunny woodlands and open plains.

Range: Mountains of se. Arizona and extreme sw. New Mexico and into Mexico.

Subspecies: Three; 1 in our range, *S. s. slevini*.

Diurnal. These shy lizards disappear into clumps of grass at the slightest threat. They feed on insects and spiders, which they catch among the grass and leaf litter.

375 Eastern Fence Lizard
(*Sceloporus undulatus*)
Subspecies: Southern Fence, Southern Prairie, White Sands Prairie, Northern Plateau, Red-lipped Prairie, Northern Prairie, Northern Fence, Southern Plateau.

Description: 3½–7½" (9.0–19.0 cm). *Dark band along rear of thigh.* Color varies geographically: gray to brown or rusty, dark or light stripes down back, sometimes vague crossbars or spots. Back and belly scales about same size. Males usually marked by black-bordered blue patches on belly and blue throat patch.

Breeding: Mates April to August. Yearling lays 1 clutch of 3–13 eggs, June to September; older females lay 2–4 clutches. Eggs hatch June to September.

Habitat: Generally sunny locations: favors rotting logs, open woodlands, open grassy dunes, prairies.

Range: Delaware to Florida and west to New Mexico and Arizona.

Subspecies: Southern Fence (*S. u. undulatus*), wavy dark crossbars on back, blue patches on belly and throat; s. South Carolina to c. Florida, west to e. Louisiana.
Southern Prairie (*S. u. consobrinus*), weak dorsolateral light stripes, throat patches often fused, female lacks belly patches; se. Arizona to extreme s. Oklahoma, through c. Texas to Mexico.
White Sands Prairie (*S. u. cowlesi*), pale, back pattern faint or absent; White Sands, New Mexico.
Northern Plateau (*S. u. elongatus*), wavy dark crossbars on back, blue patches on throat and belly; sw. Wyoming to ne. Arizona and nw. New Mexico.
Red-lipped Prairie (*S. u. erythrocheilus*), dorsolateral stripes indistinct or absent, blue throat patches meet at midline, chin yellow, lips orange during breeding season; extreme ne. New Mexico and w. Oklahoma through e. Colorado to se. Wyoming.

Northern Prairie (*S. u. garmani*), prominent dorsolateral light stripes, throat patches reduced or absent, female lacks belly patches; se. Wyoming and s. South Dakota through e. Colorado, Kansas and Oklahoma to extreme n. portion of Texas panhandle. Northern Fence (*S. u. hyacinthinus*), wavy dark crossbars on back, belly and throat patches black-bordered; e. Kansas to New Jersey, south to c. South Carolina, southwest to e. Texas, and north through e. Oklahoma. Southern Plateau (*S. u. tristichus*), faint dorsolateral light stripes, blue throat and belly patches; c. Arizona to c. New Mexico.

Diurnal. In the east this lizard is primarily arboreal, seldom far from a tree or wall up which it will flee to avoid capture. In the prairie states it is more terrestrial, sheltering under brush or in burrows. It will eat almost any insect, spider, centipede, or snail, but beetles seem a favorite food.

373 Rose-bellied Lizard
(*Sceloporus variabilis*)
Subspecies: Texas

Description: 3¾–5½" (9.5–14.0 cm). Small spiny lizard with *pocketlike indentation in tail* behind hind leg. Light tan to greenish brown; light dorsolateral stripe, line of dark spots down each side of midline. *Pink patches* on belly with dark blue border extending up to form blue spot behind foreleg.

Breeding: Mating and egg-laying occur throughout the year. 1–5 eggs are laid near base of small tree, under rotting log, in earth, or in dry humus.

Habitat: Arid regions, from sea level to 7,500′ (2,200 m). Frequents mesquite branches, cacti, and, less often, rocks.

Range: Extreme sc. Texas to Costa Rica.

Subspecies: At least 4; 1 in our range.
Texas (*S. v. marmoratus*), more than 59
dorsal scales, female has light,
unmarked belly; s. Texas into Mexico.

Diurnal. This species can be seen
foraging in the leaf litter for large
crawling insects: roaches, beetles, bugs.
It also likes grasshoppers.

374 Striped Plateau Lizard
(*Sceloporus virgatus*)

Description: 4–6⅞" (10.2–17.4 cm). *Only U.S.
spiny lizard without belly markings.*
Reddish, with prominent light stripes
along side separated by broader dark-
brown stripe. Males show patch of blue
on side of throat. Gravid females show
orange throat patches.

Breeding: One clutch of eggs is laid each year,
usually in June; hatches July to August.

Habitat: Elevations of 5,000–9,600' (1,500–
2,900 m) in mixed hardwood and
conifer forest, especially in association
with shallow rocky or sandy streams.

Range: Extreme se. Arizona and sw. New
Mexico and adjacent Mexico.

Diurnal. This lizard searches along the
ground for its insect prey, but it can
climb readily should the need arise.

376 Florida Scrub Lizard
(*Sceloporus woodi*)

Description: 3½–5⅜" (8.9–13.7 cm). A small
spiny lizard with *broad dark-brown stripe
along each side.* Gray or tan; vague
pattern of wavy dark bars across back,
reduced or absent in male. Blue patches
on sides of throat and sides of belly.
Belly patches black-bordered in males.

Breeding: Mating begins in February. Clutches of
2–8 eggs are laid April to August.

Multiple clutches may be produced by one female. Incubation lasts 6–8 weeks.

Habitat: Sand pine and rosemary scrub, especially on old sand dunes.

Range: C. and s. Florida.

Diurnal. These lizards can frequently be found resting in the shade of a rosemary bush or foraging in leaf litter for insects. Although primarily terrestrial, they will climb trees to avoid capture.

Coachella Valley Fringe-toed Lizard
(*Uma inornata*)

Description: 5–7" (12.7–17.9 cm). *Comblike fringe of pointed scales on trailing edge of toes.* Flattened body covered with velvety granular scales. Dark bands under tail. Gray to white with dark eyelike spots, rough lines over shoulders, diagonal dark lines on throat. Sides pink during breeding season. Belly sometimes has small black spots, never has black blotches.

Breeding: Clutches of 2–4 eggs are laid throughout spring and summer.

Habitat: Arid dunes of fine windblown sand and sandy flats where vegetation is sparse.

Range: Coachella Valley and San Gorgonio Pass, Riverside County, California.

Diurnal. Food consists of a variety of insects, small lizards, occasional blossoms. At night these lizards commonly sleep in the sand under shrubs.

343 Fringe-toed Lizard
(*Uma notata*)
Subspecies: Desert, Sonoran

Description: 5–7" (12.7–17.9 cm). *Comblike fringe of pointed scales on trailing edge of toes.*

Flattened body covered with velvety
granular scales. Dark bands under tail.
Gray to white with dark eyelike spots,
rough lines over shoulder, *diagonal lines
on throat.* Black blotch on each side of
belly. Orange or pink on sides.

Breeding: Clutch of 1–5 eggs, average 2, is laid
every 4–6 weeks through summer.

Habitat: Arid stretches of windblown dunes.

Range: Se. California, sw. Arizona and adjacent
Mexico.

Subspecies: Desert (*U. n. notata*), orange patch on
side of belly; extreme s. California and
adjacent n. Baja.
Sonoran (*U. n. rufopunctata*), no orange
belly patch; extreme sw. Arizona and
adjacent Mexico.

Diurnal. This species is well adapted to
living in sand. The toe fringes act like
"snowshoes" to stop the feet from
sinking. Fringe-toed Lizards "swim"
into the sand to avoid capture, also to
avoid extreme heat or cold. The setback
jaw, scaly flaps over the ear,
overlapping eyelids, and valves in the
nostrils all serve to keep out sand while
the lizard is burrowing.

344 Mojave Fringe-toed Lizard
(*Uma scoparia*)

Description: 5–7" (12.7–17.9 cm). *Comblike fringe of
pointed scales on trailing edge of toes.*
Flattened body covered with velvety
granular scales. Dark bands under tail.
Gray to white, with dark eyelike spots;
no lines over shoulders. *Dark crescents
across center of throat.* Black blotch on
each side of belly. Breeding adults have
pink on sides and yellow-green on
belly.

Breeding: Clutches of 1–4 eggs are laid in
midsummer, probably more than 1
clutch per season. Following dry
winters little food is available, so
reproduction may not occur.

Habitat: Windblown sand dunes with low-growing vegetation.

Range: Mojave Desert in California; extreme w. Yuma County, Arizona.

Diurnal. This species can move extremely fast across the loose sand of its habitat. One specimen was clocked at a speed of 23 miles (37 km) an hour.

---

380 Long-tailed Brush Lizard
(*Urosaurus graciosus*)
Subspecies: Western, Arizona

Description: 5¾–7¼" (14.6–18.4 cm). Medium-sized lizard with fold across throat; *single band of abruptly larger scales down back.* Divided frontal scale. Tail frequently twice head and body length. Gray with dark crossbars; occasionally a light stripe along side. Throat yellow or orange. Male has bluish patches on side of belly.

Breeding: Mates throughout spring. Clutch of 4 eggs is laid in May to July, hatches July to October.

Habitat: Loose sandy desert where scrub growth is plentiful, especially creosote bushes.

Range: S. Nevada, w. Arizona, se. California and adjacent Mexico.

Subspecies: Western (*U. g. graciosus*), back pattern inconspicuous; se. California, s. Nevada and w. Arizona and Mexico.
Arizona (*U. g. shannoni*), dark markings prominent on back; sc. Arizona.

Diurnal. Although active during all the daylight hours, this lizard frequently is found resting head down on the smaller branches of a bush or tree. When discovered foraging on the ground, it instantly dashes for the safety of the nearest shrub or ground hole.

382 Small-scaled Tree Lizard
(*Urosaurus microscutatus*)

Description: 3¾–4¾" (9.5–12.0 cm). Small lizard
with fold across throat. *No abruptly
larger scales down back.* Usually single
frontal scale. Tail less than twice head
and body length. Gray with darker
crossbands or blotches. Male has yellow
spot in center of blue throat. Female
lacks the blue.

Breeding: 4–8 eggs are laid June to July, hatch
August to September. Hatchlings are
approximately 2" (5 cm) long.

Habitat: Dry areas among rocks, especially
where trees are present.

Range: Extreme s. California into c. Baja
California.

Diurnal. An active rock climber, this
lizard scoots into burrows and other
underground holes to escape predators.
It feeds on a variety of insects, but
seems to prefer crawling forms. It also
relishes spiders.

369 Common Tree Lizard
(*Urosaurus ornatus*)
Subspecies: Eastern, Canyon, Lined,
Big Bend, Colorado River, Northern

Description: 4½–6¼" (11.4–15.9 cm). Has fold
across throat; band of small scales
separates *2 bands of abruptly larger scales
down middle of back.* Divided frontal
scale. Tail less than twice head and
body length. Brown to gray, with dark
crossbands and blotches. Male has
bright blue belly patches and blue to
yellow-orange throat patches.

Breeding: Up to 6 clutches of 3–13 eggs are laid
April to September. Hatchlings appear
July to late October, according to
range.

Habitat: Trees, rocks, fence posts, and buildings
in arid regions; often near streams, and
dry washes.

Range: Extreme sw. Wyoming, southeast to
sc. Texas and west to extreme se.
California.

Subspecies: Eight; 6 in our range.

Eastern (*U. o. ornatus*), inner series of
enlarged back scales twice width of
outer series; c. and s. Texas and
adjacent Mexico.

Canyon (*U. o. levis*), 3–4 rows of
enlarged back scales, belly patches not
connected; nc. New Mexico.

Lined (*U. o. linearis*), bands of enlarged
back scales separated by less than width
of enlarged band; c. and se. Arizona to
c. New Mexico and into Mexico.

Big Bend (*U. o. schmidti*), irregular
rows of enlarged back scales, inner
series less than twice width of outer
series, belly patches separate; Big Bend
region of Texas (west of Pecos River) to
extreme se. New Mexico.

Colorado River (*U. o. symmetricus*),
bands of enlarged back scales separated
by more than width of enlarged band,
belly patches separate; extreme se.
Nevada and California to w. Arizona
and n. Mexico.

Northern (*U. o. wrighti*), enlarged back
scales in 3–4 rows, belly patches
connected; extreme sw. Wyoming to n.
New Mexico and Arizona.

Diurnal. This lizard is often found in
pairs or groups. Shy and wary, it is
adept at hiding by agilely keeping a
tree trunk or branch between itself and
a pursuer. It is most commonly
encountered in the morning and late
afternoon, foraging for insects, spiders,
and centipedes.

363 Side-blotched Lizard
(*Uta stansburiana*)
Subspecies: Northern, California,
Nevada, Desert, Colorado.

Description: 4–6⅜" (10.0–16.2 cm). A small lizard
with small back scales, some larger
scales on head; a fold with granular
scales across throat. Brown; back
pattern may consist of blotches, spots,
speckles, stripes. *Single dark blue to
black spot on side behind foreleg.* External
ear openings.

Breeding: Mates throughout year in southernmost
range, in summer elsewhere. 3 clutches
of 2–6 eggs are laid. Sperm may be
stored in female's oviduct for up to 3
months, so at least 2 fertile clutches can
result from each mating.

Habitat: Arid and semiarid regions with coarse,
gravelly soil and low-growing
vegetation.

Range: C. Washington southeast to w. Texas
and Mexico; west to Pacific coast and
Baja California, north through c. and e.
California to c. Oregon.

Subspecies: Six; 5 in our range.

Northern (*U. s. stansburiana*), male has
light spots and blue specks, back scales
weakly keeled; c. Nevada, w. Utah,
nw. Arizona, and e. California.
California (*U. s. elegans*), conspicuous
dorsolateral stripes, fewer than 8 scales
between femoral pores; s. California, w.
Arizona, and into Mexico.
Nevada (*U. s. nevadensis*), back pattern
weak, breeding male has orange sides;
n. Nevada, Oregon, Washington, and
Idaho.
Desert (*U. s. stejnegeri*), dorsolateral
stripes evident toward front, more than
8 scales between femoral pores; c. New
Mexico, w. Texas, south to Mexico.
Colorado (*U. s. uniformis*), back pattern
weak or absent; e. Utah, w. Colorado,
ne. Arizona, and nw. New Mexico.

Diurnal. These lizards live on or near
the ground and are voracious consumers

of insects. In the northern parts of their range they become inactive in the winter; they are active on any warm day throughout the year in the southern regions.

ANGUID LIZARD FAMILY
(Anguidae)

Eleven genera of about 80 species; 5
species of *Gerrhonotus* and 3 species of
Ophisarus inhabit our range.
Anguid lizards are characterized by
elongate, shiny, and stiff bodies and
tails; closeable eyelids; external ear
openings; and tiny (or absent) legs and
toes. The stiffness is a result of an
abundance of bony armor (osteoderms)
in the skin. Many species are so stiff
they could not expand to breathe were
it not for a lengthwise flexible groove of
soft granular scales along the side. All
the species in our range have this
groove.
Most anguids are terrestrial or
burrowing. While some species will
bite, the primary defenses are fleeing,
smearing an attacker with excrement,
and giving up part of the tail. In some
species the tail vertebrae have fracture
planes along which the tail will readily
break, leaving the writhing tip to be
eaten by the predator while the rest of
the lizard crawls to safety. In legless
species, in which the tail may account
for more than half the total length, the
loss of a tail may give the impression
that the lizard has been broken in two.
Contrary to folklore, the parts will not
grow together again, but the body will
grow a new tail in several weeks.
Anguids are carnivores; they consume
insects, small mammals, and other
lizards. Most are egg-layers, but a few
mountain-dwellers bear live young.

---

448 Northern Alligator Lizard
(*Gerrhonotus coeruleus*)
Subspecies: San Francisco, Sierra,
Northern, Shasta

Description: 8¾–13″ (22.2–33.0 cm). Eyes dark.
Distinct fold along sides. Olive to

bluish; indistinct crossbands on back or transverse blotches of darker brown. *Dark stripes along edges of belly scales.* Young have broad light stripe down the back, no crossbands.

Breeding: Live-bearing. Mates in April at low elevations, in June in highlands. Litters of 2–15 are born in 7–10 weeks.

Habitat: Under rotten logs, rocks, or loose bark in cool, moist woodlands to about 10,500' (3,200 m).

Range: Along coast and in Sierra Nevada range, n. California to s. British Columbia, southeast into n. Idaho and w. Montana. Disjunct populations in extreme ne. California (Modoc County) and s. Oregon (Lake County).

Subspecies:  San Francisco (*G. c. coeruleus*), dark blotches on back, occasionally fused into lengthwise band, back scales strongly keeled; California coast north and south of San Francisco Bay.
Sierra (*G. c. palmeri*), dark markings on back indistinct or absent, back scales strongly keeled; Sierra Nevada mountains of California.
Northern (*G. c. principis*), nearly unmarked or black on sides with light mid-back lengthwise band, back scales weakly keeled; Cascade mountains of Oregon and Washington to Victoria Island, southeast into n. Idaho and w. Montana.
Shasta (*G. c. shastensis*), many dark blotches on back (sometimes in irregular crossbands), back scales weakly keeled; s. Oregon into Cascades and Sierra Nevada mountains of n. California.

Diurnal. This species prefers cooler temperatures than most lizards; hence its occurrence at high elevations. It remains active throughout the day, feeding on insects, millipedes, spiders, and snails.

446 Arizona Alligator Lizard
(*Gerrhonotus kingi*)

Description: 7½–12½" (19.0–31.7 cm). Eyes light
orange to pink. Distinct fold along
sides. Brown; prominent crossbands on
back. *Spots and bars on belly; no stripes.*
Black and white spots on upper lip. Young
brightly crossbanded.

Breeding: Lays 9–12 eggs at a time, June to July.

Habitat: High dry grasslands, open woods of fir,
pine, or chaparral; 5,000–7,000'
(1,500–2,100 m).

Range: Se. Arizona and sw. New Mexico,
south into Mexico along the Sierra
Madre Occidental.

Subspecies: Three; 1 in our range, *G. k. nobilis.*

Diurnal. This lizard is primarily a
ground dweller. It is active throughout
the day.

Texas Alligator Lizard
(*Gerrhonotus liocephalus*)

Description: 10–20" (25.0–50.8 cm). Eyes yellow.
Distinct fold along sides. Yellow to
reddish-brown; *light crossbands*
on back and tail. Belly gray mottled with
white, sometimes with dark lines down
center of scales. Young prominently
crossbanded.

Breeding: Throughout the year; 1 or more
clutches of 5–31 eggs. Female broods
eggs during incubation.

Habitat: Rocky slopes with some scrub
vegetation.

Range: Edwards Plateau and Big Bend regions
of Texas, south into Mexico.

Subspecies: Six; 1 in our range.
Texas (*G. l. infernalis*), dark marks on
belly, tail less than twice head and
body length.

Diurnal. This lizard is rather slow and
deliberate in its movements.

445, 449 Southern Alligator Lizard
(*Gerrhonotus multicarinatus*)
Subspecies: California, Oregon,
San Diego

Description: 10–16⅞" (25.4–42.8 cm). Eyes
yellow. Distinct fold along sides.
Reddish brown to yellowish gray;
*distinct crossbands on back and tail. Dark
stripes on center of belly scales.* Young have
light band along back.

Breeding: 2–3 clutches of 1–41 eggs (average 12)
are laid during the warm months.

Habitat: Grasslands, open woods, moist areas;
wherever cover is plentiful, but
primarily oak woodlands of foothills.

Range: S. Washington south through w.
Oregon and California into n. Baja
California.

Subspecies: Five; 3 in our range.
California (*G. m. multicarinatus*), dark
mottling on head, red blotches down
middle of back; n. California south
along coast and Cascade Mountains to
Los Angeles, also on San Miguel, Santa
Rosa, and Santa Cruz islands.
Oregon (*G. m. scincicauda*), no dark
marks on top of head, scales of upper
foreleg smooth; s. Washington and
Columbia River Basin into w. Oregon
and n. California.
San Diego (*G. m. webbi*), no dark marks
on head, 3–4 rows of keeled scales on
upper foreleg; Sierras east of the San
Joaquin Valley southwest to the coast
near Los Angeles, also San Nicolas and
Santa Catalina Islands, and into Baja
California.

Primarily diurnal. This lizard can
sometimes be found climbing bushes as
it searches for insects and other small
prey. Its stiffly prehensile tail can wrap
over small branches to aid climbing. If
grabbed roughly, it may bite or
defecate. It feeds on almost any animal,
vertebrate or invertebrate, that it can
catch and swallow.

447 Panamint Alligator Lizard
(*Gerrhonotus panamintinus*)

Description: 10–15" (25.4–38.1 cm). Eyes pale
yellow. Distinct fold along sides. Pale
yellow to tan; *broad dark crossbands.*
Belly scales mottled with light gray.
Young brightly crossbanded.

Breeding: Mates in spring, presumably lays eggs.
Little else is known about its
reproduction.

Habitat: Damp areas with some undergrowth,
beneath rocks and among rock slides;
2,800–5,100' (853–1,500 m).
Frequently found near willow-shaded
watercourses.

Range: Panamint, Nelson, and Inyo mountains
of ec. California.

Diurnal. The Panamint is quick to
escape an intruder by darting under
boulders or rocky talus. But it seems
much more reluctant than other
alligator lizards to climb trees or
shrubs.

455 Slender Glass Lizard
(*Ophisaurus attenuatus*)
Subspecies: Western, Eastern

Description: 22–42" (56–106.7 cm). Stiff legless
lizard with eyelids and external ear
openings. *White markings in center of*
back scales. Groove along sides, *dark*
stripes or speckling below groove.
Prominent dark stripe down middle of
back.

Breeding: Mates in May. Clutches of 6–17 eggs
laid June to July. Female broods eggs.

Habitat: Dry grassland and dry open woodland.

Range: Virginia to Florida, west to Texas,
Oklahoma, and Nebraska, north to s.
Wisconsin, and Illinois.

Subspecies: Western (*O. a. attenuatus*), tail less than
2½ times head and body length; s.
Wisconsin, Illinois and w. Indiana,
south to Louisiana, west to Kansas,

Oklahoma, and Texas.
Eastern (*O. a. longicaudus*), tail more
than 2½ times head and body length;
e. coastal states, Virginia to e.
Louisiana, and Tennessee and s.
Kentucky.

Diurnal. When held, this lizard
squirms and wriggles vigorously in an
attempt to escape. If the tail is grabbed
or takes a blow, it may shatter into
several pieces—hence the name glass
lizard.

454 Island Glass Lizard
(*Ophisaurus compressus*)

Description: 15–24″ (38.0–61.0 cm). Stiff legless
lizard with eyelids and external ear
openings. *White markings near edges of
back scales.* Groove along sides; *no dark
stripes or pigment below groove.* Sometimes
poorly defined dark stripe down middle
of back.

Breeding: Average clutch: 4 eggs.

Habitat: Dry grass and scrub vegetation along
coast; old dunes inland.

Range: Coast and offshore islands of South
Carolina, Georgia, peninsular Florida.

Diurnal. This species can frequently be
found under tidal debris on sandy
beaches. Unlike other U.S. glass
lizards, it lacks fracture planes in the
tail. Consequently the tail, although it
will break, does not do so easily.

453, 456 Eastern Glass Lizard
(*Ophisaurus ventralis*)

Description: 18–42⅝″ (46.0–108.3 cm). Stiff
legless lizard with eyelids and external
ear openings. *White markings near edges
of back scales.* Groove along sides; *no
dark stripes or pigment below groove;* no

dark stripe down middle of back.

Breeding: Clutch of 8–17 eggs, laid in spring or summer, hatch in 2 months. Females remain with eggs during incubation.

Habitat: Damp grassland, open woods.

Range: Coastal Plain of North Carolina, through Florida to e. Louisiana.

Diurnal. Best time to observe this lizard is early morning, when it forages in the damp grass lining highways near marshes and wet savannas. Tail shatters readily.

⊗ GILA MONSTER FAMILY
(Helodermatidae)

One genus of 2 species confined to North America.

Gila monsters are heavy-bodied, with short stout legs and a thick tail. With an abundance of food the tail becomes fat. During periods of scarcity, the tail may lose up to 20% of its bulk as the stored fat is used up. On the back and head are non-overlapping beadlike scales of shiny black, pink, or yellow. The scales contain osteoderms, a sort of bony armor.

Gila monsters are venomous carnivores that seem to rely more on taste and smell to locate their quarry than on vision. They have been observed tracking prey by tasting the ground with their long thick tongues. Their venom causes great pain, but rarely human death. It is the persistent grip of a Gila monster's jaw that is the single most important reason for handling these lizards carefully.

332 Gila Monster
(*Heloderma suspectum*)
Subspecies: Reticulate, Banded

Description: 18–24″ (45.7–61.0 cm). Large and heavy-bodied. Small beadlike scales on back. *Broken blotches, bars, spots of black and yellow, orange, or pink, with bands extending onto blunt tail.* Face black.

Breeding: Mates throughout summer months. 3–5 eggs are laid, fall to winter.

Habitat: Arid and semiarid regions of gravelly and sandy soils, especially areas with shrubs and some moisture. Found under rocks, in burrows of other animals, sometimes in holes it digs itself.

Range: Extreme sw. Utah, s. Nevada and adjacent California, south through s.

Arizona and sw. New Mexico to
Mexico.

Subspecies: Reticulate (*H. s. suspectum*), adults
mottled and blotched; c. and s. Arizona
and adjoining New Mexico:
Banded (*H. s. cinctum*), adults have
broad double crossband; sw. Utah
through s. Nevada and adjoining
California and w. Arizona.

Primarily nocturnal, although also
active on warm winter or spring days.
Gila Monsters and the related Mexican
Beaded Lizards are the only venomous
lizards. Their bite, although rarely fatal
to humans, serves to overpower animal
predators and prey. Produced in glands
lying along the lower jaw, the poison is
not injected like that of a snake but
flows into the open wound as the lizard
chews on its victim. Most prey is small
enough to be taken easily without
venom: small birds, eggs, small
rodents, and other lizards.

CALIFORNIA LEGLESS LIZARD FAMILY
(Anniellidae)

One genus of 2 species found only in California and adjacent Baja California, Mexico. Only 1 species, *Anniella pulchra,* occurs in our range.
These are long and slender lizards with movable eyelids. Although legless, they still retain hip and shoulder bones internally. They are related to the anguid lizards, from which they differ by the absence of external ear openings, and bony armor within the skin. California legless lizards burrow in loose soil and are seldom seen on the surface. They eat insects; young are born alive.

---

452 California Legless Lizard
(*Anniella pulchra*)
Subspecies: Silvery, Black

Description: 6–9¼" (15.2–23.5 cm). *Shiny legless lizard with eyelids but without external ear openings.* Usually silvery or tan; dark stripe down middle of back, dark striping on sides. Back sometimes completely dark. Belly yellow.

Breeding: Mates May to June. Litters of 1–4 young (average 2) are born alive, September to November.

Habitat: Areas of loose soil: sandy loam, sand dunes, gravelly banks of streams. Prefers some vegetation, can be found in leaf litter.

Range: W. California from San Francisco Bay south into Baja California.

Subspecies: Silvery (*A. p. pulchra*), silvery above; San Francisco Bay south through w. California into n. Baja California.
Black (*A. p. nigra*), adults dark brown to black above; found only on California coast from Monterey Bay to Morro Bay.

Nocturnal. Its small smooth scales and blunt tail make burrowing easy for this lizard. Most of its time is spent under the surface of the soil or beneath leaf litter, where it pursues small insects and insect larvae. Use of pesticides on agricultural lands has decimated some populations.

NIGHT LIZARD FAMILY
(Xantusidae)

Three genera of 18 species found only in North America and Cuba; 3 species occur in our range.

The night lizards are related to the geckos and like them have soft skin, a somewhat flattened body, and no movable eyelids. But unlike the geckos the night lizards have small round scales on the back, large rectangular scales on the belly, and large shields on the head. Their toes end in a sharp claw.

As the name implies, night lizards are mostly active at night, hiding in rocky crevices or under brush and leafy debris during the day. Their light-sensitive eyes have vertically elliptical pupils. Developing embryos are nourished by a primitive placenta while in the mother's oviduct. Young are born tail-first and alive.

404, 405, 407 **Granite Night Lizard**
(*Xantusia henshawi*)

Description: 4–5⅝" (10.0–14.3 cm). Flat-bodied. No eyelids; pupils vertical. Light-colored with large dark-brown to black spots that expand by day, contract at night. Skin is soft. Head scales large, symmetrical; back scales small and granular; belly scales large, square, and smooth, with *14 scales across midbelly.*

Breeding: Mates in late spring; 1–2 young are born alive in September.

Habitat: Arid and semiarid territory, especially in cooler areas among large peeling or flaking rocks and outcroppings.

Range: S. California into Baja California.

Nocturnal. This extremely wary and secretive lizard can be encountered only at dusk or after dark. Sometimes it can be found by prying loose the rocky flakes under which it hides. The collector should take care to restore the habitat by replacing the flakes.

408 Island Night Lizard
(*Xantusia riversiana*)

Description: 5⅜–8¼" (13.7–20.9 cm). No eyelids; pupils vertical. Gray to yellowish brown; dark mottling, sometimes a light dorsolateral stripe. Skin soft, appears loose around neck and shoulders. Back scales small and granular; head scales large, symmetrical; belly scales large, square, smooth; *16 scales across midbelly.*

Breeding: 3–9 young are born alive in September.

Habitat: Rocky beaches, areas of sparse vegetation.

Range: San Clemente, Santa Barbara, and San Nicolas islands off coast of s. California.

Primarily nocturnal. An endangered species; its restricted habitat is threatened by introduced livestock and also by human activities. Occasionally active during the day; it feeds on insects, spiders, scorpions, small crustaceans, leaves, seeds, blossoms.

403, 406 Desert Night Lizard
(*Xantusia vigilis*)
Subspecies: Desert, Arizona, Sierra, Utah

Description: 3¾–5⅟₁₆" (9.5–12.8 cm). No eyelids; pupils vertical. Olive, yellow, brown, or orange; usually many small dark spots tending to form rows. Skin soft. Head scales large, symmetrical; back scales large, square, and smooth; 12 *scales across midbelly.*

Breeding: Mates May to June. 1–3 young born

alive, tail first and upside down,
September to October.

Habitat: Arid and semiarid granite outcroppings
and rocky areas, among fallen leaves
and trunks of yuccas, agaves, Joshua
trees.

Range: S. Nevada, s. Utah, and w. and c.
Arizona through s. California into
Mexico.

Subspecies: Six; 4 in the U.S.

Desert (*X. v. vigilis*), olive brown
above, spots tend to fuse, smallest
subspecies; s. Nevada and extreme sw.
Utah and nw. Arizona into s. California
and n. Baja California; disjunct
population in sw. Arizona (Kofa
Mountains).
Arizona (*X. v. arizonae*), yellow or gray
above, largest subspecies; c. Arizona
along s. edge of Colorado Plateau
(Mohave, Pinal, and Yavapai counties.
Sierra (*X. v. sierrae*), spots fused into
netlike pattern, broad light stripe
behind eye; s. California (Kern County)
in sw. foothills of Sierra Nevada
Mountains.
Utah (*X. v. utahensis*), yellow-orange
above; sc. Utah (Garfield County).

Diurnal. Activity may continue after
dusk, hence the name. This species
feeds on termites, ants, beetles, and
flies encountered under decaying
vegetation or rocks. The tail breaks off
easily.

WHIPTAIL AND RACERUNNER FAMILY
(Teiidae)

Forty genera of about 230 species confined to the New World and most abundant in South America; 16 species of *Cnemidophorus* are native to the U.S., and 1 species of *Ameiva* has been introduced.

Teiids are long slender lizards with long whiplike tails and well-developed legs. Movements are characteristically rapid and jerky. They range from 4–48 inches (10–122 cm) in length.

Typically they have small round, non-overlapping scales on the back and large rectangular scales on the belly. There are no bony plates (osteoderms) in the skin. The large regular head shields are fused to the skull.

Whiptails are diurnal, terrestrial carnivores. Small species feed on insects and other invertebrates, while large species consume small mammals, birds and bird eggs, and other reptiles. Prey is located both by sight and by smell or taste, by means of a long protrusible, deeply-forked tongue.

All teiids are egg-layers. In most species, females produce fertile eggs only after breeding with males of the same species. However, there are true unisexual species among *Cnemidophorus*. All individuals are females, so there is no mating. A mature female lays fertile —but unfertilized—eggs that hatch into more females.

Jungle Runner
(*Ameiva ameiva*)
Subspecies: Amazon

Description: 15–25" (38.0–63.5 cm). A large lizard of streamlined appearance. Blue to brown with rows of light spots across back; sometimes a pale brown stripe along middle of back. *Fine granular scales on back;* 10 or more lengthwise rows of large smooth rectangular belly scales. More than one large preanal scale.

Breeding: Mates and lays clutch of 1–4 eggs seasonally in some areas, any part of year in others.

Habitat: Open sunny, grassy areas, usually on sand or sandy loam.

Range: Introduced into Miami, Florida, and surrounding district; native to n. South America.

Subspecies: Amazon (*A. a. petersi*), introduced in Hialeah (Miami), Florida.

Diurnal. The Jungle Runner is an extremely alert lizard whose scratch-here, poke-there foraging behavior can best be described as fidgety. The slightest provocation sends it sprinting for the nearest hole or thick brush. It feeds on insects, spiders, crustaceans, and smaller lizards.

416 Giant Spotted Whiptail
(*Cnemidophorus burti*)
Subspecies: Giant, Red-backed

Description: 11–17¾" (27.0–45.1 cm). Long slender lizard; blue-gray to gray-green with profuse light spotting and 6 or 7 light stripes, prominent in juveniles, faded or absent in large adults. Back scales small and granular. Belly uniform gray to white; 8 lengthwise rows of large, smooth rectangular belly scales. *Head and back rusty red.* Tail brown; orange in young.

Breeding: Clutches of 1–4 eggs are laid presumably in summer.

Habitat: Rocky, grassy, and brush-covered semiarid areas, often near small streams.

Range: S. Arizona and extreme sw. New Mexico, south to Mexican coast.

Subspecies: Three; 2 in our range.

Giant (*C. b. stictogrammus*), larger subspecies, large light spots, red on head and neck only; s. Arizona, sw. New Mexico and n. Sonora. Red-backed (*C. b. xanthonotus*), red on head and body, but not low on sides; sc. Arizona (Pima County).

Diurnal. Subsists on insects and their larvae, plus spiders.

Gray Checkered Whiptail
(*Cnemidophorus dixoni*)

Description: 8–12½" (20.3–31.8 cm). Slender. 10–12 light stripes over black squarish areas create effect of *light net over dark background*. Throat and belly uniform white; 8 lengthwise rows of large, smooth rectangular belly scales. Hips and base of tail russet. Juveniles more strongly striped.

Breeding: Unisexual; no mating. Eggs laid June to July, hatch in about 45 days.

Habitat: Gravelly alluvial slopes; degraded grassland, creosote bush, ocotillo.

Range: Presidio County, Texas. Hidalgo County, New Mexico.

Diurnal. These lizards can be found foraging in the vegetative litter beneath shrubs. When frightened, they flee into the nearest hole.

---

418 Chihuahuan Spotted Whiptail
(*Cnemidophorus exsanguis*)

Description: 9½–12⅜" (24.1–31.4 cm). Slender, with 6 light stripes separated by dark-brown bands; *light spots in both stripes and bands*. Small granular back scales, 5–8 scales between light stripes down middle of back. Belly uniform light gray to white; 8 lengthwise rows of large smooth rectangular belly scales.

Tail blue-gray to green. Juveniles have orange tail and light spots in interposed dark bands.

Breeding: Unisexual; no mating. Clutches of 1–6 eggs are laid June–August and hatch about a month and a half later.

Habitat: Desert, desert grasslands, and mountain woodlands, especially pine-oak.

Range: Se. Arizona and c. New Mexico south into w. Texas and Mexico.

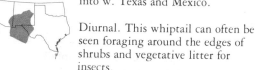

Diurnal. This whiptail can often be seen foraging around the edges of shrubs and vegetative litter for insects.

Gila Whiptail
(*Cnemidophorus flagellicaudus*)

Description: 8–12″ (20.3–30.6 cm). Slender; 6 light greenish-yellow stripes separated by dark-brown bands, a few *light spots in both stripes and dark bands.* Back scales small and granular. 3–6 scales between 2 light stripes down middle of back. Throat and belly white; 8 lengthwise rows of large, smooth rectangular belly scales. Tail olive green. Juveniles lack spots on dark bands.

Breeding: Unisexual; no mating.

Habitat: Woodlands and forest in the Gila River Basin.

Range: C. and se. Arizona, sw. New Mexico.

Diurnal. Until 1964, when it was officially described, this species was confused with both the Chihuahuan Spotted Whiptail and the Sonoran Spotted Whiptail.

Texas Spotted Whiptail
(*Cnemidophorus gularis*)

Description: 6½–11 (16.5–27.9 cm). Slender; 7–8 light stripes separated by *light-spotted*

dark brown to green bands most prominent on sides. Wide light stripe, sometimes split in two, down middle of back. Stripes faded or absent in some populations. Back scales small and round; 8 lengthwise rows of large, smooth rectangular belly scales. Male has reddish throat, bright blue belly, usually black spot on chest. Female has pale pink throat, white belly. Tail pink to brownish-orange. Juveniles are brightly striped; tail and hips reddish.

Breeding: Mates April to May. Clutch of 1–5 eggs is laid May or June, a 2nd clutch in late July.

Habitat: Semiarid prairie grasslands, open brushy areas; also arid washes and canyons, frequently in vicinity of streams.

Range: Se. New Mexico to s. Oklahoma, south through Texas into Mexico.

Subspecies: Six to eight; 1 in our range, *C. g. gularis.*

Diurnal. This lizard can be seen scratching in sand or dry leaves in search of prey. It will eat insects and spiders, but termites, grasshoppers, and caterpillars are its staple fare. When disturbed, it dashes off only a few feet before pausing to see if it is being pursued.

409 Orange-throated Whiptail
(*Cnemidophorus hyperythrus*)

Description: 6⁷⁄₁₆–8¹⁄₁₆″ (16.4–20.5 cm). Slender, with 5 or 6 light stripes separated by dark-brown to black bands with no spots. Back scales small, granular. *Throat orange.* Belly uniform light blue-gray to white; 8 lengthwise rows of large, smooth rectangular scales. Tail gray; blue in juveniles.

Breeding: Mates April to July. Clutch of 1–4 eggs is laid in June, again in July; Hatch in 50–55 days.

Habitat: Chaparral country and similar arid and semiarid areas, most frequently where sand or loose soil and rocks are present.

Range: Sw. California into Baja California and the offshore islands.

Subspecies: Eight; 1 in our range, *C. h. beldingi.*

Diurnal. This whiptail feeds on spiders and a variety of insects, especially termites. Males defend food and water sources against other males by arching the back, compressing the sides, twitching the tail tip, and pointing the snout at the ground. This threat display drives off interlopers without actual contact.

414 Little Striped Whiptail
(*Cnemidophorus inornatus*)
Subspecies: Arizona, Trans-Pecos

Description: 6½–9⅜" (16.5–23.8 cm). Slender, with 6 to 8 light stripes separated by dark reddish-brown to black bands without spots. Small granular scales on back. Throat and belly blue, intense in males, pale in females; 8 lengthwise rows of large smooth rectangular belly scales. *Tail brownish at base, remainder bright blue.*

Breeding: Mates in spring. Clutch of 2–4 eggs, laid May to July.

Habitat: Arid and semiarid grasslands with some low brush; flatlands, gentle slopes.

Range: New Mexico, extreme se. Arizona, w. Texas south into Mexico.

Subspecies: Four; 2 in our range.

Arizona (*C. i. arizonae*), interposed dark bands blue-gray, belly light blue; se. Arizona and w. New Mexico.
Trans-Pecos (*C. i. heptagrammus*), dark bands black, belly dark blue; w. Texas, se. New Mexico and n. Mexico.

Diurnal. Beetles, grasshoppers, and spiders are the chief food of this whiptail. When frightened, it tends to

seek cover under shrubs, then abandon
them in favor of burrows if the threat
persists. Populations are densest where
vegetation is scattered, neither too open
nor too thick.

Laredo Striped Whiptail
(*Cnemidophorus laredoensis*)

Description: 6–11⅝" (15.2–28.7 cm). Slender; 7
light stripes separated by *dark-green to
greenish-brown bands;* small ill-defined
light spots only in dark bands on sides.
Stripe down middle of back not as wide
as adjacent light stripes. Back scales
small and granular. Throat and belly
uniform white; 8 lengthwise rows of
large smooth rectangular belly scales.
Tail greenish-brown, striped on side.

Breeding: Unisexual; no mating. One or more
clutches of 1–4 eggs are laid June to
July; hatch July to August.

Habitat: Semiarid regions of sparse vegetation
near streambeds and hills.

Range: Webb County, Texas.

Diurnal. This all-female species is
found in the same habitat as the Texas
Spotted Whiptail.

415 New Mexican Whiptail
(*Cnemidophorus neomexicanus*)

Description: 8–11⅞" (20.3–30.2 cm). Slender; 7
light stripes on sides separated by light-
spotted dark bands. *Wavy or zigzag light
stripe down middle of back,* forked on
head. Small granular scales on back.
Throat pale blue or blue-green. Belly
uniform white or pale blue; 8
lengthwise rows of large, smooth
rectangular scales. Tail gray at base,
gray-green toward tip; bright blue in
juveniles.

Breeding: Unisexual; no mating. Clutch of 2–4

eggs, laid in summer, hatches in 50–60 days.

Habitat: Grassy areas where periodic flooding occurs: sandy arroyos, washes, playas, sandy river bottoms.

Range: Rio Grande valley of New Mexico and extreme w. Texas into Mexico.

Diurnal. This species seems to prefer disturbed areas where vegetation is sparse and ditches, fences, and trash piles abound. When frightened it flees to the shelter of these human artifacts.

417 Plateau Spotted Whiptail
(*Cnemidophorus septemvittatus*)

Description: 8–12½" (20.3–31.8 cm). Slender; 6 or 7 light stripes separated by dark brownish-green bands with light spots. Sometimes a light narrow stripe down middle of back. In adults, spotting is more pronounced and stripes faded or absent toward rear. Back scales small and granular. Throat white in males, orange in females, occasionally with scattered black scales. Belly white or blue-white; 8 lengthwise rows of large, smooth rectangular belly scales. Hips and base of tail rusty red; remainder of tail brown to gray, blue-green in juveniles.

Breeding: Eggs are laid in midsummer.

Habitat: Rocky terrain with sparse vegetation, from low desert foothills to mountains.

Range: Big Bend region of Texas, south into Mexico.

Diurnal. This lizard seems less fidgety and more deliberate than other whiptails in its hunting habits. It preys on insects.

411 Racerunner
(*Cnemidophorus sexlineatus*)
Subspecies: Six-lined, Prairie Lined

Description: 6–10½" (15.2–26.7 cm). Slender, with 6 *or 7 light stripes separated by dark-greenish brown to black bands without spots.* Sometimes a narrow light stripe down middle of back. Back scales small and granular. Throat green or blue in males, white in females. Belly white or blue-white; 8 lengthwise rows of large, smooth, rectangular belly scales. *Tail brown,* prominently striped on sides; light blue in juveniles.

Breeding: Mates April to June. Clutch of 1–6 eggs, laid June to July; 2nd clutch 3 weeks later. Eggs hatch June to September.

Habitat: Dry sunny areas; grasslands, open woodlands, usually on well-drained soils.

Range: Delaware south through Florida, west to Texas, New Mexico, Colorado, se. Wyoming, South Dakota.

Subspecies: Six-lined (*C. s. sexlineatus*), 6 light stripes; Atlantic coast to e. Texas and c. Missouri.
Prairie Lined (*C. s. viridis*), 7 light stripes, head and body bright green; c. Texas and w. Missouri to se. Wyoming, e. Colorado and New Mexico.

Diurnal. The Racerunner is most active in the morning, when it can be seen basking or hunting for insects. It avoids cool seasonal and night temperatures by burrowing into the soil.

410 Sonoran Spotted Whiptail
(*Cnemidophorus sonorae*)

Description: 8–11⅓" (20.3–28.7 cm). Slender, with 6 light-yellow stripes separated by dark brown bands, *light spots in both stripes and dark bands.* Small granular scales on back. *5–8 scales between two light stripes down middle of back.* Throat and belly white; 8 lengthwise rows of large, smooth rectangular belly scales. Tail olive. Juveniles lack spots in dark bands.

Breeding: Unisexual; no mating.

Habitat: Desert grassland and mountain oak-pine woodlands.

Range: Se. Arizona, extreme sw. New Mexico, and adjacent Mexico.

Diurnal. This species is closely related to the Gila Whiptail and the Chihuahuan Spotted Whiptail.

419 Checkered Whiptail
(*Cnemidophorus tesselatus*)

Description: 11–15½" (27.9–39.4 cm). Long and slender, with 6 faint light stripes separated by *bold black checks, bars, or spots.* Spotting faded in some populations. Back scales small and granular. Throat and belly white; small scattered black spots most prominent on chin and chest; 8 lengthwise rows of large, smooth rectangular belly scales. Tail yellow or brown, dark spots on sides. Juveniles have dark bands with a few light spots.

Breeding: Primarily unisexual; only a few males have been found. Unmated female lays clutch of 2–8 eggs in June to July, which hatch in August.

Habitat: Rocky locations on sand or gravel supporting grass or sparse brush.

Range: S. Colorado and New Mexico, w. Texas into adjacent Mexico.

Diurnal. This whiptail is a more agile rock climber than others. Food includes scorpions as well as the usual insects and spiders. Many geographically localized color forms have been identified in this all-female species.

420 Western Whiptail
(*Cnemidophorus tigris*)
Subspecies: Great Basin, Southern,
Marbled, Coastal, California, Northern.

Description: 8–12" (20.3–30.5 cm). Slender; *4–8
light stripes, often with many dark spots
and lines on light gray or tan.* Stripes and
spotting sometimes faded or absent.
Throat and belly usually white or
yellow (rarely all black), with black
spotting on chest. 8 longitudinal rows
of large smooth rectangular belly scales.
Tail gray or gray-green, usually with
black speckling on sides; bright blue in
juveniles.

Breeding: Mates April to May. Clutch of 1–4
eggs is laid in June in northern range.
In south 1st clutch is laid in May, 2nd
in July. Eggs hatch July to August.

Habitat: Arid and semiarid desert to open
woodlands; where vegetation is sparse
enough to make running easy.

Range: Baja California and California to e.
Oregon and s. Idaho, south to w. Texas
and Mexico.

Subspecies: About 15; 6 in U.S.
Great Basin (*C. t. tigris*), usually 4
light stripes, vertical dark bars on sides;
e. Oregon and sw. Idaho, south
through c. Utah and w. Arizona into
extreme nw. Sonora, e. Baja peninsula,
north through e. California.
Southern (*C. t. gracilis*), 4–6 distinct
brown stripes with many light spots in
interposed dark bands, throat and chest
black; s. Arizona to extreme sw. New
Mexico and n. Sonora.
Marbled (*C. t. marmoratus*), faded light
gray-brown stripes and spots, vertical
dark bars, chest salmon, s. New Mexico
through w. Texas into Mexico.
Coastal (*C. t. multiscutatus*), 8 ill-
defined light stripes, large black spots
on throat; coast of s. California and w.
Baja California.
California (*C. t. mundus*), 8 light
stripes, distinct large dark spots and
black spots on throat; nc. California

south to c. California coast, isolated
population in nc. Oregon.
Northern (*C. t. septentrionalis*), 6 fairly
distinct yellow stripes, small black
spots on throat; w. Colorado to nw.
New Mexico, west through s. Utah and
n. Arizona.

Diurnal. This species digs burrows
both for safe retreats and to find
underground prey. Like most whiptails,
it stalks any small moving object, even
fluttering leaves. Insects, scorpions,
spiders, and daddy-longlegs are eaten.

412 Desert-Grassland Whiptail
(*Cnemidophorus uniparens*)

Description: 6½–9⅜″ (16.5–23.8 cm). Slender; 6
*or 7 light stripes separated by dark reddish-
brown or black bands without spots.* Back
scales small, granular. Throat white or
blue-white; chin blue. Belly uniform
white; 8 lengthwise rows of large,
smooth rectangular belly scales. *Tail
olive green;* blue in juveniles.

Breeding: Unisexual; no mating. Clutch of 1–4
eggs, laid in summer, hatches in 50–
55 days.

Habitat: Arid and semiarid grassland, desert
scrub.

Range: C. Arizona to extreme w. Texas, and
south into adjacent Mexico.

Diurnal. The range of this species
seems to be expanding as more and
more grassland is degraded to desert
scrub. But this expansion means that
the grassland range of the Little Striped
Whiptail is shrinking.

413 Plateau Striped Whiptail
(*Cnemidophorus velox*)

Description: 8–10¾" (20.3–27.3 cm). Slender; *6 or 7 light stripes separated by dark-brown or black bands without spots.* Back scales small, granular. Throat white or blue-white; chin blue-green. Belly uniform white or pale blue-green. 8 lengthwise rows of large, smooth rectangular belly scales. *Tail light blue;* bright blue in juveniles.

Breeding: Unisexual; no mating. Clutch of 3–5 eggs, laid June to July, hatches in August.

Habitat: Pinon-juniper woodland and ponderosa pine forest, 5,500–6,000′ (1,600–1,800 m).

Range: W. Colorado south to c. New Mexico, west to c. Arizona, north through se. Utah.

Diurnal. While it is foraging in leaf litter beneath bushes this lizard can be approached quite closely before it takes flight. Even then, it may flee only a short distance.

TYPICAL OLD WORLD LIZARD FAMILY
(Lacertidae)

Lacertids are native to Europe, Africa, and Asia; 1 species of *Lacerta* and 1 of *Podarcis* have been introduced into the U.S.

Lacertids are characterized by slender round bodies, well-developed legs, and long tails. The back is covered with hexagonal scales; belly scales are large and rectangular. The large head shields usually contain bony plates (osteoderms) and are not fused to the skull. Most species have movable eyelids. Terrestrial species are variously colored, while arboreal species are green. Males have slightly larger heads and may be differently colored from females.

Lacertids are daytime hunters of insects, spiders, scorpions, and small vertebrates. All lacertids are egg-layers, save one north Eurasian species, *Lacerta vivipara*.

386 Green Lizard
(*Lacerta viridus*)

Description: 9–15¼" (22.8–38.7 cm). Long and slender. *Green or greenish-brown with fine dark stippling;* sometimes light dorsolateral stripes. Breeding males may be blue on side of head; juveniles are brown with stripes. Small granular scales on back; 6–8 lengthwise rows of large, smooth rectangular belly scales. Single large preanal scale.

Breeding: Mates April to May. Clutch of 5–20 eggs laid May to June, hatch July to August. Some females lay 2nd clutch about time 1st clutch is hatching.

Habitat: Dense vegetation, shrubs, open woodlands, frequently in moist areas.

Range: Introduced into Topeka, Kansas; native to Mediterranean Europe.

Subspecies: Five; 1 in our range; *L. v. viridus.*

The Green can frequently be seen basking in early morning. It is equally at home on the ground and in shrubbery. When pursued it flees into rodent burrows or thick brush. It feeds on insects, spiders, smaller lizards, small bird eggs, fruit.

---

387 Ruin Lizard
(Podarcis sicula)
Subspecies: Southern, Northern

Description: 6–9¹³⁄₁₆″ (15.2–25.0 cm). Slender; green, yellow, or light brown, usually with *dark stripe or row of spots along middle of back;* often with light dorsolateral stripes. Small granular scales on back, 6 lengthwise rows of large smooth rectangular belly scales. Single large preanal scale.

Breeding: Probably mates in spring; eggs laid in early summer.

Habitat: Grassy fields, stone walls, gardens, city parks.

Range: Introduced into Philadelphia, Pennsylvania, and Long Island, New York; native to Italy and Mediterranean islands, also Spain, Turkey.

Subspecies: 40 European; 2 in our range.

Southern (*P. s. sicula*), scattered dark spots or netlike pattern; established in West Hempstead, New York. Northern (*P. s. campestris*), dark stripe or row of spots down middle of back; introduced in Philadelphia.

Diurnal. Although primarily terrestrial, this lizard will climb to escape capture. It seems adept at surviving in urban environments. It feeds on a variety of insects, leaves, and blossoms.

SKINK FAMILY
(Scincidae)

A cosmopolitan group of 87 genera and 1,280 species found on every continent, except Antarctica, and many oceanic islands, most abundant in the tropics of southeast Asia and the Indo-Australian archipelago. Fifteen species of 3 genera, *Eumeces, Neoseps,* and *Scincella,* inhabit the U.S.

A representative skink has a long cylindrical body and tail covered with smooth sleek scales containing bony plates (osteoderms). Terrestrial skinks have small legs; those adapted for burrowing have tiny legs or none. A clear window in the lower eyelid of burrowing forms enables the animal to see when the eyelid is closed to keep out dirt. Fracture planes in the tails of many species allow the tail to break off easily when grasped by a predator. In such cases usually the tail is vividly colored to draw the attack of the predator away from the vulnerable body.

Skinks are diurnal. Most are insect eaters, but a few giant species are herbivorous. All species have thick oval tongues with a shallow notch at the tip. Most will nip the hand that catches them, and large species can inflict a painful bite. All North American species are egg-layers. Females may tend the eggs during incubation.

425, 429 Coal Skink
(*Eumeces anthracinus*)
Subspecies: Northern, Southern

Description: 5–7" (13–17.8 cm). Brown, with *4 light stripes extending from neck onto tail*. Dark band on side, more than 2 scale rows wide, separates pair of light side stripes. Upper light stripe follows edge where 3rd and 4th scale rows meet,

counting from middle of back. No
stripes on head. Upper labial scales, 7.
Male sometimes has reddish color on
head. Young have blue tail.

Breeding: Mates in spring, early summer. Clutch
of 8 or 9 eggs is laid in June, hatches in
4–5 weeks.

Habitat: Damp wooded areas with abundant leaf
litter or loose stones.

Range: Scattered populations; w. New York
through Appalachians to Gulf Coast,
Louisiana and Missouri west to c.
Kansas, Oklahoma, and Texas.

Subspecies: Northern (*E. a. anthracinus*), young
striped like adults; w. New York and c.
Pennsylvania, also scattered south
through the Appalachians to Kentucky
and North Carolina.
Southern (*E. a. pluvialis*), young black
or with only an occasional trace of light
stripes; c. Kansas and s. Missouri
through ne. Texas and nw. Louisiana,
e. Gulf coast from s. Louisiana to w.
Florida panhandle, with scattered
populations in Alabama, Georgia, and
extreme w. South Carolina.

Diurnal. Coal Skinks readily dive into
water to avoid capture. When this
happens, they can usually be located by
turning over stream-bed rocks, under
which they tend to hide.

435, 436, 438, **Mole Skink**
440 (*Eumeces egregius*)
Subspecies: Florida Keys, Peninsula,
Northern, Blue-tailed, Cedar Key.

Description: 3½–6½" (8.9–16.5 cm). Relatively
long-bodied brownish skink with 4
light stripes on head and body. Upper 2
stripes confined to 2nd scale rows
counting from middle of back. Legs
tiny; 5 toes. Ear opening partly closed.
Supraocular scales, 3. *Tail orange, red,
pink, or blue.* Breeding male has reddish
chin and belly.

Breeding: Mates September to October. Clutch of
2–9 eggs is laid in nest burrow, April
to June. Female tends eggs.

Habitat: Sandy soils of coastal dunes; inland
sandhill scrub and turkey-oak. Also,
under rocks and tidal wrack on beaches.

Range: Coastal plain of Georgia, Alabama, and
Florida including Keys.

Subspecies: Florida Keys (*E. e. egregius*), tail red or
reddish-brown; Florida Keys and Dry
Tortugas.
Peninsula (*E. e. onocrepis*), tail red to
yellow to violet, upper light stripes
widen or diverge posteriorly;
throughout Florida peninsula except for
central ridge and Everglades.
Northern (*E. e. similis*), tail red or red-
brown; Georgia, Alabama, and n.
Florida.
Blue-tailed (*E. e. lividus*), tail blue,
upper light stripes widen or diverge;
central sand ridge of Florida peninsula
(Highlands and Polk counties).
Cedar Key (*E. e. insularis*), tail dark
orange or maroon, ill-defined upper
light stripes; Cedar Keys (Levy
County), Florida.

Diurnal. Adapted for tunneling and
digging, as its name implies, this
species successfully preys on burrowing
or secretive insects, spiders, and small
crustaceans.

427, 437, 443 **Five-lined Skink**
(*Eumeces fasciatus*)

Description: 5–8$\frac{1}{16}$″ (12.7–20.5 cm). Black or
brown with 5 *broad light stripes*,
including dorsolateral stripe along 3rd
and 4th scale rows counting from
middle of back. Stripes fade with age;
adults may be uniform brown. Tail
blue to gray. Wide lengthwise row of
scales under tail. Breeding males
usually have red-orange head. Juveniles
have brilliant striping, bright blue tail.

Breeding: Mates April to May. Clutch of 4–15 eggs is laid in nest excavation, May to June; hatches June to August. Female tends eggs.

Habitat: Humid woodlands with decaying leaf litter, stumps, logs. May be seen in gardens and around houses.

Range: S. New England to n. Florida, west to e. Texas, north to Kansas, Wisconsin and s. Ontario.

Diurnal. Terrestrial; the Five-lined Skink climbs only to bask on stumps or the lower reaches of tree trunks. It feeds on insects and their larvae, spiders, earthworms, crustaceans, lizards, even small mice.

430, 434 Gilbert's Skink
(*Eumeces gilberti*)
Subspecies: Greater Brown, Variegated, Northern Brown, Western Red-tailed, Arizona

Description: 7⅛–12⅞" (18.1–32.6 cm). Yellow-brown, with *4 light stripes extending to base of tail;* upper stripes cover adjoining edges of 2nd and 3rd scale rows counting from middle of back. Striping fades with age; some adults uniform brown or speckled. *Tail blue or yellow-pink.* Supraocular scales, 4; upper labial scales, usually 8. Juveniles have prominent striping and bright tail color. Breeding adults have reddish heads.

Breeding: Clutch of 5–9 eggs is laid in midsummer in burrows or rotting logs.

Habitat: Open grassy areas, especially where rocks provide ready retreats. Also found in sunny woodlands and around streams and springs.

Range: Sierra Nevada and Coast Ranges of c. and s. California; also scattered populations in mountains of e. California, s. Nevada, c. Arizona, n. Baja California.

Subspecies:  Greater Brown (*E. g. gilberti*), juvenile tail blue, adults lack stripes; c. portions of Sierra Nevada, California.

Variegated (*E. g. cancellosus*), juvenile tail pink below and blue above, adults barred or speckled; wc. area of California, including e. Contra Costa and Alameda counties, sw. San Joaquin County, and nw. Merced County.

Northern Brown (*E. g. placerensis*), juvenile tail blue, adults retain some striping on neck; nc. California in the foothills of the n. Sierra Nevada and the counties of Placer, Sacramento, Eldorado, Amador, and San Joaquin.

Western Red-tailed (*E. g. rubricaudatus*), juvenile tail pink, adults uniform; s. Sierra Nevada and Coast Range, San Gabriel and San Bernardino mountains, and scattered populations into s. Nevada, and in n. Baja California.

Arizona (*E. g. arizonensis*), juvenile tail yellow above and pink below, adults retain striping; c. Arizona in Yavapai and Maricopa counties.

Diurnal. Gilbert's Skink can often be encountered sunning itself or foraging in a sunny grassy plot. Prey consists of insects and spiders.

426, 444 Southeastern Five-lined Skink
(*Eumeces inexpectatus*)

Description: 5½–8½" (14–21.6 cm). Black or brown skink with *5 narrow light stripes*, including dorsolateral stripe along 5th, or 4th and 5th, scale rows counting from middle of back. Stripes fade with age; adults may be uniform brown. Tail blue or gray. No wide lengthwise row of scales under tail. Breeding male usually has red-orange head. Juveniles have brilliant striping and bright blue or purple tail.

Breeding: Mates May to June. Clutch of up to 11

eggs laid in June, hatches in August.
Female tends eggs.

Habitat: Moist woods and grassy areas, but also
found in drier areas with sparse
vegetation and no permanent fresh
water.

Range: S. Maryland and Virginia, south
through Florida and the Keys, west to
Louisiana, and northeast to Kentucky.

Diurnal. Although primarily a
terrestrial species, its climbing ability
is well developed. Sometimes domestic
cats eat this skink, in which case they
may lose their sense of balance or
develop a paralysis requiring veterinary
attention.

424, 431 Broad-headed Skink
(*Eumeces laticeps*)

Description: 6½–12¾″ (16.5–32.4 cm). Large and
brown, with wide head and 5 *broad
light stripes,* including *dorsolateral stripe
along 4th scale row* counting from
middle of back. Stripes fade with age;
adult males uniform brown with red-
orange head. Tail blue to brown; wide
lengthwise row of scales under tail.
Juveniles black with brilliant striping,
bright blue tail.

Breeding: Mates April to May. Clutch of 6–16
eggs laid May to July in excavations
under logs or leaf litter; hatches June to
August, occasionally later. Female
tends eggs.

Habitat: Moist wooded areas; also open areas
where low shelter is provided by leafy
debris or piles of rubble.

Range: Se. Pennsylvania to c. Florida, along
the Gulf Coast to e. Texas, north to
Kansas and Illinois.

Diurnal. This lizard is often found
hunting insects high in trees. It has
been observed shaking the nests of
paper wasps to dislodge pupae, which it

consumes; apparently wasp stings are unable to penetrate the skink's bony scales. Throughout much of their range Broad-headed Skinks are locally called scorpions.

---

422 Many-lined Skink
(*Eumeces multivirgatus*)
Subspecies: Many-lined, Variable

Description: 5–7⅝″ (12.7–19.4 cm). Long-bodied, with *many alternating light and dark stripes,* including dorsolateral light stripe along 3rd scale row counting from middle of back. Back striping faded or absent in some populations. Tail tapers so gradually it appears swollen. Young have bright blue tail.

Breeding: Lays clutch of 5 eggs.

Habitat: Areas of rocks and small brush in open grassy plains, sand hills, and desert; mountainous wooded areas to 8,200′ (2,500 m).

Range: Sw. South Dakota through se. Wyoming, south to Arizona and New Mexico; scattered populations in w. Texas and Mexico.

Subspecies: Many-lined (*E. m. multivirgatus*), stripes persistent; sw. South Dakota, se. Wyoming, w. and c. Nebraska to ec. Colorado and n. New Mexico. Variable (*E. m. gaigeae*), stripes faded or absent in adults; extreme s. Colorado and se. Utah into Arizona, w. Texas, and Mexico.

Diurnal. This skink typically burrows under rocks, logs, trash, even dry cow chips. It feeds on insects.

432 Great Plains Skink
(*Eumeces obsoletus*)

Description: 6½–13¾" (16.5–34.9 cm). The largest skink. *Scale rows on sides oblique to rows on back.* Dark edges of brown scales may align to form indistinct lengthwise stripes. Sides yellow. Juveniles black with white spots on lips, bright blue tail.

Breeding: Mates April to May. Clutch of 7–21 eggs, laid in nest excavations under rocks, May to June, hatches July to August. Female tends eggs.

Habitat: Open rocky grasslands of the Great Plains; near permanent or semipermanent water in otherwise drier areas.

Range: Se. Wyoming, s. Nebraska, and extreme sw. Iowa through the Great Plains to c. Arizona, Mexico, and wc. Texas.

Diurnal. This husky skink feeds on insects, spiders, and small lizards. It will bite if handled.

423, 428 Prairie Skink
(*Eumeces septentrionalis*)
Subspecies: Northern, Southern

Description: 5–8⅛" (13–20.6 cm). Brown, with 4 *dark-edged light stripes extending onto tail.* Light stripes on side separated by dark band 2 scale rows wide. *Upper light stripes follow adjoining edges of 4th and 5th scale rows* counting from middle of back. Pale stripe may run down middle of back. Upper labial scales, 7. Breeding male may have orange on head. Young have bright blue tail.

Breeding: Mates May to June. Clutch of 5–18 eggs is laid May to July.

Habitat: Moist terrain with vegetation and loose soil; rocky, gravelly washes.

Range: S. Manitoba south through e. North

Dakota, Minnesota, and w. Wisconsin; south to coastal Texas.

Subspecies: Northern (*E. s. septentrionalis*), mid-back pale stripe present; s. Manitoba south to extreme ne. Oklahoma; isolated populations in Wisconsin and Illinois.
Southern (*E. s. obtusirostris*), mid-back stripe faint or absent; sc. Kansas to c. Texas and the Gulf Coast.

Active during twilight and predawn hours. Terrestrial; the Prairie Skink is found under rocks, boards, or leaf litter. When disturbed, it retreats into a burrow or disappears into the vegetation. It feeds on insects, spiders, and snails.

421, 441, 442 Western Skink
(*Eumeces skiltonianus*)
Subspecies: Western, Coronado Island, Great Basin

Description: 6½–9⁵⁄₁₆″ (16.5–23.7 cm). *4 light stripes extending well onto tail; upper 2 stripes along adjoining edges of 2nd and 3rd scale rows* counting from middle of back. Broad brown band on back between light stripes; broad dark band on side between light stripes. Tail usually gray or brown; bright blue in juveniles. Upper labial scales 7. Breeding male has orange on sides of head, tip of tail.

Breeding: Mates May to June. Clutch of 2–6 eggs is laid in burrows or under rocks, June to July; hatches July to August. Females tend eggs.

Habitat: Forest, open woodland, and grassy areas, especially where rocks are abundant. Usually found under leaf litter, logs, or rocks.

Range: N. Arizona and s. Nevada to s. British Columbia; south along the coast through California into Baja peninsula.

Subspecies: Western (*E. s. skiltonianus*), light stripe

on side has dark border along lower edge; s. British Columbia, w. Montana, n. Idaho, e. Washington, Oregon and n. California, south along coast to San Diego County, and on Santa Catalina Island.
Coronado Island (*E. s. interparietalis*), dark bands on side extend at least half length of tail; San Diego County, California into n. Baja California.
Great Basin (*E. s. utahensis*), light stripe on side lacks dark border; extreme sw. Montana, s. Idaho, Nevada, and nc. Arizona.

Diurnal. The Western Skink feeds on a variety of insects, their larvae, spiders, and earthworms.

439 Four-lined Skink
(*Eumeces tetragrammus*)
Subspecies: Four-lined, Short-lined, Mountain

Description: 5–7⅞" (12.7–20.0 cm). Has *4 light stripes that end on body;* upper 2 stripes follow 4th scale rows counting from middle of back. Broad brown band on back between light stripes; broad black band on side between light stripes. Juveniles more prominently striped than adults. Usually light Y-shaped mark on head. Tail gray or brown; bright blue in juveniles. Upper labial scales, 7. Breeding males have reddish heads.

Breeding: Clutch of 5–12 eggs is laid in shallow depression, April to July. Female tends incubating eggs.

Habitat: Arid and semiarid country; rocky ravines, grassy zones, scrub, forest, woodland, sea level to 6,500′ (1,900 m).

Range: C. and s. Texas, south into Mexico, north to sc. Arizona and extreme sw. New Mexico.

Subspecies: Four-lined (*E. t. tetragrammus*), upper

light stripe ends near hind leg, adults
lack light Y on head; c. and s. Texas
into Mexico.
Short-lined (*E. t. brevilineatus*), upper
light stripe ends near shoulder, adults
have light Y on head; extreme s. Texas
to Mexico.
Mountain (*E. t. callicephalus*), upper
light stripe ends near hind leg, adults
have light Y on head; sc. Arizona to
Mexico.

Diurnal. This skink can be found under
rocks, rotting logs, leaf litter, and
trash. It feeds on insects and spiders.

450 Sand Skink
(*Neoseps reynoldsi*)

Description: 4–5⅛″ (10.1–13.0 cm). Small but
long-bodied. Chisel-shaped snout, no
external ear opening; *1–2 toes on tiny
legs*. White to tan, with faint dark band
along sides.

Breeding: Mates March to April. Pair of eggs is
laid May to June.

Habitat: Rosemary scrub on sandhills.

Range: C. Florida (Highland, Lake and Polk
counties).

Diurnal. This skink is adept at digging
and burrowing. It can be found resting
under logs or other surface debris, but
when disturbed escapes by "swimming"
into the loose sand. It eats beetle larvae
and termites.

433 Ground Skink
(*Scincella lateralis*)

Description: 3–5⅛″ (7.6–13.0 cm). Long-tailed;
brown, with black dorsolateral stripes; no
stripes on back. Small legs, 5 toes.
Transparent window in movable lower
eyelid.

Breeding: Mates January to August. Clutch of 1–
7 eggs is laid almost monthly, April to
August, with a maximum of 5 clutches
a season. Female abandons nest after
laying.

Habitat: Humid forests, hardwood hammocks,
and forested grasslands, generally where
leaf litter is abundant.

Range: New Jersey south through Florida, west
to c. Texas, north to Nebraska and
Missouri.

Diurnal. Food consists of insects and
spiders. Occasional specimens have
been observed to bite off their own tails
and eat them.

AMPHISBAENID FAMILY
(Amphisbaenidae)

Nineteen genera of approximately 135
recognized species found in Africa, the
Mediterranean countries, South
America to Mexico, and the West
Indies. One species occurs in
Florida.

Amphisbaenids resemble earthworms in
appearance. Body scales are fused into
rings which encircle body. Specialized
for burrowing, family members lack
external ear openings, have eyes buried
under skin, and with the exception of
one genus, *Bipes,* which has short front
legs, they are limbless.

Amphisbaenids live underground much
of their lives feeding on insects and
worms. Breeding biology is poorly
known. Most seem to be egg-layers.

451 Worm Lizard
(*Rhineura floridana*)

Description: 7–16″ (17.8–40.6 cm). Legless,
opalescent-pink lizard, *resembles an
earthworm.* Depressed and chisel-like
snout facilitates burrowing; lower jaw is
countersunk. Scales are arranged in
rings around body; tail slightly
depressed and upper surface rough. No
external eyes or ear openings.

Breeding: 1–3 eggs are deposited in summer,
hatch in early fall.

Habitat: Sandy areas, primarily in high pine and
broadleaf hammock areas.

Range: N. and c. Florida.

A burrower, it lives underground
searching for earthworms and termites.
It is sometimes seen at the surface after
heavy rains have saturated the soil or
cultivation has disrupted its burrow.
Mockingbirds occasionally unearth a
worm lizard while searching for
earthworms.

Snakes
(order Squamata; suborder Serpentes)

Numbering some 2,700 species, snakes, like lizards, are found on all continents except Antarctica. In the Old World they range from the Arctic Circle in Eurasia to the Cape of Good Hope in Africa, in the New World from the Northwest Territories southward nearly to the southern tip of South America. They are absent from Ireland, Iceland, and New Zealand. Of 11 families, 5 are represented in the United States and Canada.

Snakes are distinctive in possessing an elongated scaly body without limbs, external ear openings, or eyelids. Their life styles vary; some are active by day, others at night. Some occupy terrestrial or subterranean situations, others live in the trees or in the water. They range from sites at sea level to elevations over 10,000' (3,050 m) in our western mountain ranges. All snakes are exclusively carnivorous and swallow their prey whole. They continue to increase in length throughout their lives, but the growth rate slows after maturity is reached. Snakes periodically shed their outer layer of skin, usually in one piece beginning at the tip of the snout. Like lizards, they may be active throughout the year in the tropics, but in our range they overwinter in a protected retreat.

Snakes usually mate in the fall before going into hibernation or shortly after emerging from it in spring. Males locate females by scent; after a brief courtship internal fertilization takes place. Some snakes are egg-layers, others bear their young alive.

Of the 115 species of snakes that may be seen north of Mexico, 17 species are venomous. Learn to recognize them quickly and leave them alone. Do not attempt to catch them. A bite from a venomous snake may be an extremely painful experience and, in some cases, may lead to disfigurement, damage to the nervous system, or even death.

SLENDER BLIND SNAKE FAMILY
(Leptotyphlopidae)

One genus, *Leptotyphlops,* of 95 species, distributed in Africa, southwestern Asia, the West Indies, and tropical America; 2 species extend northward into the southwestern United States.

Also called worm snakes or thread snakes, slender blind snakes are indeed blind, but possess tiny vestigial eyes that appear as black dots beneath an irregularly shaped ocular scale. These snakes are characterized by a short blunt head, short tail, and slender cylindrical body covered with close-fitting, overlapping smooth scales arranged in 14 rows around the body; they lack enlarged belly scales. Most range from 5–15" (13–38 cm) in length. Teeth are scarce and present only on the lower jaw.

These seldom seen burrowers and crevice dwellers may emerge around sundown and crawl about on the surface. They feed largely on termites and ants. Females lay small clutches of long slender eggs.

---

464 Texas Blind Snake
(*Leptotyphlops dulcis*)
Subspecies: Plains, New Mexico

Description: 5–10¾" (12.7–27.3 cm). Smooth, shiny cylindrical snake; reddish-brown, pink, or silvery tan, with blunt head and tail. Small spine on tip of tail. Eyes mere black spots beneath ocular scales; *more than 1 scale on top of head between large scale covering each eye. No enlarged belly plates.* 14 rows of scales around body.

Breeding: Clutch of 2–7 elongate, thin-shelled eggs, ⅝" (16 mm) long, is laid late June to July. Females tend incubating eggs and may share communal nesting sites in rocky fissures or earthen burrows. Hatchlings are about 2¾" (7 cm) long.

Habitat: Semiarid deserts, prairies, hillsides, mountain slopes with sandy or loamy soil suitable for burrowing; sea level to 5,000' (1,500 m).

Range: Sc. Kansas through Oklahoma and Texas to Mexico, west to s. New Mexico and se. Arizona.

Subspecies: Plains (*L. d. dulcis*), 1 upper lip scale between large scale containing eye and scale surrounding nostril; s. Oklahoma, c. Texas into Mexico.
New Mexico (*L. d. dissectus*), 2 upper lip scales between ocular and lower nasal scale; s. Kansas, w. Texas, s. New Mexico, se. Arizona and adjacent Mexico.

Nocturnal. This burrowing snake is seldom seen on the surface except in the evening following heavy summer rains. It is most frequently found in damp soil under slabs of rock, logs, or other surface debris. In farming areas it is sometimes uncovered and eaten by chickens scratching through barnyard soil for worms. It defends itself by coiling and writhing about, smearing cloacal fluid over its body. Eats termites, ants, and ant pupae.

---

457 Western Blind Snake
(*Leptotyphlops humilis*)
Subspecies: Southwestern, Desert, Trans-Pecos, Utah

Description: 7–16" (17.8–40.6 cm). Smooth, shiny cylindrical snake; brown, purplish, or silvery pink with blunt head and tail. Small spine on tip of tail. Black eye spots beneath ocular scales; *only 1 scale on top of head between large scale covering each eye. No enlarged belly scales.* 14 rows of scales around body.

Breeding: Mates in spring. Clutch of 2–6 slender

eggs, ⅝" (15 mm) long, is laid July to August. Females tend eggs, and may use communal nests. Hatchlings are about 3½" (9 cm) long.

Habitat: Deserts, grassland, scrub, canyons, and rocky foothills with moist sandy or gravelly soils suitable for burrowing; sea level to 5,000′ (1,500 m).

Range: Extreme sw. Utah, s. Nevada and California into Baja California, s. Arizona, sw. New Mexico, w. Texas, and Mexico.

Subspecies: Southwestern (*L. h. humilis*), 12 scales around tail, 7–9 darkly colored scale rows on midback; s. California (except se. corner) and s. Nevada, southeast to sc. Arizona.
Desert (*L. h. cahuilae*), 12 scales around tail, 5 deeply colored scale rows on midback; sw. Arizona and se. California into Mexico.
Trans-Pecos (*L. h. segregus*), 10 scales around tail, 7 scale rows on midback deeply colored; se. Arizona, sw. New Mexico, w. Texas into Mexico.
Utah (*L. h. utahensis*), 12 scales around tail, 7 deeply pigmented scale rows on midback, 4th middorsal scale behind rostral divided—5th expanded; se. Nevada and extreme sw. Utah.

Subterranean in habit, and capable of borrowing quickly into loose soil or sand. On warm evenings it emerges at sunset from beneath moist rocks or among roots of bushes and forages for termites and ants, following their trails by smell.

BOA AND PYTHON FAMILY
(Boidae)

Giants of the snake world, boids
include the well-known Boa Constrictor
of the American tropics, which attains a
length of 18' (5.5 m), and the
Reticulated Python of southeastern Asia
and Anaconda of South America, both
capable of exceeding 30' (9.1 cm).
Most boids, like the two that inhabit
our range, are significantly smaller. Of
59 species belonging to 20 genera, the
majority are found in tropical and
subtropical regions of the world.
Boids are the most primitive of living
snakes. They have stout muscular
bodies and short tails. Back scales are
smooth, relatively small, sometimes
iridescent, and occur in numerous scale
rows. By contrast, belly scales are large
and form transverse plates. Boid eyes
have vertical pupils and many species
possess temperature-sensitive pits in the
lip scales. Vestiges of hind limbs are
present as "spurs," usually visible on
either side of the vent.
Boids occupy a variety of habitats.
Many of the smaller species are
burrowers, favoring loose or sandy soils.
Large forms generally are terrestrial or
semiarboreal. Pythons are egg-layers,
boas live-bearers. All boids feed on
birds and mammals, which they
suffocate in their constricting coils.

472 Rubber Boa
(*Charina bottae*)

Description: 14–33" (35.6–83.8 cm). Looks
rubbery. Short broad snout and short
blunt tail give it a *two-headed
appearance*. Uniformly olive-green,
reddish-brown, or tan to chocolate-
brown. *Large scales on top of head.* Eyes
small with vertical pupils. Scales
smooth, in 32–53 rows. Anal plate

single. Adult males have well-developed anal spurs; small spurs in females usually hidden.

Breeding: Live-bearing; 2–8 young, 7" (17.8 cm) long, are born late August to September.

Habitat: Damp woodland and coniferous forest, large grassy areas, meadows, and moist sandy areas along rocky streams. Sea level to 9,200' (2,800 m).

Range: British Columbia to s. California and eastward to Montana, Wyoming, and Utah.

Crepuscular and nocturnal. An accomplished burrower, it retreats under rocks or into damp sand, hollow rotting logs, or forest litter. Also a good swimmer and climber; its prehensile tail enables it to climb shrubs and small trees. A constrictor, it preys on small mammals, birds, and lizards. It is docile and curls into ball when picked up. Captive longevity exceeds 11 years.

508, 525 Rosy Boa
(*Lichanura trivirgata*)
Subspecies: Mexican, Desert, Coastal

Description: 24–42" (61–106.7 cm). Stout, smooth, and shiny. Gray, tan, brown, or rosy-red with 3 brown stripes down body; occasionally blotched. Head and tail somewhat short and blunt. *No large symmetrical scales on top of head or under chin.* Neck nearly as wide as head. Males have clawlike spur on each side of anal plate. Scales smooth, in 35–45 rows. Anal plate single.

Breeding: Habits poorly known. Mates May to June; 6–10 young are born October to November. Gestation takes approximately 130 days; newborn are about 12" (31 cm) long.

Habitat: Desert, arid scrub, brushland, rocky chaparral-covered foothills—

particularly where moisture is available, as around springs, streams, and canyon floors; sea level to 4,000' (1,200 m).

Range: S. California into n. Baja California, sw. Arizona, and adjacent Mexico.

Subspecies: Mexican (*L. t. trivirgata*), distinct dark brown stripes, black speckles on belly; sw. Arizona (Organ Pipe Cactus National Monument) into Mexico. Desert (*L. t. gracia*), distinct rose, reddish-brown, or light brown stripes, brown speckles on belly; se. California and sw. Arizona. Coastal (*L. t. roseofusca*), ill-defined pink, rose, reddish-brown, or brown stripes; extreme sw. California into Baja.

Nocturnal. Primarily terrestrial, but occasionally climbs shrubs. A powerful constrictor, it preys on small mammals and birds. It seldom bites when handled, but when frightened may coil into a tight ball with head hidden within the coils. The anal spurs are used by the male to stroke the female during courtship. Record longevity is 18½ years.

COLUBRID SNAKE FAMILY
(Colubridae)

Largest of all snake families, the colubrids represent more than three-fourths of the world's 2,700 known snake species. They are found on all continents except Antarctica and range to the Arctic Circle in Eurasia and the Cape of Good Hope in Africa. Of the 115 snake species in our range, 92 belong to this family.

This huge assemblage of snakes displays so great a range of physical characters that biologists are forever trying to reshuffle the species into more manageable subgroups. But few distinctive characters are shared by all. In general, the colubrid head is as wide or wider than the neck with large and regularly arranged scales. Eyes are well developed with round or vertical pupils. Back scales may be smooth or keeled. Belly scales are as wide as the body; those on the underside of the tail are usually divided. Colubrids lack all vestiges of hind limbs and pelvic girdle. Teeth are present on both jaws, but there are no enlarged hollow poison-injecting fangs. Some species have grooved teeth toward the back of the upper jaw that are connected to a poison-producing gland. Most such rear-fanged snakes are small and harmless; the African Boomslang and Bird snakes are two notable exceptions capable of inflicting lethal bites. Rear-fanged species in our range, like other colubrids, are harmless to man.

Colubrid habits are as varied as their anatomy. Colubrids range from arboreal to terrestrial, from burrowing to aquatic habitats. All devour whole animals. Many prey on small mammals and birds; some eat reptiles, amphibians, or fishes; others take worms, snails, scorpions and centipedes, or insects.

Most colubrids are egg-layers, but a

large number bear living young. The
male's tail is usually longer and thicker
at its base than the female's.

566, 577, 587   Glossy Snake
(Arizona elegans)
Subspecies: Kansas, Texas, Mojave,
Desert, Arizona, California,
Painted Desert

Description: 26–70" (66–178 cm). Resembles
Gopher and Great Plains Rat snakes
but has *smooth glossy scales* rather than
keeled scales. Snout somewhat pointed;
lower jaw inset. Variable number of
black-edged tan, brown, or gray
blotches mark cream, pinkish, or light-
brown upper surfaces. Dark line runs
from angle of jaw to eye. Belly
unmarked. Scales in 25–35 rows. Anal
plate single.

Breeding: Mates in spring. Clutch of 3–23 eggs,
about 2⅜" (60 mm) long, is laid
during summer, hatches in 10–12
weeks.

Habitat: Dry, open sandy areas, coastal
chaparral, creosote-mesquite desert,
sagebrush flats, and oak-hickory
woodland; below sea level to 5,500'
(1,700 m).

Range: Se. Texas and extreme sw. Nebraska
west to central California, south into
Mexico.

Subspecies: Nine, poorly differentiated; 7 in our
range.
Kansas (*A. e. elegans*), 39–69 large dark
body blotches, 29–31 scale rows;
extreme sw. Nebraska south through
w. Texas into Mexico.
Texas (*A. e. arenicola*), 41–58 body
blotches, 29–35 scale rows; se. Texas.
Mojave (*A. e. candida*), 53–73 narrow
body blotches, 27 or fewer scale rows;
Death Valley area, s. Nevada south
through w. Mojave Desert in
California.
Desert (*A. e. eburnata*), pale-colored

with 53–83 small narrow body
blotches, 27 or fewer scale rows;
extreme sw. Utah, s. Nevada south
through center of Mojave Desert to
Gulf of California.
Arizona (*A. e. noctivaga*), body blotches
slightly wider than interspaces; 25–29
scale rows; s. and w. Arizona south to
c. Sinaloa, Mexico.
California (*A. e. occidentalis*), dark with
51–75 dark brown blotches, 27 scale
rows; San Joaquin Valley south into
Baja California.
Painted Desert (*A. e. philipi*), usually
less than 200 ventral scales, 53–80
body blotches, 27 scale rows; se. Utah,
ne. and se. Arizona, w. New Mexico.

Occasionally called the faded snake
because of its bleached appearance. It is
a capable burrower and is usually seen
on the surface in the early evening
hours during the warmer months. It
feeds chiefly on lizards and occasionally
takes small mammals. Captive
longevity exceeds 12 years.

493 Worm Snake
(*Carphophis amoenus*)
Subspecies: Eastern, Midwest, Western

Description: 8–14¾" (20.3–37.5 cm). A tiny,
glossy, cylindrical-bodied snake;
unpatterned brown, gray, or black with
a *bright reddish-pink belly*. Belly
coloration extends up onto dorsal scale
rows. Tail short and tapers to a sharp
tip. Scales smooth, in 13 rows. Anal
plate divided.

Breeding: Mates April to May and September to
October. 1–8 elongate, thin-shelled
eggs, about ⅞" (23 mm) long, are
deposited June to July and hatch about
7 weeks later. Hatchlings are 3–4"
(8–10 cm) long; mature in 3 years.

Habitat: Damp hilly woodlands, partially
wooded or grassy hillsides above

streams, farmland bordering
woodlands; sea level to 4,300′
(1,300 m).

Range: S. New England to c. Georgia, west to
se. Nebraska, e. Kansas, e. Oklahoma,
and extreme ne. Texas.

Subspecies: Eastern (*C. a. amoenus*), brown above,
prefrontal and internasal scales separate;
se. New York and s. New England
south through South Carolina and n.
Georgia, west into s. Ohio, e.
Kentucky, e. Tennessee, and nc.
Alabama.
Midwest (*C. a. helenae*), brown above,
prefrontal and internasal scales fused;
w. Ohio west to s. Illinois, southward
east of the Mississippi River to the
Gulf.
Western (*C. a. vermis*), black above,
prefrontal and internasal scales separate;
se. Nebraska and sw. and se. Iowa,
southward west of the Mississippi River
to nw. Louisiana.

Secretive; dwells in damp situations
under rocks, decaying logs, or stumps,
or in loose soil. It is more likely to be
seen in spring while habitat is still
moist. During dry and cold periods it
retreats deep into soil. Feeds on
earthworms, and is preyed upon by
milk snakes and kingsnakes. It does not
bite when handled.

595, 596, 607 **Scarlet Snake**
(*Cemophora coccinea*)
Subspecies: Florida, Northern, Texas

Description: 14–32¼″ (35.6–81.9 cm). Often
mistaken for Coral Snake, but Scarlet's
wide red bands are separated by much
narrower black-bordered yellow bands.
Bands do not encircle body. Belly plain
white or yellow. *Snout pointed and red.*
Scales smooth, in 19 rows. Anal plate
single.

Breeding: In June, lays 3–8 elongated leathery

eggs, 1–1⅜" (26–35 mm) long.
Young hatch in late summer at about
6" (15 cm).

Habitat: Hardwood, mixed, or pine forest and
adjacent open areas with sandy or loamy
well-drained soils.

Range: S. New Jersey to s. Florida, west to e.
Oklahoma and extreme e. Texas; s.
Texas.

Subspecies: Florida (*C. c. coccinea*), black bands
extend to 1st or 2nd scale row, 7 upper
lip scales, 1st black band does not
touch parietal scales; peninsular
Florida.
Northern (*C. c. copei*), similar to
Florida, 6 upper lip scales, 1st black
band touches parietals; s. New Jersey to
n. Florida west to e. Oklahoma and
extreme e. Texas.
Texas (*C. c. lineri*), black bands do not
extend below 3rd scale row, no black
along lower edges of red band; s. Texas.

A burrower, this species is rarely
encountered during the day. From May
to September it may prowl about at
night. Occasionally it is found under
rotting logs or stones or is unearthed by
plows. Eggs of other reptiles appear to
be the preferred food.

605 Banded Sand Snake
(*Chilomeniscus cinctus*)

Description: 7–10" (17.8–25.4 cm). Tiny, with a
flattened and protruding shovel-shaped
snout. Head and neck same width. Pale
yellow to reddish-orange above
patterned with 24–48 dark-brown or
black crossbands that encircle tail.
Belly white. *Scales smooth and shiny, in
13 rows.* Anal plate divided.

Breeding: Habits poorly known; lays small
clutches of eggs.

Habitat: Fine sandy areas in open desert
dominated by creosote bush; coarse
sandy areas in rocky upland washes and

arroyos with paloverde and saguaro.
Range: C. and sw. Arizona south to s. Baja
California and s. Sonora, Mexico.

Often emerges and moves on the surface
at night. Highly specialized for desert
existence; its spadelike snout,
streamlined head with nasal valves,
glossy skin, and angular-ended belly
scales enable it to literally swim
through fine sand. Serpentine-shaped
grooves in the sand between bushes are
evidence of its shallow subsurface
activity. Diet includes centipedes,
sand-burrowing cockroaches, and other
soft-bodied insects.

604, 612   **Western Shovel-nosed Snake**
(*Chionactis occipitalis*)
Subspecies: Mojave, Colorado Desert,
Nevada, Tucson

Description: 10–17" (25.4–43.2 cm). Whitish or
yellow, with *21 or more saddle-shaped
dark-brown or black crossbands,*
sometimes with intervening reddish-
orange crossbands. Snout flattened and
juts well beyond lower jaw. Scales
smooth, in 15 rows. Anal plate
divided.
Breeding: 2–4 eggs are deposited in summer.
Habitat: Arid desert land; sandy washes, dunes,
and rocky hillsides; prefers areas with
scattered mesquite-creosote bush; below
sea level to 4,700' (1,450 m).
Range: Sc. Nevada south into Baja California
and Sonora, Mexico.
Subspecies: Mojave (*C. o. occipitalis*), 25 or more
dark crossbands, most not crossing
belly, red crossbands absent; se.
California, southern tip of Nevada, wc.
Arizona.
Colorado Desert (*C. o. annulatus*), 25 or
fewer dark crossbands on body, most
cross belly; narrow red crossbands
present; se. California and sw. Arizona,
south into Mexico.

Nevada (*C. o. talpina*), brown-marked light spaces between brown crossbands appear as secondary crossbands; sc. Nevada southwestward into California. Tucson (*C. o. klauberi*), distinct narrow secondary crossbands between black primary crossbands; sc. Arizona.

The small shovel-shaped head, valved nostrils, flattened belly, and smooth scales allow this burrower to move quickly through sand. Occasionally seen during the day, but essentially nocturnal and most apt to be encountered crossing a road. May strike repeatedly when approached, but the tiny teeth can do little damage. Diet includes centipedes, scorpions, and insects.

610 Sonoran Shovel-nosed Snake
(*Chionactis palarostris*)
Subspecies: Organ Pipe

Description: 10–15½" (25.4–39.4 cm). A tiny yellow snake with alternating black and red saddle-shaped crossbands. Black crossbands—*generally fewer than 21 on body*—usually extend across underside. Snout yellow, overhangs lower jaw, is slightly convex in profile. Scales smooth, in 15 rows. Anal plate divided.

Breeding: Habits unknown; presumably lays about 4 eggs in summer.

Habitat: Saguaro-paloverde dominated upland desert in Arizona; arid mesquite-creosote bush-bur sage terrain to the south.

Range: Organ Pipe Cactus National Park, sw. Arizona south to wc. Sonora, Mexico.

Subspecies: Two; 1 in our range. Organ Pipe (*C. p. organica*).

Nocturnal. A strong burrower with habits similar to those of the Western Shovel-nosed, but appears to prefer rockier terrain. Feeds on invertebrates.

551 Kirtland's Snake
(Clonophis kirtlandi)

Description: 14–24½" (35.5–62.2 cm). Slender brown or grayish snake with 2 rows of alternating dark squarish spots (often indistinct) on either side of midline of back and a *line of round black spots along each side of red belly.* Scales keeled, in 19 rows. Anal plate divided.

Breeding: Early August to late September, female gives birth to 4–15 young, 5–6½" (13–17 cm) long.

Habitat: Vicinity of marshy meadows, woodland ponds, and open swamplands.

Range: Wc. Pennsylvania west through Ohio, S. Michigan, n. Kentucky and Indiana to Illinois.

Dramatically flattens body when frightened. Swims well but is least aquatic of the water snakes and is rarely encountered in the water. Usually seen under flat rocks in wet meadows. Eats earthworms and slugs.

468, 478, 480, 486 Racer
(Coluber constrictor)
Subspecies: Northern Black, Buttermilk, Tan, Eastern Yellow-bellied, Blue, Brown-chinned, Black-masked, Western Yellow-bellied, Mexican, Everglades, Southern Black

Description: 34–77" (86.4–195.5 cm). Large, slender, agile, and fast moving. Adults uniformly black, blue, brown, or greenish above; white, yellow, or dark gray below. Young typically gray and conspicuously marked with dark spots on sides and dark gray, brown, or reddish-brown blotches down midline of back. Scales smooth, in 17 rows (*15 rows at vent*). Anal plate divided.

Breeding: Mates April to late May in most of range, 1–2 months earlier in Deep

South. Female lays 5–28 soft leathery eggs *with a rough granular texture,* 1–1⅞" (25–48 mm) long, in rotting tree stump, sawdust pile, under rocks or in small mammal tunnel, mid-June to August. Occasionally a number of females deposit their eggs in a communal nest. Young hatch in 6–9 weeks, July to September, are 8–13" (20–33 cm) long. Mature in 2–3 years.

Habitat: Abandoned fields, grassland, sparse brushy areas along prairie land, open woodland, mountain meadows, rocky wooded hillsides, grassy-bordered streams, and pine flatwoods; sea level to ca. 7,000' (2,150 m).

Range: S. British Columbia and extreme s. Ontario; every state in continental United States, except Alaska; scattered populations through e. Mexico to n. Guatemala.

Subspecies: Eleven, poorly defined.

Northern Black (*C. c. constrictor*), slate black, upper lip scales black, some white on chin, iris of eye brown; s. Maine to e. Ohio south to n. South Carolina, n. Georgia, n. Alabama, and ne. Mississippi.

Buttermilk (*C. c. anthicus*), black, bluish-black or bluish-green with random white, gray, or yellow spots; extreme s. Arkansas, Louisiana, and adjacent e. Texas.

Tan (*C. c. etheridgei*), light tan with pale spots; extreme wc. Louisiana into adjacent e. Texas.

Eastern Yellow-bellied (*C. c. flaviventris*), pale blue, bluish-green, olive-green, gray or brown above, belly cream to bright yellow; e. Montana, w. North Dakota, and Iowa south to extreme n. Arkansas, and the Gulf coast of se. Texas and w. Louisiana.

Blue (*C. c. foxi*), pale-blue or bluish-green above, belly white or bluish-white; extreme s. Ontario and nw. Ohio west to se. Minnesota, e. Iowa, and Illinois.

Brown-chinned (*C. c. helvigularis*), slate

black, lip scales and chin tan or brown;
Apalachicola and Chipola River valleys
in Florida panhandle and adjacent
Georgia.
Black-masked (*C. c. latrunculus*), slate
gray above, belly pale grayish-blue,
black stripe behind eye; se. Louisiana.
Western Yellow-bellied (*C. c. mormon*),
green, olive-green, yellowish-brown or
reddish-brown above, belly yellow; s.
British Columbia to Baja California east
to sw. Montana, w. Wyoming, and w.
Colorado.
Mexican (*C. c. oaxaca*), midline of back
green or greenish-gray, sides lighter,
belly yellow or greenish-yellow, adults
20–40″ (51–102 cm) long; s. Texas
and Mexico.
Everglades (*C. c. paludicola*), bluish-,
greenish-, or brownish-gray above,
belly whitish with pale gray or powder-
blue markings, iris of eye usually red;
s. Florida Everglades region and Cape
Canaveral area of e. Florida.
Southern Black (*C. c. priapus*),
resembles Northern Black, more white
on chin, iris of eye usually red or
orange; coastal plain from extreme se.
North Carolina to Florida Everglades
(and lower Florida Keys) west to se.
Louisiana, Mississippi, Arkansas, se.
Oklahoma and adjacent Texas; north in
Mississippi Valley to s. Illinois and s.
Indiana.

Diurnal. May be encountered in most
any terrestrial situation except atop
high mountains and in hottest deserts.
Often observed streaking across roads.
Although agile and a good climber, it
spends most of its time on the ground.
When hunting, it holds its head high
and moves swiftly through cover. Often
hibernates in rocky hillsides in large
numbers and with other species. When
annoyed it may make a buzzing sound
like a rattler by vibrating the tail tip in
dead vegetation. If grabbed, it will bite
repeatedly and thrash about violently.

Eats large insects, frogs, lizards,
snakes, small rodents, and birds.
Despite the scientific name, it is not a
constrictor.

Black-striped Snake
(*Coniophanes imperialis*)

Description: 12–20" (30.5–50.8 cm). A light
brown snake with 3 dark brown to
purplish-black stripes; a thin one down
center of back and a broad stripe on
either side. Thin light line extends
from snout to top of eye to rear of head.
Upper lip scales white with tiny black
dots. Belly, pink, red, or orange. Scales
smooth, in 19 rows. Anal plate
divided.

Breeding: 2–10 eggs are laid late April to June.
Young develop quickly; eggs may hatch
in 40 days.

Habitat: Semiarid coastal plain; sea level to 500'
(150 m).

Range: Extreme s. Texas south to Central
America.

Subspecies: Three; 1 enters U.S., *C. i. imperialis.*

Crepuscular and nocturnal. A secretive
snake, it burrows into the soil or takes
refuge under vegetative debris during
the day. Grooved rear fangs and a mild
venom immobilize prey—small toads,
frogs, lizards, snakes, and baby mice.
Although generally harmless, a bite
may produce itching, burning
numbness, and localized swelling in
humans.

471 Sharp-tailed Snake
(*Contia tenuis*)

Description: 10–19" (25.4–48.3 cm). The short
spine-tipped tail and *alternating black
and whitish crossbars on belly* quickly
identify it. Upper surfaces brown,

yellowish- or reddish-brown, or gray often with an indistinct and lighter colored dorsolateral line. Scales smooth, in 15 rows. Anal plate divided.

Breeding: About 2–8 eggs, 1¾″ (44 mm), are presumably laid late June to July; young 3″ (7.6 cm) long, hatch in autumn.

Habitat: Near streams or moist situations: pastures, open meadows, digger pine-blue oak woodland, oak-dominated foothills, and Douglas fir-vine maple forest; sea level to 6,300′ (1,900 m).

Range: C. California, along Coast Ranges and east of the San Joaquin Valley in Sierra Nevadas, north to Willamette Valley, Oregon. Isolated populations in Pierce and Klickitat counties, Washington, and North Pinder Island, British Columbia.

Surface activity coincides with the rainy season. This snake is most commonly seen beneath logs or rocks, March to early June, after a warm rain. During dry months it retreats underground. Diet is largely restricted to slugs.

495, 496, 497, **Ringneck Snake**
498 (*Diadophis punctatus*)
Subspecies: Southern, Key, Pacific, Prairie, Northern, San Bernardino, Northwestern, Coral-bellied, Regal, San Diego, Mississippi, Monterey

Description: 10–30″ (25.4–76.2 cm). A small slender snake, with a *yellow, cream, or orange neck ring* and *bright yellow, orange, or red belly*. Back gray, olive, brownish, or black; belly frequently marked with black spots. Neck ring may be interrupted, obscure, or occasionally absent. Loreal scale present. Scales smooth, in 15–17 rows. Anal plate divided.

Breeding: Mates in spring or fall. Clutches of 1–10 elongate white or yellowish eggs, 1″

(25 mm) long, are laid June to July in communal nesting sites. Young hatch in about 8 weeks, at 4–6" (10–15 cm); mature in 2–3 years. A Florida female gave live birth to 6 young.

Habitat: Moist situations in varied habitat; forest, grassland, rocky wooded hillsides, chaparral, into upland desert along streams; sea level to ca. 7,000' (2,150 m).

Range: Nova Scotia to Florida Keys, west to the Pacific coast, south to c. Mexico.

Subspecies: Southern (*D. p. punctatus*), neck ring interrupted, single row of half-moon spots down midline of belly; coastal plain and piedmont, s. New Jersey to Mobile Bay, Alabama, south to Florida Keys.
Key (*D. p. acricus*), neck ring absent, 15 scale rows; Big Pine Key, Florida.
Pacific (*D. p. amabilis*), neck ring narrow, numerous belly spots; San Francisco Bay area, California.
Prairie (*D. p. arnyi*), neck ring complete or interrupted; numerous scattered belly spots; se. South Dakota and extreme se. Minnesota, south to nw. Arkansas, sc. Texas, and e. New Mexico.
Northern (*D. p. edwardsi*), neck ring golden, belly typically unspotted; Nova Scotia, south in the Appalachians to n. Georgia and ne. Alabama, west to se. Illinois and the Great Lakes region through Wisconsin.
San Bernardino (*D. p. modestus*), neck ring narrow, heavy black spotting on belly, 17 scale rows on neck region; n. San Diego County to Los Angeles County, east to San Bernardino Mountains.
Northwestern (*D. p. occidentalis*), neck ring wide, belly lightly spotted; extreme sw. Washington, south along coast to Sonoma County, California. Isolated populations in Idaho and Washington.
Coral-bellied (*D. p. pulchellus*), neck ring wide, belly spots few or absent;

western slopes of Sierra Nevadas, California.

Regal (*D. p. regalis*), neck ring absent, 17 scale rows; se. Idaho south through w. Utah, se. Nevada, Arizona, w. New Mexico, Trans-Pecos region of Texas, and into Mexico. Isolated populations from se. Idaho to se. California.

San Diego (*D. p. similis*), neck ring narrow, belly moderately spotted; sw. San Bernardino County south into Baja California.

Mississippi (*D. p. stictogenys*), narrow neck ring often interrupted, paired black spots down midline of belly; extreme s. Illinois, south in Mississippi Valley to the Gulf, Mobile Bay west to e. Texas.

Monterey (*D. p. vandenburghi*), neck ring wide, belly spots few and small, 17 scale rows on neck region; Ventura County to Santa Cruz County, California.

Secretive. Most often seen under flat rocks, logs, or loose bark of dead trees. When threatened, red-bellied forms tightly coil the tail and elevate it to display brightly colored underside. Rarely attempts to bite when picked up, but will void musk and foul-smelling contents of the cloaca. Partially constricts prey, which includes earthworms, slugs, small salamanders, lizards, and newborn snakes.

---

489 Indigo Snake
(*Drymarchon corais*)
Subspecies: Eastern, Texas

Description: 60–103½″ (152–263 cm). *Largest North American snake.* Heavy-bodied. Lustrous blue-black or mixed brown and black. Chin, throat, and sides of head suffused with cream, orange, or red. Scales smooth, in 17 rows. Anal plate single.

Breeding: Eastern form mates November to February. Deposits 5–12 leathery eggs, 3–4" (76–102 mm) long, April to May. Hatchlings 19–26" (48–66 cm) long appear late July to October. Texas form lays April to May.

Habitat: In southeast: pine woods, turkey oak, and palmetto stands near water, orange groves, and tropical hammocks; in Texas: dry grassland and thickets near ponds and rivers.

Range: Se. Georgia through Florida Keys; scattered populations in Florida panhandle; formerly in s. Alabama. Also, Texas south to Argentina.

Subspecies: Eastern (*D. c. couperi*), body uniformly shiny blue-black; se. Georgia and Florida.
Texas (*D. c. erebennus*), front half of body often brownish-black with trace of pattern, black lines radiate downward from eye; s. Texas to Veracruz and Hidalgo, Mexico.

Not a constrictor, the Indigo immobilizes food with its jaws. It feeds on frogs, small mammals and birds, other snakes—including venomous ones—lizards, and young turtles. When disturbed, it hisses, vibrates its tail, and flattens its neck. The Eastern Indigo Snake is vanishing in the wild. Habitat destruction, commercial collecting, and the practice of gassing tortoise burrows—the Indigo's favorite retreat—have drastically reduced its numbers. It is protected by law. Long-lived; one captive lived nearly 26 years.

559 **Speckled Racer**
(*Drymobius margaritiferus*)

Description: 30–50" (76–127 cm). Streamlined black body with greenish cast. *Dart-shaped yellow spot* occupies central area of each back scale; concealed front edge of each scale blue, exposed rear edge

black. Black stripe behind eye. Scales
weakly keeled, in 17 rows. Anal plate
divided.

Breeding: About 2–8 eggs, 1½" (38 mm) long,
are laid April to July; hatch in 2
months.

Habitat: Dense thickets near water.

Range: Extreme s. Texas south to Colombia.

Subspecies: Four; 1 in our range, *D. m.
margaritiferus*.

Rarely encountered north of the Rio
Grande. Fast-moving and difficult to
capture. Feeds on frogs.

570, 608 Corn Snake
(*Elaphe guttata*)
Subspecies: Corn, Great Plains Rat

Description: 24–72" (61–182.9 cm). Long and
slender; orange or brownish-yellow to
light gray, with large black-edged red,
brown, olive-brown, or dark-gray
blotches down middle of back. Two
alternating rows of smaller blotches on
each side, extending onto edges of belly
scales. *Large squarish black marks on
belly, becoming stripes under tail.* Dark
spear-point mark on top of head, and
dark stripe extending from eye onto
neck. Belly scales flat in middle, with
ends angled up sharply. Scales smooth
or weakly keeled, in 27–29 rows. Anal
plate divided.

Breeding: Mates March to May. Clutches of 3–21
eggs are laid late May to July, hatch
July to September. Hatchlings are
10–15" (25–38 cm) long, mature in
18–36 months.

Habitat: Wooded groves, rocky hillsides,
meadowland; along watercourses,
around springs, woodlots, barnyards,
and abandoned houses. Sea level to ca.
6,000' (1,850 m).

Range: S. New Jersey south through Florida
and s. Tennessee to Texas, Mexico, and
e. New Mexico, se. Colorado, se.

Nebraska to sw. Illinois. Separate
population in e. Utah and w.
Colorado.

Subspecies: Corn (*E. g. guttata*), red, orange, or
yellow with red, red-orange, or red-
brown blotches; s. New Jersey through
Florida to Louisiana.
Great Plains Rat (*E. g. emoryi*), brown
to light gray with olive, brown, or dark
gray blotches; Louisiana and sw. Illinois
west to Colorado and Utah, e. New
Mexico and south to Mexico.

Primarily nocturnal, but often active in
early evening. It readily climbs trees
and enters abandoned houses and barns
in search of prey: mice, rats, birds, and
bats. The name Corn Snake probably
originated not from an association with
barns and corncribs but from the
similiarity of the belly markings to the
checkered patterns of kernels on Indian
corn. It is one of the most beautiful
snakes in our range. Captive longevity
is 21¾ years.

484, 509, 524, **Rat Snake**
526, 540, 581   *(Elaphe obsoleta)*
Subspecies: Black, Baird's, Texas,
Yellow, Everglades, Gray

Description: 34–101″ (86.4–256.5 cm). Long,
powerful constrictor with 3 different
adult color patterns predominating:
plain, striped, and blotched. Plain is
black often with white showing
between scales. Striped is red, orange,
yellow, brown, or gray with 4 dark
stripes. Blotched is light gray, yellow,
or brown with dark brown, gray, or
black blotches down back. Belly
uniformly white, yellow, orange, or
gray, often with dark mottling or
checks. Belly scales flat in middle, ends
angled up sharply. Underside of tail not
striped. If present, dark stripe through
eye does not reach neck. All young

vividly blotched. Scales weakly keeled, in 25–33 rows. Anal plate divided.

Breeding: Mates April to June and in autumn. Clutch of 5–30 smooth-shelled, oblong eggs, 1½–2¼″ (38–57 mm) long, laid in rotten logs, leaf litter or under rocks June to August; hatch in 7½–15½ weeks, August to October. Hatchlings are 11–16″ (28–41 cm) long.

Habitat: Hardwood forest, wooded canyons, swamps, rocky timbered upland, farmland, old fields, barnyards; from wet to arid situations; sea level to 4,400′ (1,350 m).

Range: E. Ontario and s. Vermont south to Florida Keys, west to w. Texas and adjacent Mexico, north to sw. Minnesota, and s. Michigan.

Subspecies: Black (*E. o. obsoleta*), plain black or with traces of white between scales; s. Vermont to North Carolina coast, southwest to c. Georgia, north to sc. Illinois, south to n. Louisiana and Oklahoma, north to sw. Minnesota and s. Michigan.
Baird's, (*E. o. bairdi*), brown to orange-brown with 4 dark stripes, the upper pair darkest; c. Texas west to Big Bend region and adjacent Mexico.
Texas (*E. o. lindheimeri*), yellow or grayish, with brown to blackish blotches, often with orange showing between scales; s. Louisiana to ec. Texas.
Yellow (*E. o. quadrivittata*), tan, yellow, or yellow-orange with 4 distinct dark stripes, tongue black; coastal North Carolina, South Carolina, Georgia, and most of Florida peninsula.
Everglades (*E. o. rossalleni*), red, red-orange, or orange with 4 faint stripes, tongue red; Everglades of s. Florida.
Gray (*E. o. spiloides*), whitish to gray with brown to dark gray blotches, occasionally with 4 stripes on neck; s. Illinois and extreme sw. Indiana south to Mississippi coast and east to sw. Georgia and nw. Florida panhandle.

Active during the day in spring and fall
but becomes nocturnal in summer. A
skillful climber, it ascends trees or
rafters of abandoned buildings in search
of birds, eggs, and mice. Also eats
other small mammals and lizards.
Hawks may home in on a nest-raiding
Rat Snake when it is being heckled by
other birds. In northern areas the Rat
Snake frequently shares winter dens
with Timber Rattlesnakes and
Copperheads; thus the local names
Rattlesnake Pilot and Pilot Black
Snake. Captive longevity exceeds 20
years.

523 Trans-Pecos Rat Snake
(*Elaphe subocularis*)

Description: 34–66" (86–167.6 cm). A handsome,
"bug-eyed" rat snake; yellowish-tan or
olive-yellow, marked with a series of
dark-brown H-shaped blotches. Arms
of H's may be partly connected, thus
forming fragmented dorsolateral
stripes. Head unpatterned with *large
eyes separated from upper lip scales by row of
small scales.* Scales weakly keeled, in
31–35 rows. Anal plate divided.

Breeding: Mates in late spring. About 3–7 soft,
leathery eggs, 2⅝" (67 mm) long, are
laid in summer. Young hatch in
10½–15 weeks, at 11–14"
(28–36 cm). Mature in 2–3 years.

Habitat: Chihuahuan Desert. Agave-
creosote bush-ocotillo-dominated
slopes to rocky areas characterized by
persimmon-shinoak or cedar;
ca. 1,500–4,500' (450–1,350 m).

Range: Big Bend and Trans-Pecos regions of
Texas and s. New Mexico southward to
nc. Mexico.

Most active during early evening hours
on warm, dry nights. Spends day in
rock crevices or abandoned burrows.
This constrictor feeds on small

mammals, birds, and lizards. Record longevity is 13¾ years.

479   Green Rat Snake
(Elaphe triaspis)

Description: 24–50" (61–127 cm). Slender; plain green, greenish-gray, or olive, with an unmarked yellow-tinged white or cream belly. Elongated head with long squarish snout. *Scales faintly keeled, in 25 or more rows.* Anal plate divided.

Breeding: Habits poorly known. Presumably lays 5 or more smooth white eggs, 2" (51 mm) long, in late summer or early fall.

Habitat: Vicinity of streams in wooded rocky canyon bottoms in mountainous areas; ca. 1,500–7,000' (450–2,150 m).

Range: E. Arizona, south through Mexico into Costa Rica.

Subspecies: Three; 1 in our range, *E. t. intermedia;* Baboquivari, Pajarito, Santa Rita, and Chiricahua mountains, se. Arizona into Mexico.

Rarely seen in our area. Active on the ground during the cooler parts of the day. An excellent climber, it presumably spends much of the day in trees and bushes. It feeds on birds and small mammals.

564   Fox Snake
(Elaphe vulpina)
Subspecies: Western, Eastern

Description: 34–70½" (86.4–179 cm). Yellowish or light brown, marked with bold chocolate-brown to black blotches down midline of back and tail; 2 alternating rows of smaller blotches on sides. Belly yellow with dark squarish blotches. Dark band runs from eye to angle of mouth, a second band extends vertically from eye to mouth; *lacks lance-point design on head.* Scales usually keeled, in 25 or 27 rows. Anal plate divided.

Breeding: Mates April to July. Female lays 6–29 firm leathery eggs, 1½–2″ (38–51 mm) long, late June to early August. Young hatch late August to October at 10–13″ (25–33 cm).

Habitat: Rolling prairies, farmland, wooded stream valleys, Lake Michigan dune country, marshland bordering lakes Erie and Huron.

Range: Great Lakes region west to se. South Dakota, e. Nebraska, and n. Missouri.

Subspecies: Western (*E. v. vulpina*), average of 41 large blotches on body (not including tail); upper peninsula of Michigan south to nw. Indiana and west to e. Nebraska.
Eastern (*E. v. gloydi*), average of 34 large blotches on body; s. Ontario, e. Michigan, nc. Ohio.

Although an excellent climber, it is usually seen on the ground in fields near streams or marshes. Unfortunately, it is often mistaken for a Copperhead and killed. When excited, it may rapidly vibrate its tail in surface litter —the sound somewhat suggestive of an aroused rattler. It feeds on meadow voles, deermice, eggs, fledgling birds, and newborn rabbits.

492 Mud Snake
(*Farancia abacura*)
Subspecies: Eastern, Western

Description: 38–81″ (96.5–205.5 cm). Shiny blue-black snake with pink or red belly bars extending upward on sides. Body cylindrical. Tail short and tipped with a sharp spine. Scales smooth, in 19 rows. Anal plate usually divided.

Breeding: Mates in spring. July to August female lays 11–104 eggs, about 1⅜″ (35 mm) long, in an earthen cavity. She may remain with eggs until hatching in 7–8 weeks, August to September. Hatchlings 6¼–9½" (16–24 cm) long.

Habitat: Swampy, weedy lake margins; slow-moving, mud-bottomed streams; shallow sloughs criss-crossed with rotting logs; floodplains.

Range: Se. Virginia to s. Florida, west to e. Texas, and north in Mississippi Valley to s. Illinois. Isolated population in nc. Alabama.

Subspecies: Eastern (*F. a. abacura*), tops of red bars pointed, 53 or more (not counting tail); se. Virginia to s. Florida and west to se. Alabama.
Western, (*F. a. reinwardti*), tops of red bars rounded, 52 or fewer; w. Alabama to e. Texas, north in Mississippi Valley to s. Illinois.

Habits much like those of Rainbow Snake. Especially active on rainy nights and may be seen crossing roads in swampy areas. It does not bite when picked up, but captor may be poked with its harmless spine-tipped tail. Eats sirens and amphiumas. Longevity exceeds 18 years.

546 Rainbow Snake
(*Farancia erytrogramma*)
Subspecies: Rainbow, South Florida

Description: 35–66" (89–167.6 cm). Cylindrical-bodied, glossy black or blue-black snake with 3 narrow red stripes running length of body. Yellow or reddish-yellow stripe on sides along margins of belly scales. Underside red, with a double row of black spots, one pair on each belly scale. Tail short and tipped with a sharp spine. Males smaller than females. Scales smooth, in 19 rows. Anal plate usually divided.

Breeding: In July, female lays a clutch of 20–52 eggs, each about 1½" (38 mm) long, in a cavity in sandy soil.

Habitat: Areas of loose sandy soil near water; streams, rivers, cypress swamps, spring runs, and marshland.

Range: Coastal plain, s. Maryland south to c. Florida and west to Mississippi River.

Subspecies: Rainbow (*F. e. erytrogramma*), belly predominantly red; s. Maryland south to c. Florida, west to Mississippi River. South Florida (*F. e. seminola*), belly scales and several rows of adjacent scales with heavy black pigment; vicinity of Lake Okeechobee, Florida.

Rarely seen. Burrows in sandy soil or under wet debris and mats of vegetation along water's edge. Active at night but occasionally may be seen during the day foraging for eels, its principal prey. Hatchlings eat salamanders and tadpoles. Folk tales have it that the "stinging snake," "hoop snake," or "thunderbolt," bites its tail, rolls like a hoop, and stings a victim to death with the spine on the tail tip. In fact, the Rainbow Snake is usually docile and the spine is harmless.

547 Mexican Hook-nosed Snake
(*Ficimia streckeri*)

Description: 9–19" (22.9–48.3 cm). *Snout upturned into a hooklike point.* Pale brown to dark brown spots or short narrow crossbands from neck to tail tip. Top of head unpatterned. Scales smooth, in 17 rows. Anal plate divided.

Breeding: Egg-layer. Little is known of its reproductive biology.

Habitat: Thorn forest in Texas to cloud forest at southern edge of range; sea level to 4,900' (1,500 m).

Range: S. Texas south to n. Veracruz, Mexico.

A burrower. Usually encountered at night or after a rainstorm during spring and summer months. It feeds on spiders and centipedes.

588 Western Hook-nosed Snake
(*Gyalopion canum*)

Description: 8–14¼" (20.3–36.2 cm). Pale brown and patterned with dark-edged brown crossbands; *2 prominent crossbands on head. Snout upturned and sharp-edged.* Scales smooth, in 17 rows. Anal plate *divided.*

Breeding: One female is known to have laid a single egg, 1⅛" (29 mm) long, early in July.

Habitat: Arid regions dominated by creosote bush, mesquite, and shadescale, and juniper-grassland or piñon-juniper associations.

Range: W. Texas west to se. Arizona, and south to Zacatecas, Mexico.

Although usually thought of as a desert species, it inhabits mountainous parts of New Mexico and Texas. Most often seen at night after a light rain, from April to September. When first touched, it gyrates wildly and makes popping noises by everting and retracting the lining of its cloaca. It eats small spiders, scorpions, and centipedes.

Desert Hook-nosed Snake
(*Gyalopion quadrangularis*)

Description: 7–14" (17.8–35.6 cm). *Snout upturned and sharp-edged.* Red or reddish-brown and patterned with black blotches or crossbands; *large blotch on head fused with crossband on neck.* Scales smooth, in 17 rows. Anal plate single.

Breeding: Habits of this egg-layer are unknown.

Habitat: Arizona Sonoran desert to tropical dry forest in southern parts of range.

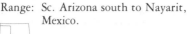

Range: Sc. Arizona south to Nayarit,
Mexico.

Secretive and rarely seen, this
burrowing species is nocturnal and
preys on spiders, centipedes, and
occasionally scorpions. "Cloacal
popping" exhibited by the Western
Hook-nosed Snake has not been
observed in the Desert Hook-nosed.

572 Western Hognose Snake
(Heterodon nasicus)
Subspecies: Plains, Dusty, Mexican

Description: 16–35¼″ (40.6–89.5 cm). *Sharply
upturned and pointed snout.* Stout body
with broad neck. Tan, brown, gray, or
yellowish-gray above with distinct or
somewhat faded series of dark blotches
down back and 2 or 3 rows of side
spots. *Belly and underside of tail distinctly
patterned with large black blotches.* Scales
keeled, in 23 rows. Anal plate
divided.

Breeding: Mates March to May. Lays 4–23
elongate, thin-shelled eggs, 1½″ (32
mm) long, in soft loamy or sandy soil,
early June to late August depending on
locality. Young, 6–7½″ (15–19 cm)
long, hatch in 7–9 weeks and reach
maturity in 2 years.

Habitat: Sand and gravelly-soiled prairie,
scrubland, river floodplains. Sea level to
8,000′ (2,450 m).

Range: Se. Alberta and nw. Manitoba, south to
se. Arizona, Texas, and into n. Mexico.
Isolated populations in Minnesota,
Iowa, Illinois, Missouri, and Arkansas.

Subspecies: Plains (*H. n. nasicus*), more than 35
midline body blotches in males, more
than 40 in females, 9 or more small
scales between prefrontal scales; se.
Alberta and sw. Manitoba south to w.
Oklahoma, Texas panhandle, and s.
New Mexico.
Dusty (*H. n. gloydi*), fewer than 32

midline body blotches in males, less than 37 in females, 9 or more small scales between prefrontal scales; se. Kansas through Texas except panhandle, Trans-Pecos Texas, and extreme s. Rio Grande Valley. Mexican (*H. n. kennerlyi*), 6 or fewer small scales between prefrontal scales; Mexico into extreme s. Texas through the Trans-Pecos region into sw. New Mexico and se. Arizona.

Primarily active during morning and late afternoon hours; burrows into loose soil to escape hot or cold conditions. Sense of smell enables it to find buried toads, lizards, snakes, and reptile eggs; also eats birds and small rodents. Defensive display parallels that of other hognose snakes, but is not nearly so well developed.

485, 563, 565 **Eastern Hognose Snake**
(*Heterodon platyrhinos*)

Description: 20–45½" (50.8–115.6 cm). *A stout-bodied snake with pointed, slightly upturned snout and wide neck.* Color extremely variable: yellow, tan, brown, gray, or reddish with squarish dark blotches on back interspaced with round dark dorsolateral blotches. All-black individuals common in some areas. Belly mottled; *underside of tail conspicuously lighter than belly color.* Scales keeled, in 23–25 rows. Anal plate divided.

Breeding: Mates spring and fall. Lays 4–61 elongate, thin-shelled eggs, about 1¼" (32 mm) long, June to July (May in Texas, to late August in northerly areas), in a shallow cavity in loose or sandy soil. Young, 6½–9½" (16.5–24 cm) long, hatch in 39–65 days.

Habitat: Prefers open sandy-soiled areas; thinly wooded upland hillsides, cultivated fields, woodland meadows. Sea level to 2,500' (750 m).

Range: Eastern-central Minnesota to extreme s. New Hampshire south to s. Florida, west to e. Texas and w. Kansas.

Commonly called puff, or spreading, adder or blow viper. Active in the daytime. Burrows deep into loose earth during cold winter months. When disturbed, it "hoods" its neck, inflates its body, hisses loudly, and strikes. If this fails to discourage a would-be predator, it rolls over and plays dead with mouth agape and tongue hanging out. It becomes limp and will remain "dead" when picked up; however, it will roll over again if placed right-side up. In captivity it loses willingness to display such behavior. Enlarged teeth on rear upper jaw are believed to inject mild venom into toads and frogs upon which it feeds. It rarely bites people.

585 Southern Hognose Snake
(Heterodon simus)

Description: 14–24" (35.5–61 cm). Short, stoutly built snake with *pointed, sharply upturned snout and wide neck.* Light brown, yellow, or grayish, and often tinged with red; a distinct row of dark blotches on back alternates with smaller dorsolateral blotches on each side. *Underside of tail not distinctly lighter than belly.* Scales keeled, usually in 25 rows. Anal plate divided.

Breeding: Little is known. Clutches of 6 and 10 eggs have been recorded.

Habitat: Open dry sandy areas, fields, dry river floodplains, and wire grass flatwoods.

Range: Coastal plain in se. North Carolina south to Lake Okeechobee, Florida, west to se. Mississippi.

A burrower. Displays death-feigning behavior exhibited by Eastern Hognose

Snake. Little else is known about this burrower's natural history.

---

586 Night Snake
(*Hypsiglena torquata*)
Subspecies: Desert, Texas, San Diego,
Mesa Verde, California, Spotted

Description: 12–26" (30.5–66 cm). Slender and
cylindrical-bodied; beige, yellowish, or
gray, patterned with numerous dark
brown or gray blotches on back and
side. Large blotch on each side of neck;
a third spot may be present on nape or
lateral blotches may be fused at
midline. *Eyes with vertical pupils.* Dark
bar behind eye; upper lip scales white.
Belly cream or white, unpatterned.
Scales smooth, in 19–21 rows. Anal
plate divided.

Breeding: Habits poorly known. Clutches of 4–6
eggs, ⅞–1¼" (22–32 mm) long, have
been deposited late April to early July.
Incubation takes 7–8 weeks.

Habitat: Semiarid and arid sandy or rocky
situations from plains and desert flats,
to heavy brush chaparral and blue oak-
Digger pine woodland; sea level to
7,000' (2,100 m).

Range: Nc. California, sc. Washington, s.
Idaho, Utah, and sw. Kansas south
through Baja California to Costa Rica.

Subspecies: About 12; 6 in our range.
Desert (*H. t. deserticola*), 3 blotches on
neck, central one greatly enlarged
posteriorly; sc. Washington, e.
Oregon, s. Idaho, Nevada, w. Utah,
and Mojave Desert region, se.
California.
Texas (*H. t. jani*), 3 elongated blotches
on neck, top of head usually rounded;
sw. Kansas, w. Oklahoma, se.
Colorado, and most of New Mexico
south through Texas into ne. Mexico.
San Diego (*H. t. klauberi*), 3 blotches
on neck, central one not enlarged
posteriorly; coastal s. California,

vicinity San Luis Obispo Bay south into
Baja California.
Mesa Verde (*H. t. loreala*), neck
blotches variable, 2 loreal scales on each
side, others with 1; e. Utah, extreme
w. Colorado, into nw. New Mexico and
extreme ne. Arizona.
California (*H. t. nuchalata*), large neck
blotch, 19 scale rows (other subspecies
have 21); slopes of Sierra Nevada and
Coast Ranges bordering Sacramento-San
Joaquin valleys.
Spotted (*H. t. ochrorhyncha*), 2 blotches
or single narrow bar across neck region,
top of head flat; Arizona, w. New
Mexico, sw. Trans-Pecos region of
Texas south to Jalisco, Mexico.

This successful, wide-ranging nocturnal
species is rarely encountered. Most of
the day it hides under rocks or plant
litter. Enlarged grooved teeth, located
near back of upper jaw, hold lizard and
frog prey while snake's mildly toxic
saliva incapacitates them.

557, 569 Prairie Kingsnake
(*Lampropeltis calligaster*)
Subspecies: Prairie, Mole Snake

Description: 30–52⅛" (76.2–132.4 cm). A slender,
variably patterned kingsnake. Tan,
grayish-brown, or yellowish-brown
above with black-edged, dark brown to
reddish-brown or greenish blotches
down back and 2 alternating rows of
smaller, less conspicuous spots on sides.
V-shaped arrowheadlike marking on
crown of head. Pattern of older
specimens may be lost or obscured by
dark pigment; some develop 4
longitudinal dusky stripes. Scales
smooth, in 21–27 rows. Anal plate
single.

Breeding: Nests June to July. Female lays 5–17
eggs, 1¾–2" (44–51 mm) long, in
earth cavity below ground surface.

Young hatch in 7–11 weeks at 8–11″ (20–28 cm), August to September.

Habitat: Open fields, cultivated farmland, barnyards, pastures, prairies, rocky hillsides, open woodland.

Range: C. Maryland to n. Florida west to se. Nebraska and e. Texas.

Subspecies: Prairie (*L. c. calligaster*), scales in 25 or 27 rows; interspaces between blotches about equal in size to blotch; belly white to yellowish, clouded or spotted with brown; w. Illinois southwest to the Gulf in sc. Louisiana, west to se. Nebraska, e. Kansas, Oklahoma, and e. Texas.

Mole Snake (*L. c. rhombomaculata*), scales in 21 or 23 rows; back blotches well separated; belly yellow with brown rectangular blotches; c. Maryland to n. Florida west to c. Tennessee and se. Louisiana.

Secretive; spends much of the day in animal burrows or under rocks or several inches of loose soil. Most frequently seen crossing roads after a rainstorm or on warm spring or summer nights. Eats small rodents, birds, frogs, lizards, and other snakes. Usually mild-tempered. One captive animal lived 11 years.

483, 522, 560, 561, 590, 592, 594

Common Kingsnake
(*Lampropeltis getulus*)

Subspecies: Eastern, California, Florida, Speckled, Black, Mexican, Desert

Description: 36–82″ (91.4–208.3 cm). A large chocolate brown to black kingsnake with a highly variable back and belly pattern. Light-centered scales may form distinct crossbands, "chainlinks," lengthwise stripes, blotches, or speckles on the back. Belly ranges from plain white to heavily blotched with dark pigment to plain black. Scales smooth, in 19–25 rows. Anal plate single.

Breeding: Mates mid-March (Florida) to June.
Clutches of 3–24 creamy white to
yellowish elongated eggs, 1¼–2¾"
(31–69 mm) long, are laid mid-May
(Florida) to August. Incubation lasts
8½–11½ weeks, depending on
temperature. Hatchlings are 9–12"
(23–30 cm) long.

Habitat: Diverse: New Jersey pine barrens to
Florida Everglades; dry rocky wooded
hillsides to river swamps and coastal
marshes, and prairie, desert, and
chaparral; sea level to 6,900'
(2,100 m).

Range: S. New Jersey to s. Florida, west to sw.
Oregon and s. California, south to s.
Baja California and Zacatecas,
Mexico.

Subspecies: Eastern (*L. g. getulus*), chocolate brown
to black with bold light chainlike
pattern, 21 scale rows; s. New Jersey to
n. Florida, west to extreme w. West
Virginia, c. Virginia, extreme w.
North Carolina, and s. Alabama.
California (*L. g. californiae*), chocolate
brown to black with bold light
crossbands or a back stripe, 23–25 scale
rows; sw. Oregon south to extreme s.
Baja California, east to s. Utah and w.
Arizona.
Florida (*L. g. floridana*), scales tipped
with chocolate brown, yellowish at
base, chainlike pattern of narrow
crossbands, often obscure, 23 scale
rows; c. and s. Florida; isolated
population in Duval and Baker
counties, ne. Florida.
Speckled (*L. g. holbrooki*), dark brown
or black scales with central light spot,
remnants of crossbands present, 21
scale rows; s. Iowa and w. Illinois south
to e. Texas, Mississippi, and c. and sw.
Alabama.
Black (*L. g. niger*), shiny black with
small light dots forming faint or broken
chainlink pattern, 21 scale rows; e.
Illinois to extreme sc. Ohio and w.
West Virginia, south to nw. Georgia
and n. Alabama.

Mexican (*L. g. nigritus*), uniformly black or black with small light spot on each side scale and some back scales, 23–25 scale rows; w. Sonora and extreme nw. Sinaloa, Mexico, into se. Arizona.

Desert (*L. g. splendida*), back dark brown or black with narrow light crossbands, scales on sides have central light spot, 23–25 scale rows; c. Texas west to se. Arizona south to c. Mexico.

Active during the day, especially early in the morning or near dusk, but becomes nocturnal in the warm summer months. It is primarily terrestrial, occasionally climbing into shrubs. A strong constrictor, it eats snakes—including rattlesnakes, Copperheads, and coral snakes—as well as lizards, mice, birds, and eggs. Record longevity exceeds 24 years.

601, 602, 603 **Gray-banded Kingsnake**
(*Lampropeltis mexicana*)

Description: 24–47½″ (61–120.7 cm). *Highly variable pattern of* white-bordered gray crossbands alternating with black-bordered reddish-orange crossbands or blotches. Head distinct from neck; eyes large. Scales smooth, in 25 rows. Anal plate single.

Breeding: Mates in spring. 4–9 eggs, 1¼–1¾″ (32–45 mm) long, are laid late May to late July, hatch in 9–11 weeks. Young are 9–11″ (23–28 cm) long.

Habitat: Arid mesquite-creosote bush desert flats, barren rocky hillsides, canyons, limestone ledges, ranging into semimoist mountainous situations; 1,200–7,500′ (350–2,300 m).

Range: Trans-Pecos region east to Balcones Escarpment of Edwards Plateau of Texas, south into Mexico.

Subspecies: Three, poorly defined; 1 enters our range, *L. m. alterna*.

Once considered rare, this handsome
snake became a favorite of collectors. A
secretive, nocturnal species it proved to
be considerably more common than
believed. Lizards are its staple diet, but
it also eats frogs, small snakes, and
rodents.

---

598 Sonora Mountain Kingsnake
(*Lampropeltis pyromelana*)
Subspecies: Arizona, Utah, Huachuca

Description: 20–41" (50.8–104.1 cm). Tricolored
kingsnake with red, black, and white,
cream, or yellow bands. *Light rings do
not widen near belly scales.* Black bands
border red and light bands and become
narrow or disappear on sides. Top of
head black; *snout light colored.* Scales
smooth, in 23–25 rows. Anal plate
single.

Breeding: Mates in spring. Clutches of 3–6
elongated eggs, 1½–2⅜" (44–60 cm)
long, are deposited June to July and
hatch in about 2½–3 months. Young
9–10" (8.8–10 cm) long.

Habitat: Chaparral woodland and pine forests in
mountainous regions; brushy rocky
canyons, talus slopes, and near streams
and springs; 2,800–9,100'
(850–2,800 m).

Range: Nc. Arizona south into Mexico; isolated
populations in Utah, Nevada, and
Arizona.

Subspecies: Four; 3 in our range.

Arizona (*L. p. pyromelana*), usually more
than 43 light rings on body, 10 lower
lip scales; c. and se. Arizona and sw.
New Mexico into nw. Mexico.
Utah (*L. p. infralabialis*), half or more
of white body rings cross belly, 9 lower
lip scales; e. Nevada, c. Nevada to
Grand Canyon area of nw. Arizona.
Huachuca (*L. p. woodini*), usually fewer
than 43 light rings on body, 10 lower
lip scales; Huachuca Mountains, s.
Arizona into Mexico.

Little is known of this handsome
species' natural history. It feeds on
lizards and presumably small rodents.
Captive longevity exceeds 18 years.
Protected in Arizona.

597, 600, 613,   **Milk Snake**
614, 615   (*Lampropeltis triangulum*)
                 Subspecies: Eastern, Louisiana,
                 Mexican, New Mexico,
                 Scarlet Kingsnake, Central Plains,
                 Pale, Red, Utah

Description: 14–78¼" (35.6–199 cm). Gray or tan
marked with a light Y-shaped or
V-shaped patch on neck and chocolate-
brown to reddish-brown, black-
bordered blotches down back and sides.
Or colorfully ringed and blotched with
red (or orange), black, and yellow (or
white). Light neck collar followed by
black-bordered red bands separated by
light rings. *Light rings widen near belly.*
Scales smooth, in 19–23 rows. Anal
plate single.

Breeding: In our range, mates in spring; deposits
clutch of 2–17 elliptical eggs often in
rotting logs, June to July; hatch
August to September. Incubation
period 6–9 weeks. Young 5½–11"
(12–28 cm) long.

Habitat: Diverse situations: semiarid to damp
coastal bottomland to Rocky Mountains
and tropical hardwood forests; pine
forests, open deciduous woodland,
meadows, rocky hillsides, prairies, high
plains, sand dunes, farmland, and
suburban areas; sea level to ca. 8,000'
(2,450 m).

Range: Se. Maine, sw. Quebec, se. and sc.
Ontario, s. Wisconsin, and c. and se.
Minnesota south through most of
United States east of the Rocky
Mountains; Mexico south to Colombia
and Venezuela.

Subspecies: Eastern (*L. t. triangulum*), 26–52" (66–
132 cm), Y- or V-shaped patch on nape

of neck; s. Maine south to n. New Jersey and in Appalachian Mountains through North Carolina, Tennessee to n. Georgia and Alabama, and westward to w. Kentucky, sw. Indiana, north half of Illinois, ne. Iowa, and sc. Minnesota.

Louisiana (*L. t. amaura*), 16–31″ (41–78.7 cm); snout mottled black and white, black border of broad red bands extends onto belly scales; Louisiana west of Mississippi River and e. Texas north into extreme se. Oklahoma and sw. Arkansas.

Mexican (*L. t. annulata*), 25–39″ (64–99 cm); snout black; edges of broad red blotches extend to belly scales, belly black; s. Texas into Mexico.

New Mexico (*L. t. celaenops*), 14–24¾″ (36–63 cm), snout mottled black and white; black and light bands expanded on center line of back and rear belly, red extends to edge of belly scales; n. and e. New Mexico and adjacent w. Texas; scattered populations in sc. New Mexico to Big Bend area, Texas.

Scarlet Kingsnake (*L. t. elapsoides*), 14–27″ (36–68.6 cm), snout red, bands usually continue across belly; North Carolina south through Florida Keys, west to the Appalachians and the Mississippi River in s. Mississippi and adjacent Louisiana, north through ne. Mississippi, c. Tennessee, and sc. and e. Kentucky.

Central Plains (*L. t. gentilis*), 18–36″ (46–91.4 cm), snout mottled black and white, black bands encroach on red bands dorsally, red bands extend onto belly—may cross it or be separated by black bar; n. Texas panhandle, w. Oklahoma, e. and w. Kansas, e. Colorado, and sc. and sw. Nebraska.

Pale (*L. t. multistrata*), 18–33½″ (46–85 cm), snout light orange with black flecks, orange often replaces red, midline area of belly white with a few scattered black marks; w. Nebraska, w. South Dakota, se. and nc. Wyoming

and se. and c. Montana.
Red (*L. t. syspila*), 21–42″ (53–106.7
cm), black border of red bands extends
to first scale rows, several side blotches
may be present; nw. Mississippi west to
ne. Oklahoma, north to sw. Indiana, c.
Illinois, ne. Iowa, and extreme se.
South Dakota.
Utah (*L. t. taylori*), 16–28¼″ (41–71.9
cm), snout black or light with black
blotch on top and tip, black bands
expanded dorsally, often fusing and
interrupting red rings, black-margined
red bands extend to first scale row or
edge of belly scales; wc. Colorado, ne.,
c., and sw. Utah, and nc. Arizona.
Note: Coast Plains Milk Snake (an
intergrade Scarlet Kingsnake) and
Eastern Milk Snake, with light collar
and reddish dorsal blotches reaching
belly scales, occur from s. New Jersey
to ne. North Carolina.

Usually discovered under rotting logs
or stumps or damp trash. Secretive and
usually not seen in the open except at
night. It eats small rodents, birds,
lizards, and snakes—including
venomous species. In the north, it is
often mistaken for the Copperhead, in
the south for the Eastern Coral Snake.
Its common name is based on the
absurd belief that it milks cows, taking
prodigious amounts in the process. The
adult size of the Milk Snake varies
geographically. North American
subspecies do not attain the great
lengths of neotropical populations. In
our range the species reaches its greatest
length in the northeastern states and
adjacent Canada. Adults in the
Southeast are smallest. Record
longevity in captivity exceeds 21 years.

599 California Mountain Kingsnake
(*Lampropeltis zonata*)
Subspecies: St. Helena, Sierra, Coast,
San Bernardino, San Diego

Description: 20–40″ (50.8–101.6 cm). One or our
most attractive snakes—ringed with
black, white, and red. Resembles
Sonora Mountain Kingsnake and
western races of the Milk Snake. Red
bands bordered by black. *Snout black.*
White bands do not widen near belly scales.
Scales smooth, in 23 rows. Anal plate
single.

Breeding: Clutches of 3–8 eggs deposited July;
hatch in 9–10 weeks. Young about 8″
(20 cm) long.

Habitat: Sierra Nevada yellow pine belt, Coast
Ranges chaparral, redwood forests south
of San Francisco Bay; sea level to
8,000′ (2,450 m).

Range: Kern County, California, north along
western slope of Sierra Nevada
Mountains into sw. Oregon, southward
in eastern portion of Coast Ranges
to San Francisco Bay area; and south in
mountains in scattered populations to
nc. Baja California. Isolated population
in sc. Washington.

Subspecies: Seven; 5 in our range.

St. Helena (*L. z. zonata*), back edge of
first white ring is behind last upper lip
scale, snout dark, more than 60% of
red bands continuous across midline of
back; Napa, Lake, Mendocino, and
Sonoma counties, California.
Intergrades with Sierra in n. California
and sw. Oregon.
Sierra (*L. z. multicincta*), first white
ring and snout resemble St. Helena,
less than 60% of red bands continuous
across back; western slopes of Sierra
Nevadas, Shasta County to Kern
County, California.
Coast (*L. z. multifasciata*), like St.
Helena, snout marked with red; Santa
Cruz and Santa Clara counties to
Ventura and Santa Barbara counties,
California.

San Bernardino (*L. z. parviruba*), back edge of first white ring on or in front of last upper lip scale, snout dark, 37 or more groups of tricolored rings (triads) around body; San Gabriel, San Bernardino, and San Jacinto mountains, s. California.

San Diego (*L. z. pulchra*), like San Bernardino except 36 or fewer triads; mountains of San Diego County, s. California.

Sometimes called Coral Kingsnake. Diurnal but becomes active at night during warm weather and sometimes can be seen crossing a road. Hides under rotting logs and stones near sunlit stretches of rocky streams. Eats lizards, snakes, and birds and their eggs. Record longevity exceeds 24 years.

606 Cat-eyed Snake
(*Leptodeira septentrionalis*)
Subspecies: Northern

Description: 18–38¾" (20.3–98.4 cm). Head conspicuously wider than neck; eyes have *vertical pupils.* Pale yellow to reddish-tan and marked with dark-brown or black saddle-shaped blotches that extend down sides close to 1st scale row. Rear edge of belly scales dark-marked. Scales smooth, in 21–23 rows. Anal plate divided.

Breeding: Information from Texas is lacking. Elsewhere, females deposit about 6–12 eggs, March to May. Hatching takes 3 months; young are about 9" (23 cm) long.

Habitat: Arid situations to rainforest; coastal plain and semidesert areas in Texas.

Range: Extreme s. Texas south through Mexico to nw. South America.

Subspecies: One in our range. Northern (*L. s. septentrionalis*).

At night, it searches for prey in bushes and amid ground litter along the margins of streams and ponds. It has grooved fangs near the back of its upper jaw. Frogs and lizards are immobilized by the mild venom. A wild solitary female imported with a banana shipment laid 3 clutches of fertile eggs during the following 5-year period. Captives have lived more than 10 years.

517 Sonora Whipsnake
(Masticophis bilineatus)
Subspecies: Sonora, Ajo Mountain

Description: 30–67" (76.2–170.2 cm). Long and slender; grayish-brown, olive, or blue-gray, becoming lighter toward tail. *2 or 3 dark-edged light stripes on each side fade before reaching tail.* Black line under eye from snout to neck. Belly cream, yellow under tail. Scales smooth, *in 17 rows.* Anal plate divided.

Breeding: Clutch of 6–13 rough, leathery eggs, 2" (52 mm) long, laid June to July.

Habitat: Thorny desert brushland to mountain pine-oak forest, generally in more open areas or near a stream, ca. 2,000–6,100' (600–1,850 m).

Range: C. and s. Arizona, extreme sw. New Mexico into Mexico.

Subspecies: Three; 2 in our range.

Sonora (*M. b. bilineatus*), uppermost light stripe 2 half-scale rows wide, chin usually unmarked; c. and s. Arizona, sw. New Mexico south into Mexico. Ajo Mountain (*M. b. lineolatus*), uppermost light stripe 1 half-scale row wide, chin spotted; Ajo Mountains, Pima County, Arizona.

Diurnal. Readily climbs into shrubs and trees in search of lizards and birds (particularly nestlings). An alert snake, it will flee at the approach of danger.

469, 491, 553, **Coachwhip**
554, 556, 558 (*Masticophis flagellum*)
　　　　　　　Subspecies: Eastern, Sonora, Baja
　　　　　　　California, Lined, Red, San Joaquin,
　　　　　　　Western

Description: 36–102″ (91.4–259 cm). Large, lithe,
　　　　　　　long-tailed and fast-moving. Western
　　　　　　　races generally yellow, tan, brown,
　　　　　　　gray, or pinkish; essentially patternless
　　　　　　　or with dark crossbars on neck. Eastern
　　　　　　　form: head and neck region dark brown
　　　　　　　to almost black, gradually fading to
　　　　　　　light brown toward rear. Occasionally
　　　　　　　all black. No pale side stripes. Scales
　　　　　　　smooth, in 17 rows (*13 at rear of body*).
　　　　　　　Anal plate divided.

Breeding: Mates in spring. Clutches of 4–16
　　　　　　　granular-surfaced eggs, 1–2¼″ (25–57
　　　　　　　cm) long, are deposited June to July,
　　　　　　　hatch in 6–11 weeks. Young 12–16″
　　　　　　　(30–41 cm) long.

Habitat: Dry, relatively open situations; pine
　　　　　　　and palmetto flatwoods, rocky hillsides,
　　　　　　　grassland prairies, desert scrub, thorn
　　　　　　　forest, and chaparral; sea level to ca.
　　　　　　　7,000′ (2,150 m).

Range: Se. North Carolina, sw. Tennessee,
　　　　　　　extreme sw. Illinois, extreme sw.
　　　　　　　Nebraska, e. Colorado, nc. New
　　　　　　　Mexico, sw. Utah, wc. and s. Nevada,
　　　　　　　and c. California, south through
　　　　　　　Florida, Texas, and California to c.
　　　　　　　Mexico.

Subspecies: Eastern (*M. f. flagellum*), head and neck
　　　　　　　area dark brown to black above, fading
　　　　　　　to light brown toward rear; se. North
　　　　　　　Carolina to s. Florida west to extreme
　　　　　　　sw. Tennessee and Mississippi River in
　　　　　　　Louisiana, and se. Kansas and s.
　　　　　　　Missouri south through w. Louisiana
　　　　　　　and e. Texas. Isolated populations in e.
　　　　　　　Tennessee and sc. Kentucky.
　　　　　　　Sonora (*M. f. cingulum*), long, dark
　　　　　　　reddish-brown bands separated by
　　　　　　　shorter, paired pale pink bands, or
　　　　　　　uniformly reddish-brown or black; sc.
　　　　　　　Arizona into Mexico.
　　　　　　　Baja California (*M. f. fulginosus*), 2

phases: yellow or light gray with zigzag pattern of black bands along body, or dark gray-brown above with lined pattern on sides; extreme s. California through Baja California.

Lined (*M. f. lineatulus*), light gray or tan, each dorsal scale on forepart of body has a dark streak down center, underside of tail salmon pink; sw. New Mexico into Mexico.

Red (*M. f. piceus*), 2 phases: pink to red above with dark crossbands on neck and forepart of body, or black above and reddish near vent and under tail; wc. and s. Nevada and sw. Utah, south through California and w. and se. Arizona and ne. Baja California and nw. Mexico.

San Joaquin (*M. f. ruddocki*), light yellow to olive-yellow above, without dark head and dark neck bands; e. and s. California.

Western (*M. f. testaceus*), light brown, olive, yellowish, or pinkish-red above, some with short dark crossbands on neck and wide crossbands on forepart of body, double row of dark spots on belly; extreme sw. Nebraska, w. Colorado, w. Kansas, w. Oklahoma, e. New Mexico through w. and c. Texas into Mexico.

Perhaps our fastest snake. Prowls about during the day in search of grasshoppers, cicadas, lizards, snakes and small rodents. When pursued, may take to a tree or disappear into a mammal burrow. If cornered, it coils, vibrates its tail, and strikes repeatedly —often at enemy's face. Contrary to popular belief, it does not chase down an adversary and whip him to death. Record longevity is 16 years, 7 months.

518 Striped Racer
(*Masticophis lateralis*)
Subspecies: California, Alameda

Description: 30–60" (76.2–152.4 cm). Long and slender; uniform black or dark brown, with *single yellow or orange stripe on each side* extending from neck to tail. Belly yellowish-white, becoming pink under tail. Scales smooth, in 17 rows. Anal plate divided.

Breeding: Mates in spring. 6–11 eggs, 2½" (54 mm) long, are laid late May to July, hatch August to October.

Habitat: Chaparral brushland, desert foothills, open hardwood-pine forest in the mountains; especially common around watercourses and ponds; sea level to 7,000' (2,150 m).

Range: Nc. California south along coast and along western slopes of Sierra Nevada, but absent from most of Great Valley, into Baja California.

Subspecies: Three; 2 in our range.
California (*M. l. lateralis*), stripe yellow, 2 half-scale rows wide; w. California, except San Francisco Bay region, and western slopes of Sierras into Baja.
Alameda (*M. l. euryxanthus*), stripe orange, 1 whole and 2 half-scale rows wide; e. San Francisco Bay area.

Active during the day. Swift, agile, and alert predator of frogs, lizards, other snakes, birds, and small mammals. Juveniles may eat insects. Like other racers, it locates its prey more by sight than smell. When hunting, it holds its head high off the ground.

521 Striped Whipsnake
(*Masticophis taeniatus*)
Subspecies: Desert, Central Texas,
Ruthven's, Schott's

Description: 40–72" (101.6–182.9 cm). Long,
slender, and fast-moving. Gray, bluish-
greenish-gray, olive, reddish-brown, or
black; typically with 2 or more
continuous or broken light lengthwise
stripes on each side. Large head scales
edged in white (except in s. Texas).
Scales smooth, *in 15 rows.* Anal plate
divided.

Breeding: Courts in early spring. May nest in
abandoned rodent burrows. Lays 3–12
eggs, June to July. Young, 14–17"
(36–43 cm) long, hatch in August.
Males mature in 1–2 years; females
in 3.

Habitat: From grassland and arid brushy flatland
to rugged mountainous terrain
dominated by piñon-juniper and open
pine-oak woodlands; sea level to 9,400'
(2,850 m).

Range: Sc. Washington southeast in Great
Basin to s. New Mexico and w. and c.
Texas, south to wc. Mexico.

Subspecies: Five; 4 in our range.

Desert (*M. t. taeniatus*), dark brown or
blackish, white side stripes divided by
thin black line; sc. Washington
southeastward in Great Basin to s. New
Mexico and adjacent extreme w. Texas
and Mexico.
Central Texas (*M. t. ornatus*), black
with well-spaced lengthwise white
patches on sides; c. and w. Texas to c.
Mexico. Isolated population in nc.
Texas.
Ruthven's (*M. t. ruthveni*), bluish-
greenish-gray, traces of stripes on neck;
extreme s. Texas to c. Mexico.
Schott's (*M. t. schotti*), bluish-greenish-
gray, 2 light side stripes, sides of neck
reddish-orange; s. Texas, Edwards
Plateau south to Falcon Reservoir
region, west into Mexico.

When surprised, this speedster quickly vanishes into brush, rocks, or mammal burrows. During the day, it hunts with head held high, watching for scurrying lizards or small mammals. Small snakes, including venomous species, are also eaten. An agile climber, it is occasionally seen high in trees, basking or searching for nesting birds.

482 Green Water Snake
(*Nerodia cyclopion*)
Subspecies: Green, Florida Green

Description: 30–74" (76.2–188 cm). Heavy-bodied; olive-green, brownish, or reddish (in south Florida), with indistinct black bars on sides alternating with crossbars on back, more distinct in juveniles. Head seemingly short, *series of small scales separating eye from upper lip scales.* Belly cream to brown, with or without light spots. Scales keeled, in 27–29 rows. Anal plate divided.

Breeding: Mates March to April; 4–101 young are born June to August; are 8¾–10¾" (22–27 cm) long.

Habitat: Marshes, swamps, ditches, canals, bayous, and estuaries. Most frequent where there is little current and dense aquatic vegetation.

Range: Coastal South Carolina through Florida, west to Louisiana and e. Texas, and north through e. Arkansas to extreme s. Illinois.

Subspecies: Green (*N. c. cyclopion*), light spots on dark belly; s. Alabama (Mobile Bay) to Texas and north to s. Illinois.
Florida Green (*N. c. floridana*), unmarked light belly; s. South Carolina through Florida to s. Alabama.

Primarily diurnal, but often active in the early evening feeding on minnows and small fishes. Most favorable habitat for western form is wooded swamp, where many individuals may be seen basking in branches that overhang water. The Florida Green prefers weed-choked marshes, is less inclined to bask, and is somewhat more nocturnal. When captured Green Water Snakes will bite or smear their captors with musk, but seem more reluctant to do so than most water snakes. Instead, they often regurgitate their last meal and attempt to flee. One of the most fecund snakes in our range.

481, 490 **Plain-bellied Water Snake**
(*Nerodia erythrogaster*)
Subspecies: Red-bellied,
Yellow-bellied, Copper-bellied,
Blotched

Description: 30–62″ (76.2–157.5 cm). Plain reddish-brown, brown, greenish, or gray, becoming lighter on sides. Some populations have dark-bordered light crossbars down back. *Belly plain red, orange, or yellow,* but occasionally belly scales have dark edges. Juveniles have vivid dark blotches down back, alternating with dark crossbars on sides. Scales keeled, in 23–27 rows. Anal plate usually divided.

Breeding: Mates April to June. 5–27 young, 7½–13″ (19–33 cm) long, are born August to October.

Habitat: River swamps and the forested edges of streams, ponds, lakes, and bayous.

Range: S. Delaware to n. Florida, west through Alabama to w. Texas and se. New Mexico, north to w. Missouri and s. Illinois and Indiana. Scattered populations in Michigan, Ohio, and e. Iowa.

Subspecies: Six; 4 in our range.
Red-bellied (*N. e. erythrogaster*), belly red, orange, or pink; Delaware to n. Florida and se. Alabama.
Yellow-bellied (*N. e. flavigaster*), belly yellow or yellow-orange; nc. Georgia to e. Texas and north to sw. Illinois.

Copper-bellied (*N. e. neglecta*), belly red or orange, frequently with dark color on edges of belly scales; w. Kentucky, se. Illinois, Indiana, Michigan, and Ohio.

Blotched (*N. e. transversa*), belly yellow, back with dark-edged light crossbars; w. Missouri, se. Kansas through Oklahoma and c. Texas, se. New Mexico, and adjacent Mexico.

Active in the early evening, Plain-bellies are often seen crossing roads on warm rainy or humid nights. Later they may take refuge and sleep high in the branches of trees overhanging streams. They have been observed to anchor themselves in the vegetation and fish with the mouth opened to the current, grabbing any small fish that happens by. They also eat frogs and tadpoles.

513, 552, 562, **Southern Water Snake**
578, 579 (*Nerodia fasciata*)
Subspecies: Banded, Gulf Salt Marsh, Mangrove, Broad-banded, Florida, Atlantic Salt Marsh

Description: 16–62½" (40.6–158.8 cm). Stout-bodied aquatic snake with dark crossbands over most of body, light back and dorsolateral stripes, or essentially patternless. Color and pattern highly variable, like Northern Water Snake; where ranges overlap, presence of *dark stripe from eye to angle of mouth* and *large squarish blotches or wormlike markings on belly scales* will identify the Southern. Some darken with age, obscuring pattern. Scales keeled, in 21–25 rows. Anal plate divided.

Breeding: Live-bearing. Mates January to February in extreme southerly parts of range. 2–57 young, 7–10½" (18–27 cm) long, are born June to August.

Habitat: Fresh- and saltwater situations;

permanent lakes, ponds, cypress and mangrove swamps, marshes, and sluggish streams; sea level to ca. 1,000′ (300 m).

Range: Coastal plain, North Carolina to Florida Keys, west to e. Texas; north in Mississippi River Valley to extreme s. Illinois.

Subspecies: Banded (*N. f. fasciata*), red, brown, or black crossbands, most darken with age, squarish spots on belly; coastal plain, North Carolina to Florida panhandle west to sw. Alabama.

Gulf Salt Marsh (*N. f. clarki*), yellowish back, dorsolateral and ventrolateral stripes, row of large white or yellow spots down midline of belly; Gulf coast, nw. Florida to se. Texas.

Mangrove (*N. f. compressicauda*), greenish with dark blotches or crossbands, occasionally striped on neck, often plain red or reddish-orange; coastal Florida, vicinity of Miami south through Keys, north to Tampa Bay area; n. Cuba.

Broad-banded (*N. f. confluens*), 11–17 broad dark crossbands, irregularly shaped yellow interspaces, large squarish blotches on belly; w. Alabama to e. Texas, north in Mississippi River Valley to extreme s. Illinois.

Florida (*N. f. pictiventris*), dark spots on side, wormlike red or black markings on belly; peninsular Florida; intergrades with Banded in panhandle and se. Alabama; introduced into Brownsville, Texas.

Atlantic Salt Marsh (*N. f. taeniata*), stripes on forepart of body, remainder light with dark blotches, broad light spots down midline of belly; coastal Florida, from near Daytona Beach to Vero Beach.

Fond of sunning, but active mostly at night after heavy rains when frogs are moving about. In cool weather it is often found under vegetative debris. Commonly mistaken for the venomous

Cottonmouth, it defends itself vigorously when disturbed. Feeds on frogs, tadpoles, and fish. Interbreeds extensively with Northern Water Snake in some parts of its range.

549 Harter's Water Snake
(*Nerodia harteri*)
Subspecies: Brazos, Concho

Description: 20–35½″ (51–90.2 cm). Greenish-brown to grayish, with *4 rows of dark spots or crossbars running down the back and sides.* Belly pink or orange, with a row of dark spots down either side in some populations. Scales keeled, in 21–25 rows. Anal plate divided.

Breeding: Mates in spring; 7–23 young are born September to October. Newborn are 8–9″ (19–23 cm) long.

Habitat: Swift rocky streams and rivers.

Range: Brazos and Concho river drainages of c. Texas.

Subspecies: Brazos (*N. h. harteri*), back and belly spots prominent; Brazos River system, Texas.
Concho (*N. h. paucimaculata*), back spots indistinct, belly spots reduced or absent; Concho and Colorado rivers, Texas.

Primarily active during the day. Found under rocks along borders of streams, or in shallow water, and amid grasses and sedges along the banks. Little is known of this species' habits. In captivity it accepts frogs, salamanders, and fishes.

574 Diamondback Water Snake
(*Nerodia rhombifera*)

Description: 30–63″ (76.2–160 cm). Heavy-bodied; greenish-brown to brown with large dark netlike pattern formed of dark blotches on back, connected with alternating dark bars on sides. Yellow belly with largest dark spots along sides. Scales keeled, in 25–31 rows. Anal plate divided. *Males have pimplelike bumps on chin scales.*

Breeding: Mates in spring; 14–62 young, 9–13" (23–33 cm) long, are born August to October.

Habitat: Margins of lakes, rivers, streams, swamps, marshes, canals, ditches, and ponds.

Range: S. Illinois and Indiana south along the Mississippi River drainage to Mississippi and sc. Alabama, west to sc. Texas and Mexico, north through Oklahoma, Kansas, and n. Missouri.

Subspecies: Three; 1 in our range, *N. r. rhombifera.*

Active in daytime and often seen basking in spring on logs and brush along the water's edge. On warm summer nights, it is frequently found feeding on fishes and frogs. It is quick to bite and capable of inflicting numerous lacerations. This aggressive behavior, coupled with its habitat, often results in Diamondbacks being misidentified as poisonous Cottonmouths.

580 **Northern Water Snake**
(*Nerodia sipedon*)
Subspecies: Northern, Lake Erie, Midland, Carolina Salt Marsh

Description: 22–53" (55.9–134.6 cm). Reddish, brown, or gray to brownish-black, with *dark crossbands on neck region,* and alternating dark blotches on back and sides at midbody. Pattern darkens with age, becoming black. Belly white, yellow, or gray, with reddish-brown or black crescent-shaped spots. *No dark line from eye to corner of mouth.* Juveniles more vivid. Scales keeled, in 21–25 rows. Anal plate divided.

Breeding: Mates April to June; 8–99 (typically

15–30) young are born August to
October. Newborn are 6½–12"
(16.5–30 cm) long.

Habitat: Found in most aquatic situations from
sea level to about 4,800' (1,450 m);
lakes, ponds, swamps, marshes, canals,
ditches, bogs, streams, rivers, even
saltmarshes of Carolina Outer Banks.

Range: Maine to coast of North Carolina, nw.
South Carolina and Georgia to s.
Alabama and e. Louisiana, west to e.
Colorado and northeast through
Minnesota to s. Ontario and Quebec.

Subspecies: Northern (*N. s. sipedon*), dark back
markings wider than spaces between
them; s. Maine to North Carolina, west
to c. Tennessee, n. Indiana, and
Illinois, west to e. Colorado, northeast
to Minnesota and s. Ontario and
Quebec.
Lake Erie (*N. s. insularum*), dark back
markings faint or absent, belly plain;
islands of Put-in-Bay Archipelago, Lake
Erie.
Midland (*N. s. pleuralis*), dark back
markings narrower than spaces between
them; n. South Carolina and Georgia
through Alabama to e. Louisiana, north
to s. Indiana and Illinois, west to e.
Oklahoma.
Carolina Salt Marsh (*N. s.
williamengelsi*), very dark, belly
markings black; Outer Bank islands
and mainland coast of Pamlico and Core
sounds, North Carolina.

Active day and night. Frequently
encountered basking on rocks or
stumps, hunting frogs in shoreline
vegetation during the day, or gorging
itself on minnows and small fishes
caught sleeping in the shallows at
night. Also eats salamanders, juvenile
turtles, crustaceans, even small
mammals. Will flee if given the
chance, but flattens body and strikes
repeatedly if cornered. Wounds caused
by the bite bleed profusely because of
the anticoagulant quality of the snake's

saliva, but there is no poison. Northern Water Snakes are often mistaken for venomous "water moccasins" and killed on sight.

567 Brown Water Snake
(*Nerodia taxispilota*)

Description: 28–69" (71.1–175.3 cm). Heavy-bodied with large head. Brown to dark brown, with large squarish dark blotches down middle of back, alternating with row of similar blotches on each side. Yellow belly has many prominent dark spots, often arranged in broken rows along sides. Scales keeled, in 25–33 rows. Anal plate divided.

Breeding: Mates April to May; 14–58 young are born June to October. Newborn are 7–11" (18–28 cm) long.

Habitat: Lakes, rivers, streams, swamps, marshes, and ponds, especially where overhanging vegetation is present.

Range: Coastal Virginia through Florida to sw. Alabama.

Primarily active during the day, but sometimes encountered foraging at night. It feeds on frogs and fishes caught among emergent vegetation along the shore. It is quite arboreal, basking and sleeping on limbs overhanging the water. When frightened it drops from its resting place into the water. It readily bites, making it a most unwelcome intruder should it accidentally drop into a boat.

477 Rough Green Snake
(*Opheodrys aestivus*)

Description: 20–45⅝" (51–115.9 cm). Slender arboreal snake; uniform pea-green with a long tapering tail. Belly white to yellowish green. Hatchlings greenish-gray. *Scales keeled, in 17 rows.* Anal plate divided.

Breeding: Mates spring and fall. Lays 3–12 smooth, rather hard, capsule-shaped eggs, 1½" (28 mm) long, June to August. Young hatch in 5–12 weeks, are 7–8⅝" (18–22 cm) long. Mature in 1–2 years.

Habitat: Vines, bushes, and trees near water; sea level to 5,000' (1,500 m).

Range: S. New Jersey west to e. Kansas, south to Florida Keys west through Texas into e. Mexico. Isolated populations in se. Iowa, ne. Missouri, w. Texas, and ne. New Mexico.

A graceful, mild-tempered tree-dweller. Abroad during the day, it moves slowly through vegetation in search of grasshoppers, crickets, caterpillars, and spiders. Swims well and may take to water when disturbed.

475, 476 Smooth Green Snake
(*Opheodrys vernalis*)
Subspecies: Eastern, Western

Description: 14–26" (35.5–66 cm). Small and streamlined; bright grass-green with long tapering tail. Belly white, tinged with pale yellow. Hatchlings bluish-gray or dark olive-green. *Scales smooth, in 15 rows.* Anal plate divided.

Breeding: Mates spring and late summer. Lays 3–11 cylindrical-shaped, thin-shelled eggs, late July to August. Young hatch in 4–23 days at 4–6½" (10–16.5 cm). A choice egg-laying site may be shared by a number of females.

Habitat: Meadows, grassy marshes, moist grassy fields along forest edge; sea level to 9,500' (2,900 m).

Range: Nova Scotia west to se. Saskatchewan, south to North Carolina and ne. Kansas; se. Idaho and Wyoming south into ne. and se. Utah and e. New

Mexico; se. Texas. Numerous scattered populations.

Subspecies: Difficult to differentiate. Eastern (*O. v. vernalis*), male has fewer than 131 belly scales, female fewer than 140; Nova Scotia west to se. Ontario, south to w. North Carolina and through Michigan; e. and n. Wisconsin, and c. and nc. Minnesota. Western (*O. v. blanchardi*), male has 131 or more belly scales, female 140 or more; nw. Indiana northwest to s. Manitoba, west through remainder of indicated range. Scattered populations in sw. Ohio, c. and sw. Indiana west.

Active during the day. A capable climber, but is largely terrestrial. Its color provides excellent camouflage as it moves through grass and low shrubs in search of insects and spiders. Large numbers of this species may overwinter together.

Mexican Vine Snake
(*Oxybelis aeneus*)

Description: 36–60″ (91.4–152.4 cm). *Extremely slender. Narrow head with elongated and pointed snout.* Generally brownish gray or gray, becoming tan or yellowish-brown on forward part of body. A thin dark-brown line runs from near nostril, through eye, to side of neck. Scales smooth, in 17 rows. Anal plate divided.

Breeding: Female deposits clutch of 3–5 eggs, about 2″ (51 mm) long, late spring to early summer. Eggs hatch in 2½ months.

Habitat: Brushy hillsides; gullies and canyons dominated by oak, sycamore, and wild grape in Arizona to tropical rainforest; near sea level to 4,000′ (1,200 m).

Range: Pajarito Mountains, extreme s. Arizona south through Mexico to Brazil.

Diurnal. It searches out lizards on the ground, in vegetation tangles, and in treetops. Grooved rear fangs inject a mild but immobilizing venom into small prey. This snake holds its head high when crawling on the ground.

---

589 Saddle Leaf-nosed Snake
(*Phyllorhynchus browni*)
Subspecies: Pima, Maricopa

Description: 12–20″ (30.5–50.8 cm). Small pink or cream snake with *fewer than 17 large brown dark-edged, saddle-shaped blotches* on body. Belly plain white. A large *free-edged, triangular-shaped patchlike scale (rostral) curves back over tip of snout* and separates internasal scales. Pupils vertical. Scales smooth or faintly keeled, in 19 rows. Anal plate single.

Breeding: Poorly known habits. 2–5 eggs presumably laid in summer.

Habitat: Upland rocky or sandy desert dominated by mesquite, creosote bush, saltbush, paloverde, and saguaro; ca. 1,000–3,000′ (300–900 m) in Arizona.

Range: Sc. Arizona to nw. Mexico.

Subspecies: Four; 2 in our range.

Pima (*P. b. browni*), dark saddles much wider than light inner spaces; Pinal County, Arizona, south into Mexico. Maricopa (*P. b. lucidus*), dark saddles and light inner spaces approximately equal in size; Maricopa County, Arizona, southwest into Mexico.

Usually seen crossing highways before midnight on warm humid summer nights. Feeds chiefly on lizards. Although harmless, it assumes a striking coil like a rattlesnake when annoyed, vertically flattens its neck, and hisses while striking.

571, 583 Spotted Leaf-nosed Snake
(*Phyllorhynchus decurtatus*)
Subspecies: Clouded, Western

Description: 12¾–20″ (32.4–50.8 cm). Stout-bodied like Saddle Leaf-nosed Snake; bears similar *triangular-shaped patchlike scale curved back over tip of snout.* Pink, gray, or tan with *more than 17 dark blotches* from neck to tail along midline of back. Belly white. Vertical pupils. Scales smooth or keeled, in 19 rows. Anal plate single.

Breeding: Poorly known. Clutches of 2–4 large eggs, approximately 1⅜″ (35 mm) long, are laid primarily June to July. Hatchlings about 7–8″ (18–20 cm).

Habitat: Open, sandy, or gravelly creosote bush desert; sea level to 3,000′ (900 m).

Range: S. Nevada south through se. California and Arizona to tip of Baja California and Sinaloa, Mexico.

Subspecies: Six; 2 in our range.

Clouded (*P. d. nubilis*), 42–60 blotches on midline of back, equal or wider than inner spaces; Pima County, Arizona, south into Sonora, Mexico.
Western (*P. d. perkinsi*), 24–48 blotches, narrower than inner spaces; s. Nevada south into Baja California and nw. Sonora, Mexico.

Nocturnal. A secretive, adept burrower, this snake is most commonly encountered in the evening on roads after heavy rains. It feeds on banded geckos and their eggs.

488, 537, 573, **Pine-Gopher Snake**
575, 591 (*Pituophis melanoleucus*)
Subspecies: Northern Pine, Sonora
Gopher, San Diego Gopher, Pacific
Gopher, Great Basin Gopher, Black
Pine, Florida Pine, Santa Cruz Gopher,
Louisiana Pine, Bullsnake

Description: 48–100″ (122–254 cm). Large and
powerfully built; small head. Light-
colored with black, brown, or reddish-
brown blotches on back and sides.
Snout somewhat pointed, with enlarged
rostral scale extending upward between
internasal scales. Usually *4 prefrontal
scales*. Scales keeled, in 27–37 rows.
Anal plate single.

Breeding: Mates in spring. Clutches of 3–24
cream to white eggs, 2–4¼″ (50–109
mm) long, are laid in burrows in sandy
soil or below large rocks or logs, June
to August; hatch in 64–79 days.
Young are 12–18″ (30–46 cm) long.

Habitat: Dry, sandy pine-oak woodlands and
pine flatwoods, cultivated fields,
prairies, open brushland, rocky desert,
chaparral; sea level to 9,000′ (2,750 m).

Range: S. New Jersey, w. Virginia, s.
Kentucky, Wisconsin, sw.
Saskatchewan, s. Alberta, and sc.
British Columbia south to s. Florida,
ec. and wc. Mexico, and tip of Baja.

Subspecies: Fifteen; 10 in our range.

Northern Pine (*P. m. melanoleucus*),
white, pale gray, or yellowish with
black blotches near front of body, often
brownish toward rear; s. New Jersey;
w. Virginia, s. Kentucky, Tennessee,
n. Alabama, n. Georgia, sw. and se.
North Carolina and South Carolina.
Sonora Gopher (*P. m. affinis*), blotches
brown or reddish brown on forepart of
body, distinctly darker on rear; extreme
sc. Colorado, w. New Mexico, extreme
w. Texas, c. and s. Arizona, and se.
California south into Mexico.
San Diego Gopher (*P. m. annectans*),
black blotches on forepart of body fuse
together; coastal s. California into Baja.

Pacific Gopher (*P. m. catenifer*), dark brown or black blotches on forepart of body separated; interspaces between side blotches suffused with gray; w. Oregon south through w. and c. California to Santa Barbara County.

Great Basin Gopher (*P. m. deserticola*), wide blotches on forepart of body usually black, connected with side blotches, and creating isolated light blotches on back; sc. British Columbia south through e. Washington, Nevada, se. California and eastward through Idaho, Utah, n. Arizona, to Wyoming, w. Colorado, and nw. New Mexico.

Black Pine (*P. m. lodingi*), almost uniformly dark brown to black, some with traces of blotching; sw. Alabama west to extreme e. Louisiana.

Florida Pine (*P. m. mugitus*), grayish in front to rusty brown toward rear, blotches indistinct, especially on forepart of body; s. South Carolina, s. Georgia, se. Alabama, and Florida.

Santa Cruz Gopher (*P. m. pumilis*), dwarf race resembling San Diego, 24–32″ (61–81 cm) long, 27–29 scale rows; Santa Cruz Island, California.

Louisiana Pine (*P. m. ruthveni*), obscure dark brown blotches on forepart of body, distinct lighter brown or reddish-brown blotches on rear portion of body; wc. Louisiana and adjacent e. Texas.

Bullsnake (*P. m. sayi*), yellowish with 41 or more black, brown, or reddish-brown body blotches, dark line from eye to angle of jaw; s. Alberta to s. and c. Wisconsin and w. Indiana, south through c. and w. Texas to ne. Mexico.

Generally diurnal, but may be active at night during hot weather. This snake takes refuge in mammal or tortoise burrows or under large rocks or logs. Noted for its consumption of rodents. When confronted, the Pine-Gopher Snake hisses loudly, sometimes flattening its head and vibrating its tail, and then lunges at the intruder.

510 Striped Crayfish Snake
(*Regina alleni*)

Description: 13–25¾" (33–65.4 cm). Iridescent
brown snake with wide yellow or
orange stripe on lower side of body. 3
indistinct dark stripes on back; 1 down
midline, others running dorsolaterally.
Belly yellow to orange-brown, plain or
marked with scattered dark smudges or
with a well-defined row of spots along
midline. Scales smooth (keeled on top
of tail and in anal region), in 19 rows.
Anal plate divided.

Breeding: Live-bearing; 6–34 young, 6–7" (15–
18 cm) long, are born June to August.

Habitat: Freshwater marshes, sloughs, canals,
shallow lakes, sphagnum bogs, and
waters choked by water hyacinth.

Range: S. Georgia and peninsular Florida.

Highly aquatic, it forages amid
hyacinth roots and floating mats of
vegetation. Occasionally seen crossing
roads or lawns on rainy nights. It feeds
on hard-shelled crayfish.

519 Graham's Crayfish Snake
(*Regina grahami*)

Description: 18–47" (45.7–119.4 cm). Plain brown
snake with broad yellow stripe on lower
side of body. Narrow black stripe, often
zigzag or irregular, follows lower edge
of yellow stripe. Dark-bordered pale
stripe occasionally runs down midline
of back. Belly dull yellow, plain or
marked with a series of faint dark spots
along midline. Scales keeled, in 19
rows. Anal plate divided.

Breeding: Mates April to May; 6–39 young,
7–10" (18–25 cm) long, are born late
July to mid-September. Males mature
in a year, females in 3.

Habitat: Sluggish streams, ponds, lakes, and
ditches where crayfish are abundant.

Range: Iowa and Illinois south through

Louisiana and e. Texas. Isolated population in nc. Mississippi.

Secretive, it often seeks shelter under debris, or in crayfish burrows. Largely nocturnal during hot summer months. Occasionally seen basking in branches overhanging water. Diet consists mainly of freshly molted crayfish; also eats frogs, snails, and shrimp.

474 Glossy Crayfish Water Snake
(Regina rigida)
Subspecies: Glossy, Delta, Gulf

Description: 14–31⅜″ (36–79.7 cm). A stout, small-headed snake; shiny brown or olive-brown with 2 faint dark stripes on back and 2 somewhat more detectable stripes on sides. Belly yellow or cream with 2 rows of half-moonshaped spots. Scales keeled, in 19 rows. Anal plate divided.

Breeding: Poorly known. 7–14 young, about 7″ (18 cm) long, are born in summer.

Habitat: Mucky situations along streams or around edges of ponds, lakes, swamps, freshwater tidal marshes, rice fields, flatwood ponds, floodplains.

Range: Coastal plain, ne. North Carolina to c. Florida, west to e. Texas. Isolated population in e. Virginia.

Subspecies: Glossy (*R. r. rigida*), narrow dark lines on sides of throat, 2 preocular scales; coastal plain, Virginia to c. Florida. Delta (*R. r. deltae*), no lines on throat, 1 preocular scale; Mississippi Delta. Gulf (*R. r. sinicola*), no lines on sides of throat, 2 preocular scales; Gulf coastal plain, c. Georgia to e. Texas.

Seldom seen. Usually observed at night crossing roads or found amid mats of aquatic vegetation. Sometimes called "stiff snake" because of its rigid body tone. Eats crayfish, dwarf sirens, fish, frogs, and dragonfly nymphs.

503 Queen Snake
(*Regina septemvittata*)

Description: 16–36¾" (40.6–93.3 cm). Tan to
olive-brown or chocolate-brown, to
almost black, with a yellow stripe on
lower side of body. *Belly yellow with 4
distinct brown stripes;* 2 near midline, 2
along sides. Sometimes 3 faded,
indistinct stripes on back. Scales
keeled, in 19 rows. Anal plate divided.

Breeding: Mates April to May. Live-bearing;
5–23 young, 7½–9" (19–23 cm) long,
are born July to early September.

Habitat: Streams and small rivers with rocky
margins and bottoms, and clear sandy-
bottomed creeks in coastal plain.

Range: S. Great Lakes region and se.
Pennsylvania south to Gulf Coast.
Isolated populations in n. Michigan and
sw. Missouri and nw. Arkansas.

Active day and night. Highly aquatic
and an excellent swimmer. Drops into
water when disturbed. Feeds almost
entirely on crayfish, particularly those
recently shed and soft-bodied.

462, 465 Pine Woods Snake
(*Rhadinaea flavilata*)

Description: 10–15⅞" (25.4–40.2 cm). Tiny
golden-brown or reddish-brown snake
with white to yellow belly. Upper lip
scales yellowish, some with dark
specks. Top of head darker than body;
dark stripe runs from snout through eye
to corner of mouth. Faint narrow back
and side stripes may be present. Scales
smooth; 17 rows. Anal plate divided.

Breeding: May to August, female deposits 2–4
eggs, 1" (25 mm) long. Hatchlings are
about 6½" (16.5 cm) long.

Habitat: Low marshy areas, damp pine flatwoods
and hammocks, and coastal islands.

Range: Coastal plain, North Carolina to s.
Florida and west to e. Louisiana.

The secretive "yellow-lipped snake" is most often seen during spring months when the water table is high. It often burrows into the centers of damp rotting pine logs and stumps or under forest litter or loose soil. Although harmless to man, its saliva is mildly toxic to its prey—frogs and lizards.

593, 609 Long-nosed Snake
(*Rhinocheilus lecontei*)
Subspecies: Western, Texas

Description: 22–41″ (55.9–104.1 cm). A tricolored snake with a tapered, pointed snout protruding beyond lower jaw. *Most scales under tail in a single row.* Light-bordered, cream-flecked black saddle-shaped blotches extend down sides to edge of belly scales. Spaces between blotches pink or reddish with black spotting on sides. Scales smooth, in 23 rows. Anal plate single.

Breeding: Mates in spring. June to August, female lays 4–9 eggs in underground nest. Young, 8–10″ (20–25 cm) long, hatch in 2–3 months.

Habitat: Dry open prairie, desert brushland, coastal chaparral to tropical habitat in Mexico; sea level to 5,400′ (1,600 m).

Range: Sw. Kansas, se. Colorado, and New Mexico, south into Mexico, and northwest to Arizona, w. Utah, Nevada, and c. California.

Subspecies: Western (*R. l. lecontei*), tip of snout not distinctly tilted upward; c. California, Nevada, and w. Utah south into Baja and w. and s. Arizona. Isolated populations in Utah and sw. Idaho. Texas (*R. l. tessellatus*), snout sharp with distinct upward tilt at tip; sw. Kansas and se. Colorado south through New Mexico and Texas into Mexico.

A good burrower. Active at night; hides amid rocks or in underground burrows during day. When first captured it exhibits an unusual defense reaction: it tries to hide its head, then coils its body, vibrates its tail, and discharges a bloody fluid and anal gland secretions. Diet includes small rodents, lizards and their eggs, small snakes.

514 Big Bend Patch-nosed Snake
(Salvadora deserticola)

Description: 22–40″ (56–101.6 cm). Slender snake with tan or brownish-orange back stripe, bordered by narrow black or dark brown dorsolateral stripes. *Narrow dark line on 4th scale row.* Belly peach. Rostral scale enlarged, triangular, and curved back over snout. *9 upper lip scales, 2 reach eye.* Scales smooth, in 17 rows. Anal plate divided.

Breeding: Little is known. Mates in spring. About 5–10 eggs are laid in summer. Hatchlings are about 9″ (23 cm) long.

Habitat: From flatlands, creosote bush desert and mesquite-dominated washes into foothills and mesas, ca. 2,000–5,000′ (600–1,500 m).

Range: Big Bend region of Texas to se. Arizona, south into nw. Mexico.

Like the Western Patch-nosed Snake, this species can tolerate higher temperatures than most other snakes. Thus it is able to search out lizards active during parts of the day when other snakes are in retreats.

516 Mountain Patch-nosed Snake
(Salvadora grahamiae)
Subspecies: Mountain, Texas

Description: 22–47″ (56–119.4 cm). Triangular-shaped patchlike scale curved back over snout. Broad white or yellow back stripe bordered by dark brown or blackish dorsolateral stripe.

Occasionally a dark line on 3rd scale row. *8 upper lip scales, 2 touch eye.* Scales smooth, in 17 rows. Anal plate divided.

Breeding: About 6–10 smooth-shelled eggs, 1–1¼" (24–33 mm) long, are laid in spring. Young 8½–10" (22–25 cm) long hatch in 3–4 months.

Habitat: Western form prefers open woodland and forested mountain slopes above 4,000' (1,200 m); eastern subspecies, prairie and brushland to rocky canyons, creek beds, and rugged hillsides; sea level to 6,500' (2,000 m).

Range: Se. Arizona to nc. Texas, south into Mexico.

Subspecies: Mountain (*S. g. grahamiae*), faint narrow stripe sometimes appears on 3rd scale row; Trans-Pecos region of Texas to nc. New Mexico and se. Arizona. Texas (*S. g. lineata*), dark side stripe and distinct narrow dark stripe on 3rd scale row; nc. Texas south into Mexico.

Diurnal and terrestrial. Fast-moving and agile, it quickly retreats to mammal burrows or rock shelters when disturbed. Eats small lizards.

---

527 Western Patch-nosed Snake
(*Salvadora hexalepis*)
Subspecies: Desert, Mojave, Coast

Description: 22–45" (56–114.3 cm). Slender gray or grayish-tan snake with wide triangular-shaped rostral scale curved back over snout. Broad beige or yellow back stripe bordered by dark side stripes. *9 upper lip scales, 1 sometimes reaches eye.* Scales smooth, in 17 rows. Anal plate divided.

Breeding: Not well known. Presumed to mate April to June; lays 4–10 eggs in summer. Young hatch in 2–3 months.

Habitat: Barren creosote bush desert flats, sagebrush semidesert, and chaparral; sea level to 7,000' (2,150 m).

Range: S. California, w. and s. Nevada, and

extreme sw. Utah, south into Mexico.

Subspecies: Four; 3 in our range.

Desert (*S. h. hexalepis*), top of head
gray, back stripe 3 scale rows wide, 1
lip scale reaches eye; c. and sw. Arizona
west to se. California, south into ne.
Baja California and nw. Mexico.
Mojave (*S. h. mojavensis*), top of head
brown, back stripe obscure, often
broken by narrow crossbars, lip scales
do not reach eye; nw. Arizona and sw.
Utah west into n. Mojave Desert in
California, north into Nevada.
Coast (*S. h. virgultea*), top of head
brown; back stripe narrow; one lip scale
reaches eye. Coastal California, San Luis
Obispo south into nw. Baja California.

This fast-moving, agile species is active
much of the day. After warming in the
morning sun, it searches for lizards,
young snakes, pocket mice, reptile eggs.

487, 494   Swamp Snake
(*Seminatrix pygaea*)
Subspecies: North Florida, South
Florida, North Carolina

Description: 10–18½" (25.4–47 cm). Glossy black,
red-bellied snake. *Black dorsal color
extends onto ends of belly scales.* Scales
smooth, in 17 rows; last 3–5 rows bear
light lines resembling keels. Anal plate
divided.

Breeding: Live-bearing; 2–13 young, 4¼–5½"
(11–14 cm) long, are born August to
October.

Habitat: Swamps, cedar and cypress ponds,
canals, and drainage ditches, especially
areas overgrown with water hyacinth.

Range: Coastal North Carolina to peninsular
Florida, west to extreme se. Alabama.
Isolated population in South Carolina.

Subspecies: North Florida (*S. p. pygaea*), 118–124
belly scales; coastal Georgia to c.
Florida, west to se. Alabama.
South Florida (*S. p. cyclas*), 117 or

653

fewer belly scales; s. Florida.
North Carolina (*S. p. paludis*), 127 or
more belly scales; coastal plain of North
and South Carolina.

Often found hiding amid mats of
aquatic vegetation or seen at night after
heavy rains. Eats leeches, small fish,
frogs, tadpoles, Dwarf Sirens. It is
timid and does not bite when handled.

459, 499, 555, **Ground Snake**
611 (*Sonora semiannulata*)

Description: 8–19″ (20.3–48.3 cm). Tiny glossy
snake; grayish, brownish, or reddish
with great variation in back pattern.
Some essentially patternless; others
with a wide red, orange, or beige back
stripe; others with crossbanding
ranging from a single neck band to
evenly spaced, saddle-shaped blotches
to bands with red interspaces encircling
body. Small dark blotch on back scales.
Scales smooth and shiny, in 13–15
rows. Anal plate divided.

Breeding: Mates spring and fall. Up to 6 eggs
deposited early June to late August
depending on locality. Young 4–5″
(10–13 cm) long hatch in 7–10 weeks.

Habitat: Dry open areas with loose sandy soil;
rocky wooded or prairie hillsides,
mesquite thickets along river beds,
sand hummocks, vacant lots, brushy
desert; sea level to 6,000′ (1,800 m).

Range: Sw. Idaho, se. Oregon south through
Nevada, se. California, and Arizona
into Baja California and n. Mexico, east
to e. Texas, and north through
Oklahoma, se. Colorado, s. Kansas,
and sw. Missouri.

Until recently this secretive burrower
was considered to be two species: *Sonora
episcopa* with 2 subspecies and *S.
semiannulata* with 5 subspecies. Plain-
colored, striped, and crossbanded

individuals may be found in the same area. It preys on scorpions, spiders, centipedes, insects.

584 Short-tailed Snake
(*Stilosoma extenuatum*)

Description: 14–25¾" (35.6–65.4 cm). Slender, cylindrical body; *tail about a tenth of total length.* Upper surface grayish, patterned with many dark-brown blotches separated by yellow, orange, or red areas. White belly with heavy brownish-black blotching. Scales smooth, in 19 rows. Anal plate single.
Breeding: Lays eggs. Habits unknown.
Habitat: Dry, high pineland and upland hammocks; to 100' (30 m).
Range: West of St. Johns River, Florida.

A rarely observed burrowing snake. A constrictor, it can subdue and eat small snakes and lizards. When annoyed, it vibrates its tail, hisses, and strikes.

550 Brown Snake
(*Storeria dekayi*)
Subspecies: Northern, Marsh, Texas, Florida, Midland

Description: 10–20¾" (25.4–52.7 cm). Small; gray, yellowish-brown, brown, or reddish-brown, with *2 parallel rows of small dark spots bordering an indistinct wide light back stripe.* Belly pale yellow, brown, or pinkish with small black dots along edges. Young have yellowish collar. Scales keeled, in 15–17 rows. Anal plate divided.
Breeding: Live-bearing. Mates spring and fall; 3–31 young, 3¼–4½" (8–11.4 cm) long, are born June to September.
Habitat: Moist upland woodland to lowland freshwater and saltwater marshes; margins of swamps, bogs, and ponds;

vacant lots, gardens, golf courses.

Range: S. Maine, s. Quebec, and s. Minnesota, south to lower Florida Keys, and through Texas and Mexico to n. Honduras.

Subspecies: Eight; 5 in our range. Wide zones of intergradation occur between races.

Northern (*S. d. dekayi*), vertical or diagonal dark bar on temporal scale on side of head usually extends through 6th and 7th lip scales; s. Maine and s. Quebec to Virginia. Intergrades with Midland from Michigan to Carolinas. Marsh (*S. d. limnetes*), horizontal dark bar on temporal scale, 6th and 7th lip scales unmarked; coastal marshes; Colorado Co., Texas, east through Louisiana, Mobile Bay and Pensacola. Texas (*S. d. texana*), no bar on temporal scale; large blotch on nape of neck extends downward to belly scales; Minnesota to Texas, south into Mexico. Florida (*S. d. victa*), 15 scale rows (all others 17), light band across head; se. Georgia to Florida Keys. Midland (*S. d. wrightorum*), similar to Northern, except parallel spots fused by narrow crossbands; Illinois and Indiana south to s. Mississippi, s. Alabama, and sw. Georgia. Intergrades with Texas in c. Wisconsin, e. Illinois, w. Missouri, Arkansas, and Louisiana.

Diurnal, but nocturnal in warm weather. Hides under flat rocks, logs, or trash. Usually found near water or damp places. Feeds on earthworms, slugs and snails. Large numbers may congregate to hibernate together.

501, 505, 506 Red-bellied Snake
(*Storeria occipitomaculata*)
Subspecies: Northern, Florida, Black Hills

Description: 8–16" (20.3–40.6 cm). A small snake; plain brown, gray, or black, with a single broad light stripe, or 4 faint narrow dark stripes, or all 5, down back. Belly red, orange, or yellow; occasionally jet black. *3 light spots on nape of neck,* which sometimes fuse to form collar. Scales keeled, in 15 rows. Anal plate divided.

Breeding: Mates in spring or fall; 1–21 young, 2¾–4″ (7–10 cm) long, are born June to September; mature in 2 years.

Habitat: Mountainous or hilly woodland; sphagnum bogs. Sea level to 5,600′ (1,700 m).

Range: Extreme se. Saskatchewan to Nova Scotia, south to c. Florida and west to e. Texas. Isolated population in w. South Dakota and e. Wyoming.

Subspecies: Northern (*S. o. occipitomaculata*), 3 neck spots distinct; all of range except Florida and South Dakota.
Florida (*S. o. obscura*), neck spots fused to form collar, top of head black; s. Georgia to c. Florida.
Black Hills (*S. o. pahasapae*), neck spots small or absent; Black Hills of w. South Dakota and e. Wyoming.

When startled or captured, this snake curls up its upper lip on one or both sides. In the South, it is often found with the Smooth Earth Snake. Hides under lumber or debris around houses. Eats slugs, earthworms, insects.

Mexican Black-headed Snake
(*Tantilla atriceps*)

Description: 5–9⅝″ (12.7–24.5 cm). Brown or gray-brown, occasionally with faint dark stripe down midback. Belly reddish-orange, especially toward tail. Distinct black headcap extends downward but does not reach corner of mouth, ends abruptly on neck, ½–2 scales behind head shields. Sometimes a narrow line or few light scales across neck; light collar usually absent; no

dark spots behind collar. Scales smooth,
in 15 rows. Anal plate divided.

Breeding: 1–3 eggs laid in May to June.

Habitat: Moist or humid spots in semiarid
regions, from brushy desert to
mountain woodland.

Range: S. Arizona and adjacent Sonora east to
w. Texas and Coahuila.

Nocturnal. Hides beneath slabs of rock,
logs, loose boards, leaf litter. It has 2
grooved fangs at rear of each side of
upper jaw. Although totally harmless
to humans, they probably have a mildly
toxic saliva that helps subdue
invertebrate prey.

466 Southeastern Crowned Snake
(*Tantilla coronata*)

Description: 5¼–13″ (13.4–33.0 cm). Back tan to
reddish-brown; belly white. Distinct
black headcap extends downward to
corner of mouth, ends abruptly on
posterior tips of head shields. Chin
black. Broad light collar borders
headcap and is followed by black band
2–3 scales wide. Scales smooth, in 15
rows. Anal plate divided.

Breeding: Mates April and May. Clutch of 1–3
eggs is laid May and June. Females
seem to ovulate larger numbers of
oviducal eggs but resorb all but the
1–3 eventually laid.

Habitat: Pine flatwoods, oak-hickory forests, and
sandhills, where soil is moist.

Range: Sc. Virginia south through c. Georgia
to the Gulf coast in the Florida
panhandle and e. Louisiana and north
to extreme s. Indiana.

Nocturnal. Found in rotten logs, under
leaf litter and rocks. Feeds on
earthworms, slugs, and insect larvae
found during its subterranean
wanderings.

463 Flat-headed Snake
(Tantilla gracilis)

Description: 7–9⅝″ (17.8–24.5 cm). Slender light-brown or reddish-brown snake with a slightly darker head. Sometimes has dark-gray headcap ending abruptly with concave border on neck, 1–2 scales behind head shields. Belly pink, color extending onto side of head. Only *Tantilla* with 6 upper lip scales. Scales smooth, in 15 rows. Anal plate divided.

Breeding: Mates in May. Clutch of 1–4 eggs is laid June and July, hatches in September.

Habitat: Rocky prairie and wooded hillsides. Sea level to 2,000′ (600 m).

Range: Extreme sw. Illinois to e. Nebraska south through Oklahoma and Arkansas to e. Texas and adjacent Coahuila.

Nocturnal. A secretive species, found in rotting stumps and logs and under slabs of rocks, where it feeds on insect larvae, centipedes, slugs, and sowbugs.

460 Plains Black-headed Snake
(Tantilla nigriceps)
Subspecies: Plains, Texas

Description: 7–14¾″ (17.8–37.5 cm). Uniform tan to brownish-gray. Belly white, with pink or orange midline. Distinct black headcap extends downward but does not reach corner of mouth, ends abruptly with convex border on neck, 3–5 scales behind head shields. Scales smooth, in 15 rows. Anal plate divided.

Breeding: Presumably lays 1–3 eggs in spring or early summer.

Habitat: Rocky and grassy prairie; hillsides where soil is moist.

Range: S. Nebraska south through w. Kansas and e. Colorado to s. Texas, New Mexico, se. Arizona and into Mexico.

Subspecies: Plains (*T. n. nigriceps*), 146 or more belly plates; Nebraska to w. Texas, se. Arizona and adjacent Mexico. Texas (*T. n. fumiceps*), usually 145 or fewer belly plates; sw. Oklahoma through c. Texas to Coahuila.

Nocturnal. Found under surface litter, in small burrows, occasionally wandering in basements. Specimens have been uncovered 8′ (2.4 m) beneath the surface in January.

Rim Rock Crowned Snake
(*Tantilla oolitica*)

Description: 5¾–9¾″ (14.7–24.6 cm). Tan with white belly. Distinct black headcap extends downward to below corner of mouth, ends abruptly on neck, 3–4 scales behind head shields. Chin black. Light collar absent or represented by row of light spots. Scales smooth, in 15 rows. Anal plate divided.

Breeding: Clutch of 1–3 eggs probably laid in early summer. Sexual maturity is probably reached at 3 years of age.

Habitat: Pineland and tropical hammock on oolitic limestone.

Range: E. Dade and Monroe counties, Florida.

Nocturnal. Most often discovered under loose palm fronds, boards, or rocks.

---

500 Western Black-headed Snake
(*Tantilla planiceps*)
Subspecies: California, Desert, Utah

Description: 7–15″ (17.8–38.1 cm). Back uniformly tan, brown, or olive-gray; occasionally with faint dark stripe down midback. Belly white, becoming orange toward tail. Distinct black headcap extends downward to or below corner of mouth, ends abruptly on neck; 1–2 scales behind head shields. Light collar, ½–1½ scales wide, usually borders headcap, followed by a few dark spots. Scales smooth, in 15 rows. Anal plate divided.

Breeding: Lays 1–3 eggs in May or June.

Habitat: Arid and semiarid regions, from Pacific coast to about 5,000' (1,500 m). Desert grassland to open mountain woodland, frequently in hilly areas and near streams.

Range: Wc. Colorado through s. Utah and Nevada to c. and w. California, south into Baja, Mexico.

Subspecies: Four, 3 in our range.

California (*T. p. eiseni*), light collar prominent, 161–175 belly plates in males, 165–187 in females; w. California south along coast to Baja. Desert (*T. p. transmontana*), light collar faint; 175–185 belly plates in males; 187–198 in females; Riverside and San Diego counties, California. Utah (*T. p. utahensis*), light collar absent; 153–165 belly plates in males, 162–174 in females; wc. Colorado through s. Utah and Nevada to se. California.

May be seen wandering on surface at night. Eats worms, burrrowing insect larvae, centipedes.

458, 461 **Florida Crowned Snake**
(*Tantilla relicta*)
Subspecies: Peninsula, Central, Coastal Dunes

Description: 5–7⅔" (12.4–19.4 cm). Back light brown or reddish-brown; belly white. Distinct black headcap usually extends downward to corner of mouth, ends abruptly on posterior tips of head shields. Chin black. Light collar on neck borders headcap and is followed by black band, 2½–4 scales wide; collar sometimes absent. Scales smooth, in 15 rows. Anal plate divided.

Breeding: Lays 1–3 eggs spring or summer.

Habitat: Sandhills, sand pine scrub, coastal dunes, and moist hammocks.

Range: C. Florida peninsula south of Suwannee River to Charlotte County on west coast and Palm Beach on east.

Subspecies: Peninsula (*T. r. relicta*), light collar broken by black at midline; c. ridge of Florida peninsula from Marion County south to Polk County, separate populations in Levy, Pinellas, Sarasota, and Charlotte counties.
Central (*T. r. neilli*), no light collar, cap ends 3–4 scales behind head shields; c. Florida, Suwannee River to Hillsborough and Polk counties, and east to St. Johns River.
Coastal Dunes (*T. r. pamlica*), broad light collar and light snout; Florida east coast from Cape Canaveral to Palm Beach County.

An efficient burrower; may bask while submerged beneath the sand by poking its head up into the sunlight. It sometimes lives together with Mole Skinks in the sandy mounds thrust up by Pocket Gophers.

Big Bend Black-headed Snake
(*Tantilla rubra*)
Subspecies: Devils River, Blackhood

Description: 8½–21¾" (21.6–55.25 cm). Light brown to gray-brown above; belly white. Distinct black headcap extends downward to lower lip scales and chin, ends abruptly on neck, 0–5 scales behind head shields. Light collar, 1–2 scales wide, usually borders headcap, is followed by broad dark band 3–5 scales wide. White spot on upper lip scales. Scales smooth, in 15 rows. Anal plate divided.

Breeding: Clutch of 1–2 eggs is laid in July.

Habitat: Arid rocky grasslands and pine-oak

forests, to at least 4,000' (1,200 m).
Frequents hillsides.

Range: Davis and Chisos mountains and Pecos
and Devils rivers in sw. Texas.

Subspecies: Three, 2 in U.S.
Devils River (*T. r. diabola*),
light collar continuous across neck,
often with light spot on snout; vicinity
of Devils and Pecos rivers, sw. Texas.
Blackhood (*T. r. cucullata*), light collar
absent or broken in middle by black;
light snout spot frequently absent;
Davis and Chisos mts., sw. Texas.

Most specimens have been collected at
night on the surface after light rains.
Others have been found in moist areas
along spring runs.

Chihuahuan Black-headed Snake
(*Tantilla wilcoxi*)
Subspecies: Huachuca

Description: 7–14" (17.8–35.6 cm). Light brown,
with some spotting on sides. Belly
white, becoming orange toward tail.
Distinct black headcap extends
downward to or below corner of mouth,
ends abruptly on rear tips of head
shields. Broad light collar borders
headcap and is followed by dark spots
or dark band, 1–1½ scales wide. Back
scales smooth, in 15 rows. Anal plate
divided.

Breeding: Lays 1–3 eggs in spring and summer.

Habitat: Rocky slopes, desert grassland to
evergreen forest, 3,000–5,000' (900–
1,500 m).

Range: Huachuca and Patagonia mountains of
extreme se. Arizona south into Mexico.

Subspecies: Two; 1 in our range.
Huachuca, *T. w. wilcoxi*.

Nocturnal; usually found under rocks,
logs, or dead leaves. Eats spiders,
burrowing insect grubs, centipedes.

Yaqui Black-headed Snake
(*Tantilla yaquia*)

Description: 7–12¾" (17.8–32.5 cm). Uniformly
tan to brown above, occasionally with
faint dark stripe down midback.
Underside white, becoming pinkish-
orange under tail. Distinct brown to
black headcap extends downward to
below corner of mouth, ends abruptly
on neck; 2–3 scales behind head
shields. White area on side of head
contrasts sharply with headcap. Light
collar, 1–1½ scales wide, borders
headcap and may be followed by several
dark spots. Scales smooth, in 15 rows.
Anal plate divided.

Breeding: Clutch of 1–4 eggs, presumably
deposited late spring or summer.

Habitat: Evergreen forest above 3,300' (1,000
m); woodland near water; coastal plain;
tropical semiarid and dry forests;
deciduous short-tree forest of Sierra
Madre Occidental.

Range: Se. Arizona south to Nayarit, Mexico.

A secretive burrower occasionally seen
under flat rocks and logs. It eats small
softbodied arthropods. In turn, it is
preyed upon by other snakes, lizards,
and carnivorous mammals.

Short-headed Garter Snake
(*Thamnophis brachystoma*)

Description: 14–22" (35.6–55.9 cm). A small
version of the Eastern Garter Snake.
Head short and not distinct from neck. Side
stripe occupies 2nd and 3rd scale rows
above belly scales. Stripes tend to be
bordered by fine black lines. *No double
rows of black spots between stripes.* Scales
keeled, in *17 rows.* Anal plate single.

Breeding: Mates April to May. Late July to mid-
September female gives birth to 5–14
young, 5–6" (12.7–15.2 cm) long.

Habitat: Old fields, meadows, pastures;

900–2,400' (275–730 m).

Range: Upper Allegheny River drainage in nw.
Pennsylvania and extreme sw. New
York. Introduced populations in Erie,
Allegheny, and Butler counties, Pa.

Unlike the Eastern Garter Snake, the
Short-headed avoids woodlands. It lives
in stone piles and under debris in open
areas near water; feeds on earthworms
and small amphibians.

---

529 Butler's Garter Snake
(*Thamnophis butleri*)

Description: 15–27¼" (38.1–69.2 cm). Easily
confused with the Short-headed. *Check
number of scale rows.* Olive-brown to
black, with yellow or orange side
stripes occupying 3rd scale row and
part of 2nd and 4th rows. 2 rows of
dark spots may be present between side
stripes and yellow back stripe. *Scales
keeled, in 19 rows.* Anal plate single.
Breeding: Mates March to April. 4–16 young,
5–7" (13–18 cm) long, are born in
summer. Females mature in 2–3 years.
Habitat: Wet meadows, pastures, margins of
marshes and streams in open country;
ca. 500–1,500' (150–460 m).
Range: Extreme s. Ontario, e. Michigan, e.
Indiana, and w. Ohio; se. Wisconsin.

When frightened, it moves frantically
from side to side but makes little
forward progress. Mild-tempered. Eats
worms, frogs, leeches, salamanders.

---

504 Western Aquatic Garter Snake
(*Thamnophis couchi*)
Subspecies: Sierra, Aquatic, Santa Cruz,
Giant, Two-striped, Oregon

Description: 18–57" (46–144.8 cm). Extremely
variable in color and markings; 3

stripes, or spotted, or blotched. Back stripe usually faint, sometimes absent. Light side stripe, if present, occupies 2nd and 3rd scale rows. *8 upper lip scales; 6th and 7th not enlarged.* Internasal scales pointed in front. Scales keeled, in 19–21 rows. Anal plate single.

Breeding: Live-bearing; about 10–25 young approximately 7–10″ (17.5–25 cm) long are born in late summer.

Habitat: Highly variable—brackish coastal marshes, ponds, and lakes; clear, swift streams and rivers high in mountains; sea level to 8,000′ (2,450 m).

Range: Sw. Oregon to n. Baja California; coastal California eastward into w. Nevada.

Subspecies: Sierra (*T. c. couchi*), narrow back stripe dull yellow, usually disappears on posterior part of body, checked with dark blotches, 11 lower lip scales; Sierra Nevadas in Calif. and Nevada. Aquatic (*T. c. aquaticus*), conspicuous wide yellow or orange back stripe, side stripes distinct, throat yellow, belly blotched with yellow-orange; San Francisco Bay area north in Coast Ranges into Mendocino Natl. Forest. Santa Cruz (*T. c. atratus*), back stripe conspicuous, side stripes less apparent, throat lemon yellow, belly bluish; San Francisco Bay south to Monterey. Giant (*T. c. gigas*), indistinct back and side stripes checkered with black, belly brownish; San Joaquin Vy., California. Two-striped (*T. c. hammondi*), no back stripe, middle of back plain olive; side stripes usually present and bordered with dark spots; Monterey to Baja. Oregon (*T. c. hydrophilus*), back stripe dull yellow, back gray with prominent dark blotching; sw. Oregon south along coastal n. California and Sacramento Vy.

Aquatic. Primarily active during the day, but may forage during early evening hours on warm days. Feeds on amphibians and their larvae, fishes, earthworms, and leeches. Record longevity is close to 8 years.

536 Black-necked Garter Snake
(*Thamnophis cyrtopsis*)
Subspecies: Western, Eastern

Description: 16–43" (40.6–109.2 cm). Olive-gray or olive-brown, with 2 large black blotches on neck separated by back stripe. Back stripe may be wavy; orange and expanded in neck region, yellow or cream toward rear. Pale side stripes occupy 2nd and 3rd scale rows; often wavy in appearance because of intrusion of bordering black spots. 2 alternating rows of black spots between back and side stripes. Top of head gray. Scales keeled, in 19 rows. Anal plate single.

Breeding: Live-bearing. About 7–25 young, 8–10" (20–25 cm) long, are born late June through August.

Habitat: Mesquite-dominated desert flats to pine-fir forests; prefers canyon and mountain streams and spring seepages; sea level to 8,750' (2,700 m).

Range: S. Utah, and s. Colorado south through e. Arizona, New Mexico, and Trans-Pecos, Big Bend, and Edwards Plateau regions of Texas.

Subspecies: Three recognized; 2 in our area. Western (*T. c. cyrtopsis*), small alternating black spots begin in neck region; se. Utah and s. Colorado south to c. Mexico.
Eastern (*T. c. ocellatus*), single large black spots in neck region; Texas, Edwards Plateau west to the Big Bend.

Active during the day; may be observed basking along rocky or heavily vegetated streams in the morning. During summer rainy period, it may travel away from water. Swims on the surface of the water rather than below it. Frogs, toads, and tadpoles are its most common prey.

511 Western Terrestrial Garter Snake
(*Thamnophis elegans*)
Subspecies: Mountain, Klamath, Coast, Wandering

Description: 18–42" (45.7–106.7 cm). Variable color and markings; may resemble Western Aquatic Garter Snake. Side stripe occupies 2nd and 3rd scale rows, but back stripe is usually well-defined. Space between stripes marked with dark spots or with scattered light specks. *8 upper lip scales; 6th and 7th enlarged.* Internasal scales not pointed in front. Scales keeled, in 19–21 rows. Anal plate single.

Breeding: Live-bearing. Mates in spring; 4–19 young 6½–9" (16.5–23 cm) long, are born July to September.

Habitat: Moist situations near water; margins of streams, ponds, lakes, damp meadows; open grassland to forest; sea level to 10,500' (3,200 m).

Range: Sw. Manitoba and s. British Columbia southward into Mexico, extreme sw. South Dakota and extreme w. Oklahoma west to Pacific coast.

Subspecies: Six; 4 in our range.

Mountain (*T. e. elegans*), conspicuous stripes, back stripe yellow or orange, belly pale, plain, or with light spotting, no red markings; extreme w. Nevada, Sierra Nevada, and e. slope of n. Coast Ranges of California, north in Cascade Mountains of w. Oregon.
Klamath (*T. e. biscutatus*), conspicuous broad yellow back stripe, belly light gray washed with slate or black; sc. Oregon and extreme ne. California.
Coast (*T. e. terrestris*), wide bright yellow back stripe, red or orange flecks on sides and belly; along coast, extreme sw. Oregon south to Santa Barbara County, California.
Wandering (*T. e. vagrans*), narrow back stripe dull yellow or brown, fades on tail; light areas between stripes marked with small dark spots, sometimes absent or enlarged, fused and filling

space between stripes; sw. Manitoba,
sw. South Dakota, and extreme w.
Oklahoma west to coastal British
Columbia, w. Washington, c. Oregon
and ec. California.

Diurnal. Occasionally seen basking
during morning hours in the open.
When disturbed, it often takes to
water. A generalized feeder, it eats
slugs, worms, tadpoles, frogs, fish,
mice, and small birds.

528 Mexican Garter Snake
(*Thamnophis eques*)

Description: 18–40" (45.7–101.6 cm). Stout-
bodied. Brown or greenish-brown with
yellow-white back stripe and another
stripe down 3rd and 4th scale rows on
each side of forward part of body;
alternating dark spots on sides. *Large
paired black blotches on back of head*
separated from corner of mouth by light
crescent; 8–9 upper lip scales. Belly
green, gray, or blue, with dark spots.
Scales keeled, in 19–21 rows. Anal
plate divided.

Breeding: Presumably mates in spring in
northerly portion of range. Up to 25
young are born in June, and probably
into August, and still later in southerly
areas. Newborn, 9½" (24 cm) long.

Habitat: In or near water in mountain pine-oak
forest, mesquite grassland, and desert;
ca. 2,000–8,500' (600–2,600 m).

Range: C. and se. Arizona, wc. New Mexico
south to Central America.

Subspecies: Three; 1 in our range, *T. e. megalops*.

Diurnal. An aquatic garter snake, it is
often found foraging along streams,
irrigation ditches, and lakes for its
principal prey, frogs.

515 Checkered Garter Snake
(*Thamnophis marcianus*)

Description: 18–42½" (45.7–108 cm). Brown,
olive or tan, with *bold checkered pattern of
large squarish black blotches on sides.*
Uppermost blotches intrude into yellow
back stripe. Light side stripe confined
to 3rd scale row on neck; 2nd and 3rd
rows farther back on body. Large paired
black blotches at back of head separated
from corners of mouth by light
crescent; 8 upper lip scales. Scales
keeled, in 21 rows. Anal plate single.

Breeding: 6–18 young, 8–9¼" (20.3–23.5 cm)
long, are born June to August.

Habitat: Arid and semiarid grassland near
streams, springs, ponds, and irrigation
sites; sea level to ca. 5,000' (1,500 m).

Range: Extreme se. California, s. Arizona, e.
and sw. New Mexico to e. Texas, north
to sw. Kansas, south to Costa Rica.

Subspecies: Two; 1 in our range, *T. m. marcianus.*

Active during the day in the more
northerly portions of its range. On
warm summer nights it can be found
foraging for frogs, fishes, and crayfish.

512 Northwestern Garter Snake
(*Thamnophis ordinoides*)

Description: 15–26" (38.1–66 cm). Brown,
greenish, bluish, or black; usually with
distinct red, orange, or yellow stripe
down middle of back. Side stripe down
2nd and 3rd scale rows, may be faint or
absent. Sides occasionally dark-spotted.
Belly yellow or gray, often with red
blotches. *7 upper lip scales.* Scales
keeled, in 17 rows. Anal plate single.

Breeding: Mates spring and fall. 3–15 young,
6–7" (15–18 cm) long, are born June
to August.

Habitat: Moist meadows, open grassy patches,
pastures, and along the edge of
thickets, but usually not in dense

forest; sea level to 4,000' (1,200 m).

Range: Vancouver Island and coastal s. British Columbia, w. Washington, w. Oregon, and extreme nw. California.

Diurnal, most active on sunny days. May be found near water, but when danger approaches, it usually flees to the safety of underbrush. Feeds on frogs, salamanders, slugs, worms.

531, 544 Western Ribbon Snake
(*Thamnophis proximus*)
Subspecies: Western, Arid Land, Gulf Coast, Red-striped

Description: 19–48½" (48.3–123.2 cm). Resembles Eastern Ribbon Snake; slim-bodied, 3 well-defined light stripes contrast sharply with dark back and sides. Side stripe involves 3rd and 4th scale rows. Dark stripe bordering belly scales is narrow or absent. Lip scales and belly unmarked; *2 fused spots on crown of head.* Tail long. Scales strongly keeled, in 19 rows. Anal plate single.

Breeding: Mates April to June. Female gives birth to 4–27 young, 8–11" (20–28 cm) long, July to September. Young mature in 2–3 years.

Habitat: Weedy margins of lakes, ponds, cattle tanks, marshes, ditches, streams, rivers; sea level to 8,000' (2,400 m).

Range: S. Wisconsin, Indiana, and Mississippi Valley, west to e. Nebraska, se. Colorado, e. New Mexico, and south through Texas to Costa Rica.

Subspecies: Six; 4 in our range.

Western (*T. p. proximus*), black back, narrow orange back stripe; no dark stripe bordering belly scales; Indiana, s. Wisconsin, and e. Nebraska south to s. Louisiana, and ne. Texas.
Arid Land (*T. p. diabolicus*), olive-gray to olive-brown back, orange back stripe, narrow dark stripe bordering belly scales; se. Colorado and Kansas

south through w. Texas into Mexico.
Gulf Coast (*T. p. orarius*), olive-brown
back, wide gold back stripe, no dark
stripe bordering belly scales; along
coast, extreme s. Mississippi to s. Texas.
Red-striped (*T. p. rubrilineatus*), olive-
brown to olive-gray back, bright red or
orange back stripe, narrow stripe may
border belly scales; c. Texas.

Forages amid vegetation along water's
edge for frogs, tadpoles, and small fish.
Young ribbon snakes, in turn, often fall
prey to bullfrogs, fish, and birds.

534, 543 Plains Garter Snake
(*Thamnophis radix*)
Subspecies: Eastern, Western

Description: 20–40" (51–102 cm). Distinct bright
yellow or orange back stripe. Cream to
yellow side stripes occupy 3rd and 4th
scale rows above belly. Double row of
squarish black spots between side and
back stripes, and a *row of black spots
below side stripe*. Black vertical bars on
lips. Scales keeled, in 21 rows. Anal
plate single.

Breeding: Mates April to May and in fall. 5–60
or more young, 6½–7½" (16–19 cm)
long, are born July to September.

Habitat: Wet meadows, open boggy areas,
vacant lots, parks, and open prairies
along margins of lakes, streams, and
marshes; 500–6,500' (150–
2,000 m).

Range: Nw. Indiana northwest and southwest
through the Great Plains to the Rockies
from s. Alberta to ne. New Mexico.
Isolated populations in nc. Ohio, nw.
Arkansas, sw. Illinois and se.
Missouri.

Subspecies: Eastern (*T. r. radix*), belly scales 154 or
fewer, usually 19 scale rows on neck;
nw. Indiana to c. Iowa and w. and s.
Wisconsin to ne. Missouri and c.
Illinois; isolated populations in c.

Ohio, sw. Illinois, and se. Missouri. Western (*T. r. haydeni*), belly scales 155 or more, 21 scale rows on neck; Minnesota, w. Iowa, and nw. Missouri west to the Rockies; nw. Arkansas.

A common species through much of its range, including urban areas. Spends warm days basking or searching for frogs, salamanders, and small rodents.

548 Narrow-headed Garter Snake
(*Thamnophis rufipunctatus*)

Description: 18–34″ (46–86.4 cm). Olive or brown garter snake with back and side stripes absent or faint vestiges on neck. *Distinct dark brown spots on back.* 8 upper lip scales. Scales keeled, in 21 rows. Anal plate single.

Breeding: Live-bearing; young born in July.

Habitat: Clear, rocky streams in piñon-juniper and Ponderosa pine forests.

Range: Mountains of c. and e. Arizona and adjacent sw. New Mexico; and Sierra Madre of nw. Mexico.

Highly aquatic, seen in quiet rocky areas of streams. Very excitable. Basking individuals usually do not fully expose themselves and are quick to retreat into rocks or to bottom of stream when disturbed. Eats fish.

520, 532 Eastern Ribbon Snake
(*Thamnophis sauritus*)
Subspecies: Eastern, Blue-striped, Peninsula, Northern

Description: 18–40″ (45.7–101.6 cm). A slender, streamlined garter snake. *3 bright, well-defined stripes* usually contrast sharply with dark back and sides. Side stripe involves 3rd and 4th scale rows. A dark, brownish stripe runs along margin of belly scales. Lip scales and belly unmarked. Tail very long, about a third of snake's total length. Scales strongly keeled, in 19 rows. Anal plate single.

Breeding: Mates in spring. Live-bearing. 3–26 young, 7–9" (18–23 cm) long, are born July to August, mature in 2–3 years.

Habitat: Wet meadows, marshes, bogs, ponds, weedy lake shoreline, swamps, and shallow, meandering streams.

Range: East of the Mississippi River: Michigan, s. Ontario, and s. Maine, south to the Florida Keys and se. Louisiana. Isolated colonies inhabit ne. Wisconsin and c. Nova Scotia.

Subspecies:  Eastern (*T. s. sauritus*), reddish-brown back, yellow side stripes, yellow- or green-tinged orange back stripe; s. Indiana, s. and e. Pennsylvania, se. New York, and s. New Hampshire, south to n. side of Lake Pontchartrain, Louisiana, Florida panhandle and South Carolina.
Blue-striped (*T. s. nitae*), velvety black or dark brown back, pale blue side stripes; Gulf coast of Florida, Wakulla County to Withlacoochee River.
Peninsula (*T. s. sackeni*), tan or brown back; light narrow side stripes; lustrous tan back stripe; extreme s. South Carolina, se. Georgia, and peninsular Florida.
Northern (*T. s. septentrionalis*), velvety black or dark brown back, yellow side stripes, yellow back stripe often masked with brown pigment; s. Ontario, Michigan and s. Maine to c. New Hampshire, c. Pennsylvania, and s. Indiana.

Semiaquatic; almost always encountered in low wet places. Likes to bask in bushes. When startled it takes to water. Unlike water snakes, which dive, ribbon snakes glide swiftly across the water's surface. They feed on frogs, salamanders, and small fish.

530, 533, 535, **Common Garter Snake**
538, 539, 541, (*Thamnophis sirtalis*)
542, 545, 576 Subspecies: Eastern, Texas,

Red-spotted, Valley,
California Red-sided, New Mexico,
Maritime, Red-sided, Puget Sound,
Chicago, Blue-striped, San Francisco

Description: 18–51⅛" (45.7–131.1 cm). Most
widely distributed snake in North
America. Coloration highly variable,
but back and side stripes usually well-
defined. *Side stripe confined to 2nd and
3rd scale rows* (except Texas). Red
blotches or a double row of alternating
black spots often present between
stripes. *Usually 7 upper lip scales.* Scales
keeled, in 19 rows. Anal plate single.

Breeding: Live-bearing. Mates mostly late March
to early May, occasionally in fall. 7–85
young born late June to August, earlier
in Florida, to early October in the
North. Young are 5–9" (13–23 cm)
long; mature in 2 years.

Habitat: Near water—wet meadows, marshes,
prairie swales, irrigation and drainage
ditches, damp woodland, farms, parks;
sea level to 8,000' (2,450 m).

Range: Atlantic to Pacific coasts; except desert
regions of Southwest.

Subspecies: Eastern (*T. s. sirtalis*), stripes normally
yellow, occasionally brownish,
greenish, or bluish, double row of
alternating spots usually between
stripes; some specimens lack stripes,
some are all black, others with red
between dorsal scales; s. Ontario, e.
Minnesota, e. Iowa, e. and s. Missouri,
Arkansas, and e. Texas to Atlantic coast
from s. Newfoundland to s. Florida.
Texas (*T. s. annectans*), broad orange
back stripe, side stripes on scale row 3
and on half of rows 2 and 4; Texas-
Oklahoma border south through ec.
Texas; isolated population in Texas
panhandle.
Red-spotted (*T. s. concinnus*), black
with well-defined narrow back stripe,
black pigment extends onto belly, top

of head red; nw. Oregon and extreme sw. Washington.

Valley (*T. s. fitchi*), brown or dark gray with well-defined back stripe, top of head black; c. and n. California (except area noted for California Red-sided), nw. Nevada, sw. and e. Oregon, Idaho, nc. Utah, w. Montana, and Washington east of the Cascades through British Columbia to se. Alaska.

California Red-sided (*T. s. infernalis*), resembles Red-sided, red blotches on sides, side stripe indistinct, top of head red; coastal California, Humboldt County to San Diego County.

New Mexico (*T. s. dorsalis*), resembles Red-sided but red markings between back and side stripes reduced and largely confined to skin between scales; Rio Grande Valley, extreme s. Colorado through New Mexico to extreme w. Texas.

Maritime (*T. s. pallidula*), resembles . Eastern, but back stripe faint or absent and spotting is well developed; status and range not resolved; Canadian Maritime provinces, Quebec, adjacent areas of New England.

Red-sided (*T. s. parietalis*), red or orange bars between back and side stripes variable, top of head olive; ec. and se. British Columbia, Alberta, and adjacent extreme sc. MacKenzie, c. Saskatchewan and s. Manitoba south through Great Plains to Oklahoma-Texas border.

Puget Sound (*T. s. pickeringi*), resembles Red-spotted, except back stripe largely confined to 1 scale row instead of 2 and top of head dark; Vancouver Island, adjacent coastal sw. British Columbia and w. Washington.

Chicago (*T. s. semifasciatus*), black vertical bars break side stripes in neck region, ne. and nc. Illinois, se. Wisconsin, and extreme nw. Indiana.

Blue-striped (*T. s. similis*), dark brown with dull yellowish or tan back stripe and bluish side stripes on 2nd and 3rd scale rows; nw. peninsular Florida, Wakulla County to Withlacoochee Riv. San Francisco (*T. s. tetrataenia*), red markings between back and side stripes form continuous stripe, belly greenish-blue, top of head red; San Mateo County, California.

The most commonly encountered snake in many parts of its range. Active during the day and most frequently seen amid moist vegetation where it searches for frogs, toads, salamanders, and earthworms. Occasionally it takes small fish and mice. This species is able to tolerate cold weather and may be active all year in the southerly part of its range. It hibernates in great numbers in community dens in northerly range. Ill-tempered when first captured, it will bite or expel musk, but it tames quickly and soon becomes docile. Record longevity is 10 years.

568, 582 Lyre Snake
(*Trimorphodon biscutatus*)
Subspecies: Sonora, California, Texas

Description: 24–47¾" (61–121.2 cm). A slimly built, "cat-eyed" snake; *broad head bears a chevron- or lyre-shaped mark.* Light brown to gray with darker brown or gray saddle-shaped blotches with light centers on back; smaller dark blotches on sides and belly scales. Scales smooth, in 21–27 rows. Anal plate single or divided.

Breeding: Little known. A California female laid 12 eggs in September, which hatched in 79 days. Young are about 8–9" (20–23 cm) long.

Habitat: Rocky hillsides, slides and canyons, boulder-strewn mountain slopes; arid rocky coastal areas; desert to evergreen forest; sea level to 7,400' (2,250 m).

Range: S. California east to Big Bend region of Texas, south to Costa Rica.

Subspecies: Six; 3 in our range.

Sonora (*T. b. lambda*), distinct chevron-shaped head mark, more than 22 body blotches, anal plate divided; s. Nevada and extreme sw. Utah south through se. California, Arizona and sw. New Mexico to Sonora.

California (*T. b. vandenburghi*), distinct head marking; about 36 body blotches; anal plate *usually* single. S. California into Baja California.

Texas (*T. b. vilkinsoni*), head pattern obscure, usually less than 23 body blotches, anal plate divided; Big Bend region of Texas, northwest to sw. New Mexico, and west into Chihuahua, Mexico.

Emerges at night from its rocky retreat to explore crevices and trees for lizards, small birds, and mammals. Enlarged grooved teeth toward back of upper jaw introduce a mild venom into prey.

---

507 Lined Snake
(*Tropidoclonion lineatum*)
Subspecies: Northern, Central, New Mexico, Texas

Description: 7½–21″ (19.1–53.3 cm). Resembles garter snakes. Dark or light olive-gray with 3 distinct stripes; back stripe whitish, pale gray, yellow, or orange; side stripe on 2nd and 3rd scale rows. Belly white or yellow with *2 rows of dark spots down midline.* 5–6 upper lip scales. Scales keeled, in 19 rows. Anal plate single.

Breeding: Mates before and after hibernation. 2–12 young, about 4–5″ (10–13 cm) long, are born in August and reach sexual maturity in 2 years.

Habitat: Open prairie hillsides, edges of woodland, and vacant suburban lots; sea level to ca. 5,300′ (1,600 m).

Range: Se. South Dakota, south to sc. Texas; west through Oklahoma panhandle to

c. Colorado and ne. New Mexico.
Isolated populations in Illinois, Iowa,
Missouri, and New Mexico.

Subspecies: Extremely difficult to differentiate.
Northern (*T. l. lineatum*), se. Dakota,
w. Iowa, e. Nebraska, ne. Kansas.
Central (*T. l. annectens*), ne. Texas,
Kansas, Oklahoma, Missouri, Illinois,
and se. Iowa.
New Mexico (*T. l. mertensi*), e. New
Mexico. Texas (*T. l. texanum*), Texas.

Crepuscular and nocturnal; during the
day it hides under rocks or debris. May
be found around trash dumps. Like its
relatives, the garter and water snakes,
the Lined Snake often voids excrement
and anal secretions when first handled;
it usually does not bite. Eats worms.

470, 473   Rough Earth Snake
(*Virginia striatula*)

Description: 7–12¾" (18–32.4 cm). Plain reddish-
brown, brown, or gray with distinctly
pointed snout. Belly cream to
yellowish. Light band behind head of
young. 5 upper lip scales. Scales
keeled, in 17 rows. Anal plate usually
divided.
Breeding: Mates March to April. 2–13 young,
3–4¾" (8–12 cm) long, born late June
to mid-September.
Habitat: Dry coastal plain, woodland, exposed
rocky wooded hillsides, and heavily
timbered uplands and valleys.
Range: E. Virginia to n. Florida, west to e.
Texas, north to s. Missouri.

Active from March to late October, but
very secretive and rarely seen on the
surface. It hides under leaves or
decaying logs, sun-warmed rocks, or in
a gardener's compost pile. Eats
earthworms, slugs, snails, and small
frogs. In turn, is prey to kingsnakes,
Eastern Coral Snake, and shrews. Often

seen with the Copperhead and
Southeastern Crowned Snake.

---

467 Smooth Earth Snake
(*Virginia valeriae*)
Subspecies: Eastern, Western,
Mountain

Description: 7–13¼" (17.7–33.7 cm). No
distinctive markings. Gray, brown,
reddish-brown, or yellowish-brown,
with widely scattered small dark flecks.
Occasionally faint light stripe down
back, or dark pigment between eye and
nostril. Belly unmarked; grayish, white,
or yellowish. 6 upper lip scales. Scales
smooth, in 15–17 rows. Anal plate
divided.

Breeding: 2–14 young, 3¼–4½" (8–11.5 cm)
long, born August to September.

Habitat: Damp deciduous forest; moist, rocky
timbered hillsides; wooded residential
areas. Sea level to ca. 2,000' (600 m).

Range: New Jersey to n. Florida, west to e.
Kansas, e. Oklahoma, e. Texas.

Subspecies:  Eastern (*V. v. valeriae*), gray or
brownish-gray, with small dark specks
on back, 15 scale rows on neck and
midbody; New Jersey to n. Florida, and
west to s. Ohio, and w. Alabama.
Western (*V. v. elegans*), scales in 17
rows, faintly keeled, reddish-brown to
gray; se. Iowa, s. Illinois, sw. Indiana
south to w. Alabama and sc. Texas.
Mountain (*V. v. pulchra*), reddish-
brown to gray, scales weakly keeled, 15
rows in neck region, 17 rows at
midbody; w. Pennsylvania through ne.
West Virginia and adjacent Maryland
into Highland County, Virginia.

Occasionally seen on the surface after a
downpour. May congregate in small
numbers prior to hibernation in pockets
of woodland debris or under large
rocks. Often found with Ringneck,
Worm, and Rough Earth snakes.

⊗ CORAL SNAKE FAMILY
(Elapidae)

This family includes the highly venomous cobras, kraits, mambas, taipans, coral and sea snakes. Mainly tropical and subtropical in distribution, some 250 species are represented in Australia, the Indian Ocean, Asia, Africa, and tropical America. Only two species occur in our range.

Coral snakes are strikingly colored with broad alternating bands of red and black, separated by narrow yellow or white rings. The head is black with a yellow ring behind the eye, and the snout is blunt. Unlike the vipers and pit vipers, the enlarged grooved fangs are fixed in position on the front part of the upper jaw and cannot be folded back.

Several harmless snakes have color patterns resembling that of the coral snakes. It is important to recognize the differences as coral snake venom is strongly neurotoxic and bites can be fatal. Do not handle!

---

616 Arizona Coral Snake
(*Micruroides euryxanthus*)

Description: 13–21″ (33–53.3 cm). Blunt-snouted and glossy, with alternating wide red, wide black, and narrow yellow or white rings *encircling the body. Head uniformly black to angle of jaw.* Scales smooth, in 15 rows. Anal plate divided.

Breeding: Habits poorly known; presumably lays clutch of 2 or 3 eggs in late summer.

Habitat: Rocky areas, plains to lower mountain slopes; rocky upland desert especially in arroyos and river bottoms; sea level to 5,900′ (1,800 m).

Range: C. Arizona to sw. New Mexico south to Sinaloa, Mexico.

Subspecies: Three; 1 in our range, *M. e. euryxanthus.*

Do not handle! Venom is highly dangerous. This snake emerges from a subterranean retreat at night, usually during or following a warm shower. When disturbed by a predator, it buries its head in its coils, raises and exposes the underside of its tail, and may evert its cloacal lining with a popping sound. Eats blind snakes, other small snakes.

617, 618 Eastern Coral Snake
(Micrurus fulvius)
Subspecies: Eastern, Texas

Description: 22–47½" (55.9–120.7 cm). *Body encircled by wide red and black rings* separated by narrow yellow rings. *Head uniformly black from tip of blunt snout to just behind eyes.* Red rings usually spotted with black. Scales smooth and shiny, in 15 rows. Anal plate divided.

Breeding: Reportedly lays 3–12 eggs in June; young hatch in September, at 7–9" (18–23 cm).

Habitat: Moist, densely vegetated hammocks near ponds or streams in hardwood forests; pine flatwoods; rocky hillsides and canyons.

Range: Se. North Carolina to s. Florida and Key Largo, west to s. Texas and Mexico.

Subspecies: Eastern (*M. f. fulvius*), black dotting in red rings fuses into pair of spots on back into a single spot on belly; se. North Carolina to s. Florida, west to Mississippi River.

Texas (*M. f. tenere*), random black spots in red rings; s. Arkansas, w. Louisiana, s. Texas into ne. Mexico.

Do not confuse this venomous species with its harmless mimics—the Scarlet Snake and Scarlet Kingsnake. Usually seen under rotting logs or leaves or moving on surface in early morning or late afternoon. Feeds on small snakes or lizards.

⊗ **PIT VIPER FAMILY**
(Viperidae)

Occurring worldwide, this family of dangerously poisonous snakes includes about 290 species. Seventeen of the 19 venomous snakes in our range belong to the Viperidae. They are represented by 3 genera—*Agkistrodon,* the copperheads and cottonmouths, and *Crotalus* and *Sistrurus,* the rattlesnakes.

Most of those in our area are stout-bodied, with heads distinctly wider than the neck, and patterned with blotches or crossbands. They have recurved, retractable hollow fangs situated near the front of the upper jaw. Normally folded back along the jaw, the fangs are rapidly swung forward as the mouth is opened to strike. A heat-sensitive pit, used to locate warm-blooded prey, is present on each side of the head between the eye and nostril. Viperids also have eyes with vertical pupils and an undivided row of scales under the tail. Rattlesnakes bear a distinctive rattle on their tail. This unique structure is a series of flattened, interlocking dry horny segments that produces a buzzing noise when shaken vigorously. A new segment is added each time the snake sheds its skin, normally 2 to 4 times each year. Viperids appear to be the most highly evolved snakes, organized for capturing, killing, and ingesting relatively large warm-blooded prey. Their venom is a complex mixture of proteins, which acts primarily on a victim's blood tissue. Extreme caution is advised—even "road kills" have been known to bite. Most viperids are nocturnal, and most bear their young alive.

649, 650, 651, Copperhead
652, 655 (*Agkistrodon contortrix*)
Subspecies: Southern, Broad-banded,
Northern, Osage, Trans-Pecos

Description: 22–53″ (55.9–134.6 cm). Stout-
bodied; copper, orange, or pink-tinged,
with bold chestnut or reddish-brown
crossbands constricted on midline of
back. Top of head unmarked. Facial pit
between eye and nostril. Scales weakly
keeled, in 23–25 rows. Anal plate
single.

Breeding: Live-bearing. Mates spring to fall, peak
April to May. 1–14 young, 7–10″
(18–25 cm) long, are born August to
early October; mature in 2–3 years.

Habitat: Wooded hillsides with rock outcrops
above streams or ponds; edges of
swamps and periodically flooded areas
in coastal plain; near canyon springs
and dense cane stands along Rio
Grande; sea level to 5,000′ (1,500 m).

Range: Sw. Massachusetts west to extreme se.
Nebraska south to Florida panhandle
and sc. and w. Texas.

Subspecies: Southern (*A. c. contortrix*), hourglass-
shaped crossbands, narrow across

midline of back, 2 halves often fail to
meet; e. North Carolina to Florida
panhandle, west through s. Alabama,
Mississippi, Louisiana, and e. Texas,
north through se. Oklahoma, Arkansas,
w. Tennessee, s. Missouri to sw. Illinois.
Broad-banded (*A. c. laticinctus*), dark
crossbands much wider than light
interspaces, nearly as wide across
midline of back as on sides; s. Kansas
through c. Oklahoma to sc. Texas.
Northern (*A. c. mokeson*), dark
hourglass-shaped crossbands, wide
portion on sides, narrow across midline
of back, small dark spots between
bands; sw. Massachusetts to sw.
Illinois, south to ne. Mississippi, n.
Alabama, n. and c. Georgia, and
piedmont of South Carolina.
Osage (*A. c. phaeogaster*), resembles
Northern, crossbands darker, no spots

between bands; ne. Oklahoma, e.
Kansas, c. and w. Missouri, se. Iowa,
Nebraska.
Trans-Pecos (*A. c. pictigaster*), dark
crossbands like those of Broad-banded
with pale area at base of each band;
Davis Mts. and Big Bend region, Texas.

It basks during the day in spring and
fall, becoming nocturnal as the days
grow warmer. Favored summer retreats
are stonewalls, piles of debris near
abandoned farms, sawdust heaps, and
rotting logs, and large flat stones near
streams. It feeds on small rodents,
lizards, frogs, large caterpillars, and
cicadas. The young twitch their yellow-
tipped tail to lure prey. In fall,
Copperheads return to their den site,
often a rock outcrop on a hillside with a
southern or eastern exposure.
Copperhead bites are painful, but rarely
pose a serious threat to life.

654, 656, 657 Cottonmouth
(*Agkistrodon piscivorus*)
Subspecies: Eastern,
Florida, Western

Description: 20–74½" (50.8–189.2 cm). A dark,
heavy-bodied water snake; broad-based
head is noticeably wider than neck.
Olive, brown, or black above;
patternless or with serrated-edged dark
crossbands. Wide light-bordered, dark-
brown cheek stripe distinct, obscure, or
absent. Head flat-topped; eyes with
vertical pupils (not visible from directly
above as are eyes of harmless water
snakes); facial pit between eye and
nostril. Young strongly patterned and
bear bright yellow tipped tails. Scales
keeled, in 25 rows.

Breeding: Live-bearing. Mates spring and fall.
August to September females give birth
to 1–15 young, 7–13" (18–33 cm)
long. Females mature in 3 years and

give birth every other year.

Habitat: Lowland swamps, lakes, rivers, bayheads, sloughs, irrigation ditches, canals, rice fields, to small clear rocky mountain streams; sea level to ca. 1,500' (450 m).

Range: Se. Virginia south to upper Florida Keys, west to s. Illinois, s. Missouri, sc. Oklahoma and c. Texas. Isolated population in nc. Missouri.

Subspecies: Three; broad zones of intergradation. Eastern (*A. p. piscivorus*), cheek stripe not well defined, snout tip lacks vertical markings; coastal plain, se. Virginia south through Carolinas, west through c. Georgia into Alabama. Florida (*A. p. conanti*), cheek stripe distinct, 2 vertical dark marks on snout tip; se. Alabama, s. Georgia, Florida. Western (*A. p. leucostoma*); head and body markings obscure or absent; sw. Kentucky, w. Tennessee, and w. Alabama, west through s. Missouri to sc. Oklahoma and c. Texas.

Do not disturb or attempt to handle! Its bite is far more serious than that of the Copperhead and can be fatal. When annoyed, the Cottonmouth tends to stand its ground and may gape repeatedly at an intruder, exposing the light "cotton" lining of its mouth. Also called trap jaw or water moccasin. Unlike other water snakes, it swims with head well out of water. Although it may be observed basking during the day, it is more active at night. Preys on sirens, frogs, fishes, snakes, and birds.

624 Eastern Diamondback Rattlesnake
(*Crotalus adamanteus*)

Description: 36–96" (91.4–244 cm). *Our largest rattler*. Heavy-bodied with large head sharply distinct from neck. Back patterned with dark diamonds with light centers and prominently bordered

by a row of cream to yellow scales.
Prominent light diagonal lines on side
of head. Vertical light lines on snout.
Scales keeled, in 27–29 rows.

Breeding: Habits poorly known; 7–21 young,
12–14" (30–36 cm) long, are born July
to early October.

Habitat: Sandhill or longleaf pine and turkey oak
country, dry pine flatwoods, abandoned
farmland; sea level to 500' (150 m).

Range: Lower coastal plain, se. North Carolina
to Florida Keys, west to s. Mississippi,
and extreme e. Louisiana.

Give it a wide berth; most dangerous
snake in North America! Venom highly
destructive to blood tissue.
Stumpholes, gopher tortoise burrows,
and dense patches of saw palmetto often
serve as retreats. Their numbers have
been substantially reduced by extensive
land development and by rattlesnake
hunters. Eats rabbits, squirrels, birds.

---

639   **Western Diamondback Rattlesnake**
(*Crotalus atrox*)

Description: 34–83⅛" (86.4–213 cm). *Largest
western rattlesnake.* Heavy-bodied with
large head sharply distinct from neck.
Back patterned with light-bordered
dark diamonds or hexagonal blotches;
blotches often obscured by randomly
distributed small dark spots, which
give back a mottled or dusky look. 2
light diagonal lines on side of face;
stripe behind eye meets upper lip well
in front of angle of jaw. *Tail encircled by
broad black and white rings.* Scales
keeled, in 25–27 rows.

Breeding: Mates late March to May and in fall;
4–25 young, 8½–13" (21.5–33 cm)
long, are born in late summer. Females
mature in 3 years.

Habitat: Arid and semiarid areas from plains to
mountains; brushy desert, rocky
canyons, bluffs along rivers, sparsely

vegetated rocky foothills; sea level to 7,000' (2,100 m).

Range: Se. California eastward to c. Arkansas south into n. Mexico.

The "coon tail rattler" is capable of delivering a fatal bite! When disturbed it usually stands its ground, lifts its head well above its coils, and sounds a buzzing warning. Take heed! Active late in the day and at night during hot summer months. Eats rodents and birds. Record longevity is nearly 26 years.

634, 647 Sidewinder
(*Crotalus cerastes*)
Subspecies: Mojave Desert, Sonora, Colorado Desert

Description: 17–32⅜" (43.1–82.4 cm). A rough-scaled rattler with a *prominent triangular, hornlike projection over each eye*. Scales keeled, in 21–23 rows.

Breeding: Mates April to May, sometimes in fall. Female gives birth to 5–18 young, about 6½–8" (17–20 cm) long, late summer to early fall.

Habitat: Arid desert flatland with sandy washes or mesquite-crowned sand hammocks; below sea level to 5,000' (1,500 m).

Range: S. Nevada and adjacent California, and extreme sw. Utah, south into Mexico.

Subspecies: Mojave Desert (*C. c. cerastes*), bottom segment of rattle brown, 21 scale rows; extreme se. Utah, s. Nevada, and Mojave Desert region of California. Sonora (*C. c. cercobombus*), bottom segment of rattle black, 21 scale rows; sc. Arizona to w. central Sonora. Colorado Desert (*C. c. laterorepens*), bottom segment of rattle black, 23 scale rows; sw. Arizona and se. California and adjacent Mexico.

Travels quickly over shifting surfaces by "sidewinding," a locomotion process in which the snake makes use of static friction to keep from slipping when crossing soft sandy areas. It leaves a trail of parallel J-shaped markings behind it. Primarily nocturnal, it is usually encountered crossing roads between sundown and midnight in spring. During the day it occupies mammal burrows or hides in shelters beneath bushes. Eats pocket mice, kangaroo rats, and lizards.

619, 620, 653 Timber Rattlesnake
(*Crotalus horridus*)

Description: 35–74½″ (88.9–189.2 cm). Northern forms range from yellow through brown or gray to black, with dark back and side blotches on front of body and blotches fused to form crossbands on rear of body. Head unmarked. Southern forms yellowish-, brownish- or pinkish-gray, with tan or reddish-brown back stripe dividing chevronlike crossbands; dark stripe behind eye. Both forms have black tail. Scales keeled, in 23–25 rows.

Breeding: Mates in autumn and shortly after emergence from hibernation. Female gives birth every other year to 5–17 young, 10–13″ (25–33 cm) long, late August to early October. Females mature in 4–5 years.

Habitat: Remote wooded hillsides with rock outcrops in the North; unsettled swampy areas, canebrake thickets, and floodplains in the South; sea level to 6,600′ (2,000 m).

Range: Extreme sw. Maine south to n. Florida, west into se. Minnesota and c. Texas.

Active April to October; in the daytime in spring and fall, at night during the summer. In northern areas, Timber Rattlesnakes congregate in large numbers about rocky den sites and may overwinter with rat snakes and

Copperheads. Often encountered coiled
up waiting for prey—squirrels, mice,
chipmunks, small birds; when
approached, remains motionless.
Record longevity exceeds 30 years.
Until recently, southern populations
were recognized as *C. h. atricaudatus,*
the Canebrake Rattlesnake.

636, 640 Rock Rattlesnake
(*Crotalus lepidus*)
Subspecies: Mottled, Banded

Description: 16–32⅝" (40.6–82.9 cm). Small-
headed and slender; greenish-gray,
bluish-gray, or pinkish-tan, with *widely
spaced, irregularly bordered narrow black or
brown crossbands.* Dusky spotting
between bands may be sparse or so
heavy as to form secondary crossbands
or lend a speckled appearance. Scales
keeled, in 23 rows.

Breeding: 2–8 young, 6¾–8¾" (17–22 cm)
long, are born July to August.

Habitat: Chiefly rocky mountainous areas; talus
slopes, gorges, rimrock, limestone
outcrops, rocky streambeds; 1,500–
9,600' (450–2,900 m).

Range: Se. Arizona, wc. and se. New Mexico
southeast to Jalisco, and through the
Trans-Pecos region to sc. Texas.

Subspecies: Four; 2 in our area.
Mottled (*C. l. lepidus*), spaces between
crossbands heavily spotted, dark stripe
from eye to angle of mouth; se. New
Mexico through Trans-Pecos region to
sc. Texas, south to San Luis Potosí.
Banded (*C. l. klauberi*), crossbands
distinct, no dark stripe from eye to
angle of mouth; se. Arizona to extreme
w. Texas, south to Jalisco, Mexico.

May be observed during the day
sunning itself among rocks. Feeds on
lizards, small snakes, and small
newborn rodents. Record longevity is
23 years.

635, 646   Speckled Rattlesnake
(*Crotalus mitchelli*)
Subspecies: Southwestern, Panamint

Description:
23–52" (58.4–132.1 cm). Pattern and color vary greatly; generally has a sandy, speckled appearance. Back marked with muted crossbands or hexagonal- to diamond-shaped blotches formed by small cluster of dots. *Large scale above eye pitted, creased, or rough-edged; or rostral scale separated from preanals by row of tiny scales.* Scales keeled, in 23–27 rows.

Breeding:
July to August, female gives birth to 2–11 young, 8–12" (20–30 cm) long.

Habitat:
Prefers rugged rocky terrain, rock outcrops, deep canyons, talus, chaparral amid rock piles and boulders, rocky foothills; sea level to 8,000' (2,450 m).

Range:
Extreme sw. Utah, s. Nevada and s. California south into nw. Sonora and throughout Baja California.

Subspecies:
Southwestern (*C. m. pyrrhus*), rostral scale separated from prenasals by small scales; sw. Utah, w. Arizona, extreme s. tip of Nevada, s. California. Panamint (*C. m. stephensi*), large scale above eye pitted, creased, or roughly edged; s. Nevada and adjacent Calif.

Active during the day in spring and fall, at night in summer. Eats ground squirrels, kangaroo rats, white-footed mice, birds, and lizards. Record longevity exceeds 16 years.

626   Black-tailed Rattlesnake
(*Crotalus molossus*)
Subspecies: Northern

Description:
28–49½" (71.1–125.7 cm). Greenish, yellowish, or grayish, with irregular light-edged, light-centered crossbands and *sharply contrasting black tail.* Individual scales are monochromatic. Scales keeled, in 27 rows.

Breeding: About 3–6 young, 9–12″ long, are born in summer.

Habitat: Most common in rocky mountainous areas; among rimrock and limestone outcrops, wooded stony canyons, chaparral, rocky streambeds; near sea level to ca. 9,000′ (2,750 m).

Range: Arizona east to c. Texas, south through c. Mexico.

Subspecies: Three; 1 in our range. Northern (*C. m. molossus*).

Generally considered an unaggressive rattlesnake; little is known of its natural history. It is seen at night and basking during cooler periods of the day. Presumably feeds largely on small rodents. Record longevity is 15½ years.

637 Twin-spotted Rattlesnake
(*Crotalus pricei*)

Description: 12–26″ (30.5–66 cm). Slender, with *2 rows of paired or slightly alternating scalloped-edged brown spots down middle of back*. Light-bordered stripe extends from eye to angle of jaw. Scales keeled, in 21 rows.

Breeding: Late July to August female gives birth to 3–8 young, 6¼–8¼″ (16–21 cm) long.

Habitat: Arid rocky areas high in the mountains; open yellow pine and oak woodland into evergreen forest; 6,300–10,000′ (1,900–3,000 m).

Range: Se. Arizona south to s. Durango, Mexico.

Subspecies: Two; 1 in our range, *C. p. pricei*.

The buzz of its rattle, like that of other small rattlesnakes, sounds like a cicada and can be heard only within a few yards. Often encountered in canyons sunning itself on rock slides or exposed slopes. In Arizona it shares its habitat with the Rock Rattlesnake. Eats lizards and small mammals.

644 Red Diamond Rattlesnake
(*Crotalus ruber*)

Description: 29–64" (73.7–162.5 cm). Stoutly
built; tan to brick-red, with diamond-
shaped blotches down midline of back.
Blotches usually light-edged but may
be indistinct. Black and white rings
encircle tail. Scales keeled, in 29 rows.

Breeding: Mates February to April. Female gives
birth to 3–20 young, 11¾–13½"
(30–34 cm) long, in August.

Habitat: Cool coastal zone into the foothills and
over the mountains into the desert;
prefers dense chaparral in foothills,
brush-covered boulders, cactus patches;
sea level to 5,000' (1,500 m).

Range: Sw. California south through Baja.

Subspecies: Five; 1 in our range, *C. r. ruber*.

Most often encountered in the spring,
coiled in a shelter in partial sun or
crossing the road at night. Eats rabbits,
ground squirrels, and birds. Record
longevity is 14½ years.

622 Mojave Rattlesnake
(*Crotalus scutulatus*)

Description: 24–51" (61–129.5 cm). Uniformly
white scales surround brown diamonds
marking midline of back. Greenish-
gray, olive-green, greenish-brown, or
occasionally yellow above. Black and
white rings encircle tail; white rings
significantly larger. Light stripe behind
eye extends backward above angle of
mouth. Scales keeled, in 25 rows.

Breeding: Female gives birth to 2–11 young,
about 9–11" (23–28 cm) long, July to
August.

Habitat: Upland desert flatland supporting
mesquite, creosote bush, and cacti; also
arid lowland with sparse vegetation,
grassy plains, Joshua tree forests, and
rocky hills; sea level to 8,300'
(2,500 m).

Range: S. Nevada, adjacent California and
extreme sw. Utah southeastward
through c. Mexico.

Subspecies: Two; 1 in our range, C. *s. scutullatus*.

Usually encountered on mild nights,
crossing a road, or before the heat of
the day, partially exposed under the
bank of a dry wash or along a mesquite-
bordered streambed. Its venom is
extremely toxic and causes more
respiratory distress than that of any
other North American pit viper.

628 Tiger Rattlesnake
(*Crotalus tigris*)

Description: 20–36″ (50.8–91.4 cm). *Numerous gray
or brownish crossbands* mark pale gray,
buff, lavender, or pinkish-gray upper
surface. Crossbands composed of tiny
dots, often poorly defined. Head
proportionally small, rattle large. Scales
keeled, in 23 rows.

Breeding: Live-bearing. Young reportedly about
9″ (23 cm) at birth.

Habitat: Arid rocky foothills and canyons,
primarily in ocotilla-mesquite-creosote
bush and saguaro-paloverde
associations; sea level to 4,800′
(1,450 m).

Range: C. Arizona south to s. Sonora, Mexico.

Natural history poorly known.
Reportedly active day and night and
occasionally encountered crossing roads
after warm showers. Occasionally
confused with larger Speckled
Rattlesnake which enters its range.
Feeds on lizards and small newborn
mammals. Record longevity exceeds
15 years.

621, 623, 627, **Western Rattlesnake**
629, 630, 631, (*Crotalus viridis*)
648 Subspecies: Prairie, Grand Canyon,
Arizona Black, Midget Faded,
Southern Pacific, Great Basin, Hopi,
Northern Pacific

Description: 16–64" (40.6–162.6 cm). Size and
color vary greatly. Brownish blotches
down midline of back, generally edged
with dark brown or black and often
surrounded by light border; begin as
oval, squarish, diamondlike, or
hexagonal markings and tend to narrow
into inconspicuous crossbands near tail.
More than 2 internasal scales touch rostral
scale. Scales keeled, in 25–27 rows.

Breeding: Mates March to May and in fall; 4–21
young, 6–12" (15–30 cm) long, are
born August to October.

Habitat: Great Plains grassland to brush-covered
sand dunes on Pacific coast, and to
timberline in the Rockies and the
coniferous forests of the Northwest;
rocky outcrops, talus slopes, stony
canyons, and prairie dog towns; sea
level to 11,000' (3,350 m).

Range: Extreme w. Iowa, south into Mexico
and west to s. Alberta, sw.
Saskatchewan, sc. British Columbia,
Washington, Oregon, and coastal
California, and south into Mexico.

Subspecies: Nine; 8 in our range.
Prairie (*C. v. viridis*), greenish or
brownish above, well-defined brown
blotches; extreme w. Iowa to the
Rockies—s. Alberta to n. Mexico.
Grand Canyon (*C. v. abyssus*), reddish
above, blotches obscure; Grand
Canyon, Arizona.
Arizona Black (*C. v. cerberus*), dark
gray, brown, or almost black above,
large dark blotches partially bordered
by white above; Arizona south of the
Colorado Plateau, from Hualapai
southeastward to Blue Mountains and
into extreme w. New Mexico.
Midget Faded (*C. v. concolor*), pale
yellow, blotches obscure or absent;

extreme sw. Wyoming, w. Colorado, and e. Utah.

Southern Pacific (*C. v. helleri*), resembles Arizona Black, diamond-shaped blotches completely bordered by light scales, poorly defined tail rings; San Luis Obispo and Kern counties, California, south to s. Baja California.

Great Basin (*C. v. lutosus*), light brown or gray above; blotches narrow (roughly equal to interspaces); se. Oregon, s. Idaho, ne. California, Nevada, w. Utah, extreme nw. Arizona.

Hopi (*C. v. nuntius*), pinkish- or reddish-brown, blotches well defined; ne. Arizona.

Northern Pacific (*C. v. oreganus*), resembles Southern Pacific, blotches oval or hexagonal, well-defined tail rings; sc. British Columbia, Washington, wc. Idaho, Oregon (except se. corner), west of Sierra Nevadas in California south to Kern County.

Western counterpart of the Timber Rattlesnake but much more excitable and aggressive. In northerly areas or at high elevations, large numbers may overwinter together at a common den site. In southerly areas or those lacking large rocky retreats, individuals may seek shelter in mammal burrows. Active April to October over much of range, and becomes crepuscular and nocturnal during hot summer months. Adults prey chiefly on small mammals; young like lizards and mice. Record longevity is 19½ years.

---

643 **Ridge-nosed Rattlesnake**
(*Crotalus willardi*)
Subspecies: Arizona, Animas

Description: 15–24″ (38.1–61 cm). *Prominent ridge along upper edge of snout.* Light gray, brown, or reddish above, marked with

dark-bordered narrow white crossbands.
White flash marks along side of face
present or absent. Scales keeled, in
25–29 rows.

Breeding: August to September, female gives
birth to 4–9 young, 6½–8¼″
(17–21 cm) long.

Habitat: Pine-oak woodland at high elevations,
5,500–9,000′ (1,700–2,750 m).

Range: Se. Arizona and sw. New Mexico,
south in scattered populations to s.
Durango and nw. Zacatecas, Mexico.

Subspecies: Five; 2 in our range.
Arizona (*C. w. willardi*), conspicuous
white flash marks on head, little dark
spotting on head; Santa Rita and
Hauchuca mountains, Arizona, south
into extreme n. Sonora, Mexico.
Animas (*C. w. obscurus*), lacks
prominent white flash marks on head,
has abundant dark spotting on head;
Animas Mountains, New Mexico.

A shy species that usually lies
undetected amid leaf litter.
Occasionally seen basking on sunlit
rocky slopes. It feeds on lizards, small
rodents, and birds. Vulnerable to
overcollection and habitat destruction,
it is protected by law. Record longevity
is 21 years.

632, 633, 638 Massasauga
(*Sistrurus catenatus*)
Subspecies: Eastern, Desert, Western

Description: 18–39½″ (45.8–100.3 cm). Unlike
other rattlers, has 9 *enlarged scales on top
of head*. Tail stocky with moderately
developed rattle. Rounded dark
blotches on back and sides; interspaces
narrow. Light-bordered dark bar
extends from eye to rear of jaw. Dark
bars (often lyre-shaped) on top of head
extend onto neck. Scales keeled, in
23–25 rows. Anal plate single.

Breeding: Mates April–May. Litter of 2–19

young, 6½–9½" (16.5–24 cm) long,
are born July to early September.
Females mature in 3 years.

Habitat: Sphagnum bogs, swamps, marshland,
and flood plains to dry woodland in the
East; grassy wetland, rocky hillsides,
sagebrush prairie, into desert grassland
in the West.

Range: S. Ontario and nw. Pennsylvania south
to ne. Mexico and extreme se. Arizona.
Isolated populations in c. New York.

Subspecies: Eastern (*S. c. catenatus*), brownish-gray
with dark brown or black blotches;
belly black with scattered light
markings; 25 scale rows; s. Ontario and
nw. Pennsylvania west to e. Iowa and
ne. Missouri; isolated populations in c.
New York.
Desert (*S. c. edwardsi*), smaller, faded
version of Western Massasauga; belly
whitish and often unmarked; 23 scale
rows; se. Colorado and sw. Kansas to
w. and s. Texas, s. New Mexico and
extreme se. Arizona.
Western (*S. c. tergeminus*), light gray
with dark-brown blotches; belly white
or cream with a few dark markings; 25
scale rows; sw. Iowa, nw. Missouri, e.
Kansas, w. Oklahoma, nc. and se.
Texas.

Massasauga means "great river mouth"
in the Chippewa language and probably
alludes to the snake's habitat in
Chippewa country—swampland
surrounding mouths of rivers. It may
be encountered sunning on mild days;
becomes crepuscular or nocturnal
during hot summer months. Eats
lizards, small rodents, and frogs.

625, 641, 642, **Pigmy Rattlesnake**
645 (*Sistrurus miliarius*)
Subspecies: Carolina, Dusky, Western

Description: 15–30⅞" (38–78.5 cm). Small rattler;
slender tail is tipped with a tiny rattle.

Gray to reddish, with brown to black blotches along midline of back; 1–3 rows of spots on sides. Narrow reddish back stripe sometimes present. Reddish-brown to black bar extends from eye to rear of jaw; usually bordered below with white line. *Top of head has 9 enlarged scales.* Scales keeled in 21–25 rows. Anal plate single.

Breeding: Female gives birth to 2–32 (usually 4–8) young, 4–7½" (10–19 cm) long, July to September.

Habitat: Everglades prairie, palmetto-pine flatwoods, sandhills, mixed pine-hardwood forest, borders of cypress ponds, and vicinity of lakes and marshes.

Range: E. North Carolina to Florida Keys, west to e. Oklahoma and e. Texas.

Subspecies: Carolina (*S. m. miliarius*), pale gray to reddish above with prominent markings, 1 or 2 rows of spots on sides, usually 23 scale rows; e. North Carolina to s. South Carolina, west to n. Alabama.
Dusky (*S. m. barbouri*), dark gray above with heavy black stippling obscuring pattern, 3 rows of spots on sides, 23–25 scale rows; Extreme s. South Carolina to Florida Keys, west to se. Mississippi.
Western (*S. m. streckeri*), pale grayish-brown above with blotches forming crossbars, 2 rows of spots on sides, usually 21 scale rows; sw. Kentucky, wc. Alabama, Louisiana west to e. Oklahoma and e. Texas.

Called "ground rattler" in parts of range. The tiny rattle makes a buzzing sound audible only for a few feet. Usually encountered in the summer, quietly sunning itself or crossing a road late in the day. Some are pugnacious and strike with little provocation, others appear lethargic. They eat lizards, small snakes, mice, and occasionally insects. Record longevity exceeds 15 years.

Part III
Appendices

GLOSSARY

Aestivation Dormancy during summer or dry season.

Amplexus The sexual embrace of a male frog or toad; the clasping of the female's body around the back by the male's forelimbs until eggs are laid and fertilized.

Anterior Located near or toward the head.

Boss A rounded knob on top of the head between the eyes of certain toads.

Bridge The part of a turtle's shell which connects the carapace and plastron.

Carapace Upper part of a turtle's shell.

Cloaca The chamber into which the digestive, urinary, and reproductive systems empty, opening to the outside through the vent.

Costal grooves Vertical grooves on a salamander's side.

Cranial crests Bony ridges on the head of some toads.

Crepuscular Active at twilight.

Cryptic Serving to conceal.

Cusp A pointed toothlike projection on the upper jaw of some turtles.

Dewlap "Throat fan;" vertical, loose flap of skin on the throat of some iguanid lizards.

Diurnal Active during daytime hours.

Dorsal Pertaining to the back or upper surface of the body.

Dorsolateral Pertaining to the area at the juncture of the back and the side.

Ectotherm A "cold-blooded" animal; an animal that regulates its temperature behaviorally by means of outside sources of heat (e.g., amphibians, reptiles).

Femoral pores A series of small openings on the underside of the thighs of some lizards.

Fossorial Adapted to digging or burrowing.

Gills External organs located on the sides of the neck and used in underwater respiration.

Gill slit An opening or hole in salamanders at the base of the external gills.

Granular scales Tiny grainlike or pebblelike scales that do not overlap one another.

Gravid Bearing eggs or developing young; pregnant.

Hemipenis (*pl.* hemipenes) One of the paired copulatory organs of snakes and lizards.

Hibernation Dormancy during winter.

Hybrids Offspring of two different varieties, races, species, or genera.

Intergrades Animals of related and adjoining subspecies that may resemble either form or exhibit a combination of their characteristics.

Keel A ridge on individual dorsal scales of

some snakes; longitudinal ridge on the carapace or plastron of turtles; the raised edge along the upper edge of the tail in some salamanders.

Larva(e) A post-hatching immature stage that differs in appearance from the adult and must metamorphose before assuming adult characters (e.g., a tadpole).

Lateral Pertaining to the side.

Metamorphosis A period of transformation from larval to adult form; (e.g., the transition of a tadpole to a frog).

Mental gland An organ on the chin of male lungless salamanders that produces a secretion sexually stimulating to females.

Middorsal Pertaining to the center of the back.

Midventral Pertaining to the center of the belly.

Nasolabial groove A hairline groove running from the nostril to the edge of the upper lip in lungless salamanders.

Neoteny A condition in which salamanders fail to metamorphose but become sexually mature and reproduce while retaining larval features.

Nocturnal Active at night.

Ocular The large scale covering the eye-spot of blind snakes.

Oviparous Producing young by means of eggs that hatch after laying (e.g., turtles).

Ovoviviparous Producing young by means of membranous eggs retained within the body of the female until hatching (e.g. "live-bearing" reptiles).

Parotoid gland A large glandular structure on each side of the neck or behind the eyes of toads and some salamanders.

Parthenogenesis   Reproduction by the development of an unfertilized egg (e.g., species of whiptail lizards that produce only one sex—females—and reproduce by means of unfertilized eggs.)

Plastron   The lower part of a turtle's shell.

Posterior   Located at or toward the rear end of the body.

Prehensile   Adapted for grasping or wrapping around.

Reticulation   A network of lines.

Scute   A large scale; horny shields or plates covering a turtle's shell.

Spermatophore   A cone-shaped jellylike mass topped with a sperm cap, deposited by male salamanders during courtship. The sperm cap is picked up by the cloacal lips of the female.

Spicule   A tiny pointed structure.

Tadpole   The larva of a frog or toad.

Transformation   *See* metamorphosis.

Tubercle   A small knob or a rounded protuberance in the skin.

Tympanum   The eardrum.

Vent   Anus; opening of the cloaca to the outside of the body.

Ventral   Pertaining to the underside or lower surface of the body.

Vocal sac   An expandable pouch on the throat of male frogs and toads that becomes filled with air and acts as a resonating chamber when they vocalize during courtship; the sac collapses at the end of the call.

PICTURE CREDITS

The numbers in parentheses are plate numbers. Some photographers have pictures under agency names as well as their own. Agency names appear in boldface. Photographers hold copyrights to their works.

Ronn Altig (7)

Animals Animals
Tom Brakefield (196, 247, 431)
W. Griffin (384) Breck P. Kent (14, 281, 282, 302, 311) Zig Leszczynski (32, 41, 43, 47, 48, 51, 62, 101, 108, 112, 120, 173, 182, 236, 248, 251, 256, 257, 277, 283, 295, 299, 300, 308, 314, 345, 379, 386, 404, 405, 406, 421, 445, 453, 462, 486, 489, 490, 501, 504, 526, 530, 532, 561, 591, 598, 599, 608, 609, 620, 622, 639, 640, 644, 649) Bruce MacDonald (448) Ray A. Mendez (169, 305) Robert W. Mitchell (1, 2, 4) Perry D. Slocum (385) Lynn M. Stone (179, 261) Jack Wilburn (207, 430)

Ray E. Ashton, Jr. (25, 39, 116, 125, 210, 363, 376, 391, 542, 581)
Ralph W. Axtell (368, 371, 373, 417)
R. W. Barbour (267, 268, 275, 292, 293, 303, 315, 325, 411, 426, 557, 562)
John Behler (259, 306)
W. Frank Blair (228, 238, 239, 332)
Arden H. Brame, Jr. (75)
Alvin L. Braswell (440)
E. Brodie (10, 15, 28, 34, 167)
Nathan W. Cohen (358)
C. J. Cole (412, 414, 418, 419, 420, 595, 616)

Robert W. Mitchell (3, 5, 11, 140, 168, 220, 479)
R. Mount (549)
John Murphy (249, 276, 341, 344, 361, 374, 377, 605, 646)
Ray Pawley (570)

Photo Researchers, Inc.
Joel Arlington (641) Robert J. Ashworth (226) Charles R. Belinky (322) Henry Bunker IV (296) John M. Burnley (71, 117) James H. Carmichael, Jr. (556) Dell O. Clark (328, 362) Joseph T. Collins (12, 38, 50, 57, 83, 93, 94, 126, 143, 165, 195, 208, 209, 269, 280, 290, 297, 326, 346, 360, 366, 375, 381, 403, 407, 408, 416, 423, 424, 428, 429, 444, 455, 460, 463, 464, 467, 471, 476, 480, 497, 506, 510, 512, 517, 519, 527, 531, 534, 543, 544, 558, 565, 571, 572, 577, 585, 588, 597, 600, 611, 613, 615, 631, 638) Robert J. Erwin (24, 218, 653) Michael P. Gadomski (201) Gilbert Grant (329) D. Hiser (339) H. R. Hungerford (370) Stephen J. Kraseman (365) John R. MacGregor (278) Tom McHugh (22, 23, 31, 221) Sturgis McKeever (99, 141, 147, 156, 161, 175, 176, 191, 241, 250, 576) Anthony Mercieca (559) Tom Myers (392) George Porter (157, 184, 200, 205, 206, 212, 217, 225, 237) Bucky Reeves (502, 522) Leonard Lee Rue III (187, 294, 484) C. W. Schwartz (427) Jeff Simon (422) Alvin E. Staffan (85, 163, 190, 270, 580) K. H. Switak (105, 106, 148, 447, 452) L. West (393)

Louis Porras (390)
P. C. H. Pritchard (262, 263, 264, 324)
Alan Resetar (335, 337)
Alan H. Savitsky (64, 82, 84)
Dr. Robert S. Simmons (6, 9, 16, 42, 45, 46, 52, 54, 87, 97, 100, 104, 107, 124, 128, 130, 149, 150, 151,

INDEX

Numbers in boldface type refer to plate numbers. Numbers in italic refer to page numbers. Circles preceding English names of reptiles and amphibians make it easy for you to keep a record of those you have seen.

A

Acris
crepitans, **153**, *400*
c. blanchardi, *401*
c. crepitans, *401*
gryllus, **161**, *401*
g. dorsalis, **162**, *401*
g. gryllus, *401*
g. paludicola, *402*

Adder, *615*

Agkistrodon
contortrix, *683*
c. contortrix, **650**, *683*
c. laticinctus, *683*
c. mokeson, **651**, *683*
c. phaeogaster, **649**, **655**, *683*
c. pictigaster, **652**, *684*
piscivorus, *684*
p. conanti, **654**, **657**, *685*
p. leucostoma, **656**, *685*
p. piscivorus, *685*

Alligator
Allegheny, *270*
○ American, **256**, **259**, *429*

Alligator Lizard
○ Arizona, **446**, *541*
○ California, *542*
○ Northern, **448**, *539*
○ Oregon, *542*

○ Panamint, **447**, *543*
○ San Diego, *542*
○ San Francisco, *539*
○ Shasta, *539*
○ Sierra, *539*
○ Southern, **445**, **449**, *542*
○ Texas, *541*

Alligator mississippiensis,
256, **259**, *429*

Ambystoma
annulatum, **47**, *288*
californiense, **43**, *289*
cingulatum, **42**, *289*
gracile, *290*
g. decorticatum, *291*
g. gracile, **57**, *291*
jeffersonianum, **56**, *291*
laterale, **58**, *292*
mabeei, **60**, *292*
macrodactylum, *293*
m. columbianum, **52**, *293*
m. croceum, **50**, *293*
m. krausei, *294*
m. macrodactylum, **49**, *293*
m. sigillatum, **53**, *294*
maculatum, **51**, **54**, *294*
opacum, **44**, **45**, *295*
platineum, **59**, *296*
talpoideum, **61**, *296*
texanum, **62**, *297*
tigrinum, **22**, **38**, **40**, *298*

t. diaboli, 299
t. mavortium, 46, 48, 298
t. melanostictum, 39, 298
t. nebulosum, 298
t. stebbinsi, 299
t. tigrinum, 37, 298
tremblayi, 55, 299

Ambystomidae, 288

Ameiva
ameiva, 553
a. petersi, 554

Amphibia, 265

Amphibians, 265

Amphisbaenidae, 580

Amphisbaenid Family, 580

Amphiuma
○ One-toed, 286
○ Three-toed, 286
○ Two-toed, 15, 18, 285

Amphiuma Family, 285

Amphiuma
means, 15, 18, 285
pholeter, 286
tridactylum, 286

Amphiumidae, 285

Aneides
aeneus, 85, 302
ferreus, 83, 303
flavipunctatus, 304
f. flavipunctatus, 142, 304
f. niger, 304
hardii, 81, 305
lugubris, 106, 305

Anguidae, 539

Anguid Lizard Family, 539

Anniella
pulchra, 452, 548
p. nigra, 548
p. pulchra, 548

Anniellidae, 548

Anole
○ Bahaman, 501
○ Bark, 499
○ Brown, 384, 501
○ Crested, 390, 498
○ Cuban, 501
○ Florida Bark, 499
○ Green, 383, 385, 497
○ Green Bark, 499
○ Knight, 389, 500
○ Large-headed, 499

Anolis
carolinensis, 383, 385, 497
cristatellus, 390, 498
cybotes, 499
distichus, 499
d. dominicensis, 500
d. floridanus, 500
equestris, 389, 500
sagrei, 384, 501
s. ordinatus, 501
s. sagrei, 501

Arizona
elegans, 566, 587, 590
e. arenicola, 590
e. candida, 590
e. eburnata, 590
e. elegans, 577, 590
e. noctivaga, 591
e. occidentalis, 591
e. philipi, 591

Ascaphidae, 359

Ascaphus truei, 165, 359

B

Batrachoseps
aridus, 306
attenuatus, 76, 307
major, 75, 307
nigriventris, 307
pacificus, 73, 308
relictus, 77, 309
simatus, 309
stebbinsi, 74, 310
wrighti, 78, 310

Black-headed Snake
○ Big Bend, 661
○ California, 500, 659
○ Chihuahuan, 662
○ Desert, 659
○ Devils River, 661
○ Huachuca, 662
○ Mexican, 656
○ Plains, 460, 658
○ Texas, 658
○ Utah, 659
○ Western, 659
○ Yaqui, 663

Blind Snake
○ Desert, 584
○ New Mexico, 464, 583
○ Plains, 583
○ Southwestern, 584
○ Texas, 583
○ Trans-Pecos, 584
○ Utah, 584
○ Western, 457, 584

Boa
○ Coastal Rosy, 587
○ Desert Rosy, 508, 587
○ Mexican Rosy, 587
○ Rosy, 525, 587
○ Rubber, 472, 586

Boas, 586

Boidae, 586

Box Turtle
○ Desert, 305, 469
○ Eastern, 306, 468
○ Florida, 308, 468
○ Gulf Coast, 309, 468
○ Ornate, 307, 469
○ Three-toed, 304, 468
○ Western, 469

Box Turtles, 446

Buck Knob, 339

Bufo
alvarius, 225, 386
americanus, 237, 387
a. americanus, 387
a. charlesmithi, 388

a. copei, 388
boreas, 240, 388
b. boreas, 226, 388
b. halophilus, 243, 389
b. nelsoni, 389
canorus, 227, 244, 389
cognatus, 247, 389
debilis, 254, 255, 390
d. debilis, 391
d. insidior, 391
exsul, 245, 391
hemiophrys, 239, 391
h. baxteri, 392
h. hemiophrys, 392
houstonensis, 228, 392
marinus, 242, 393
microscaphus, 223, 235, 394
m. californicus, 394
m. microscaphus, 394
punctatus, 234, 394
quercicus, 250, 395
retiformis, 251, 396
speciosus, 238, 396
terrestris, 236, 241, 397
valliceps, 246, 397
woodhousei, 224, 398
w. australis, 249, 398
w. fowleri, 248, 399
w. velatus, 398
w. woodhousei, 398

Bufonidae, 386

○ Bullfrog, 187, 190, 372

○ Bullsnake, 573, 644

○ Burrowing Toad, Mexican, 217, 220, 361

Burrowing Toad Family, 361

C

○ Caiman, Spectacled, 257, 260, 430

Caiman
crocodilus, 257, 260, 430
c. crocodilus, 430

California Legless Lizard
Family, 548

Callisaurus
draconoides, 362, 502
d. myurus, 502
d. rhodostictus, 502
d. ventralis, 502

Caretta
caretta, 265, 475
c. caretta, 476
c. gigas, 476

Carphophis
amoenus, 591
a. amoenus, 592
a. helenae, 592
a. vermis, 493, 592

Caudata, 267

Cemophora
coccinea, 595, 607, 592
c. coccinea, 593
c. copei, 596, 593
c. lineri, 593

Charina bottae, 472, 586

Chelonia
mydas, 267, 476
m. agassizi, 477
m. mydas, 477

Chelonidae, 475

Chelydra
serpentina, 322, 323, 435
s. osceola, 324, 436
s. serpentina, 436

Chelydridae, 435

Chilomeniscus cinctus, 605,
593

Chionactis
occipitalis, 604, 594
o. annulatus, 594
o. klauberi, 612, 595
o. occipitalis, 594
o. talpina, 595
palarostris, 595
p. organica, 610, 595

Chirping Frog
○ Cliff, 168, 422
○ Rio Grande, 167, 421
○ Spotted, 421

○ **Chorus Frog**, 415
○ Boreal, 415
○ Brimley's, 175, 412
○ Illinois, 185, 414
○ Mountain, 163, 411
○ New Jersey, 415
○ Northern, 413
○ Ornate, 176, 177, 413
○ Southern, 183, 413
○ Spotted, 180, 412
○ Strecker's, 414
○ Upland, 415
○ Western, 179, 415

Chrysemys
alabamensis, 292, 446
concinna, 287, 447
c. concinna, 448
c. texana, 448
floridana, 448
f. floridana, 288, 448
f. hoyi, 449
f. peninsularis, 449
nelsoni, 296, 449
picta, 450
p. belli, 293, 450
p. dorsalis, 297, 450
p. marginata, 450
p. picta, 294, 450
rubriventris, 295, 451
r. bangsi, 451
r. rubriventris, 451
scripta, 452
s. elegans, 286, 452
s. gaigeae, 453
s. scripta, 289, 452
s. troosti, 452

○ **Chuckwalla**, 331, 518
○ Arizona, 518
○ Glen Canyon, 518
○ Western, 518

Clemmys
guttata, 290, 453
insculpta, 302, 454

marmorata, 303, 455
m. marmorata, 455
m. pallida, 455
muhlenbergi, 301, 456

Clonophis kirtlandi, 551, 596

Cnemidophorus
burti, 416, 554
b. stictogrammus, 555
b. xanthonotus, 555
dixoni, 555
exsanguis, 418, 555
flagellicaudus, 556
gularis, 556
g. gularis, 557
hyperythrus, 409, 557
h. beldingi, 558
inornatus, 414, 558
i. arizonae, 558
i. heptagrammus, 558
laredoensis, 559
neomexicanus, 415, 559
septemvittatus, 417, 560
sexlineatus, 561
s. sexlineatus, 411, 561
s. viridis, 561
sonorae, 410, 561
tesselatus, 419, 562
tigris, 563
t. gracilis, 563
t. marmoratus, 420, 563
t. multiscutatus, 563
t. mundus, 563
t. septentrionalis, 564
t. tigris, 563
uniparens, 412, 564
velox, 413, 565

○ **Coachwhip,** *628*
○ Baja California, *628*
○ Eastern, 469, *628*
○ Lined, 558, *628*
○ Red, 491, 553, *628*
○ San Joaquin, *628*
○ Sonora, *628*
○ Western, 554, 556, *628*

Coleonyx
brevis, 393, 394, 490

reticulatus, 490
variegatus, 392, 491
v. abbotti, 491
v. bogerti, 395, 491
v. utahensis, 491
v. variegatus, 491

○ **Collared Lizard,** 355, 356, *503*
○ Baja, *504*
○ Chihuahuan, *503*
○ Desert, 359, *504*
○ Eastern, *503*
○ Mojave, *504*
○ Reticulate, 360, *505*
○ Sonoran, *503*
○ Western, *503*
○ Yellow-headed, *503*

Coluber
constrictor, 596
c. anthicus, 597
c. constrictor, 486, 597
c. etheridgei, 597
c. flaviventris, 480, 597
c. foxi, 597
c. helvigularis, 597
c. latrunculus, 598
c. mormon, 478, 598
c. oaxaca, 598
c. paludicola, 598
c. priapus, 468, 598

Colubridae, 589

Colubrid Snake Family, 589

Coniophanes
imperialis, 599
i. imperialis, 599

Contia tenuis, 471, 599

○ **Cooter,** 448
○ Eastern River, 447
○ Florida, **288,** 448
○ Peninsula, 448
○ River, 287, 447
○ Texas River, 447

Cophosaurus
texanus, 361, *503*

t. scitulus, 503
t. texanus, 503

○ **Copperhead**, 683
○ Broad-banded, 683
○ Northern, 651, 683
○ Osage, 649, 655, 683
○ Southern, 650, 683
○ Trans-Pecos, 652, 683

○ **Coqui, Puerto Rican**, 169, 418

Coral Snake
○ Arizona, 616, 680
○ Eastern, 618, 681
○ Texas, 617, 681

Coral Snake Family, 680

○ **Cottonmouth**, 684
○ Eastern, 684
○ Florida, 654, 657, 684
○ Western, 656, 684

Crayfish Snake
○ Graham's, 519, 646
○ Striped, 510, 646

Cricket Frog
○ Blanchard's, 400
○ Coastal, 401
○ Florida, 162, 401
○ Northern, 153, 400
○ Southern, 161, 401

○ **Crocodile, American**, 258, 261, 431

Crocodiles, 427

Crocodylia, 427

Crocodylus acutus, 258, 261, 431

Crotalus
adamanteus, 624, 685
atrox, 639, 686
cerastes, 687
c. cerastes, 687
c. cercobombus, 647, 687
c. laterorepens, 634, 687
horridus, 619, 620, 653, 688

h. atricaudatus, 689
lepidus, 689
l. klauberi, 640, 689
l. lepidus, 636, 689
mitchelli, 690
m. pyrrhus, 635, 690
m. stephensi, 646, 690
molossus, 626, 690
m. molossus, 691
pricei, 637, 691
p. pricei, 691
ruber, 644, 692
r. ruber, 692
scutulatus, 622, 692
s. scutulatus, 693
tigris, 628, 693
viridis, 694
v. abyssus, 648, 694
v. cerberus, 621, 694
v. concolor, 694
v. helleri, 627, 695
v. lutosus, 623, 695
v. nuntius, 631, 695
v. oreganus, 630, 695
v. viridis, 629, 694
willardi, 695
w. obscurus, 696
w. willardi, 643, 696

Crotaphytus
collaris, 355, 356, 503
c. auriceps, 504
c. baileyi, 504
c. collaris, 504
c. fuscus, 504
c. nebrius, 504
insularis, 359, 504
i. bicinctores, 505
i. vestigium, 505
reticulatus, 360, 505

Crowned Snake
○ Central Florida, 461, 660
○ Coastal Dunes, 660
○ Florida, 660
○ Peninsula, 458, 660
○ Rim Rock, 659
○ Southeastern, 466, 657

Cryptobranchidae, 269

Cryptobranchus
alleganiensis, 24, 269
a. alleganiensis, 270
a. bishopi, 270

Ctenosaura pectinata, 333,
506

D

Deirochelys
reticularia, 285, 457
r. chrysea, 457
r. miaria, 457
r. reticularia, 457

Dermochelyidae, 481

Dermochelys
coriacea, 263, 481
c. coriacea, 481
c. schlegelii, 482

Desmognathus
aeneus, 94, 311
auriculatus, 63, 312
brimleyorum, 91, 312
fuscus, 313
f. conanti, 313
f. fuscus, 89, 313
imitator, 134, 314
monticola, 65, 315
m. jeffersoni, 315
m. monticola, 315
ochrophaeus, 95, 112, 120,
124, 137, 316
quadramaculatus, 64, 316
welteri, 92, 317
wrighti, 66, 318

Devil Dog, 270

Diadophis
punctatus, 498, 600
p. acricus, 601
p. amabilis, 496, 601
p. arnyi, 497, 601
p. edwardsi, 601
p. modestus, 601
p. occidentalis, 601
p. pulchellus, 601
p. punctatus, 495, 601

p. regalis, 602
p. similis, 602
p. stictogenys, 602
p. vandenburghi, 602

Dicamptodon
copei, 300
ensatus, 23, 41, 300

Dipsosaurus
dorsalis, 345, 507
d. dorsalis, 507

Drymarchon
corais, 602
c. couperi, 489, 603
c. erebennus, 603

○ **Dusky Salamander**, 313
○ Black Mountain, 92, 317
○ Mountain, 95, 112, 120,
124, 137, 316
○ Northern, 89, 313
○ Ouachita, 91, 312
○ Southern, 63, 312
○ Spotted, 313

Dyrmobius
margaritiferus, 559, 603
m. margaritiferus, 604

E

Earless Lizard
○ Bleached, 370, 510
○ Eastern, 510
○ Greater, 361, 503
○ Keeled, 365, 511
○ Lesser, 510
○ Northern, 366, 510
○ Northern Spot-tailed, 348,
509
○ Southern Spot-tailed, 509
○ Southwestern, 503
○ Speckled, 346, 510
○ Spot-tailed, 509
○ Texas, 503
○ Texas Keeled, 511
○ Western, 510

Earth Snake
○ Eastern, 679
○ Mountain, 679

○ Rough, 470, 473, 678
○ Smooth, 679
○ Western, 467, 679

Eft, **29**, **30**, 275–280

Elaphe
guttata, 570, 608, *604*
g. emoryi, *605*
g. guttata, *605*
obsoleta, 484, *605*
o. bairdi, 509, *606*
o. lindheimeri, *606*
o. obsoleta, *606*
o. quadrivittata, 526, 540, *606*
o. rossalleni, *606*
o. spiloides, 524, 581, *606*
subocularis, 523, *607*
triaspis, 479, *608*
t. intermedia, *608*
vulpina, *608*
v. gloydi, 564, *609*
v. vulpina, *609*

Elapidae, *680*

Eleutherodactylus
coqui, 169, *418*
planirostris, 171, *419*

Emydidae, *446*

Emydoidea blandingi, 291, *458*

○ **Ensatina**, *319*
○ Large-blotched, 108, *319*
○ Monterey, *319*
○ Oregon, 104, *319*
○ Painted, 105, *319*
○ Sierra Nevada, 107, *319*
○ Yellow-blotched, *319*
○ Yellow-eyed, *319*

Ensatina
eschscholtzi, *319*
e. croceator, *319*
e. eschscholtzi, *319*
e. klauberi, 108, *319*
e. oregonensis, 104, *320*
e. picta, 105, *320*
e. platensis, 107, *320*

e. xanthoptica, *320*

Eretmochelys
imbricata, 266, *478*
i. bissa, *478*
i. imbricata, *478*

Eumeces
anthracinus, 425, *568*
a. anthracinus, *569*
a. pluvialis, 429, *569*
egregius, 436, *569*
e. egregius, 435, *570*
e. insularis, *570*
e. lividus, 438, 440, *570*
e. onocrepis, *570*
e. similis, *570*
fasciatus, 427, 437, 443, *570*

Eumeces
gilberti, 430, *571*
g. arizonensis, 434, *572*
g. cancellosus, *572*
g. gilberti, *572*
g. placerensis, *572*
g. rubricaudatus, *572*
inexpectatus, 426, 444, *572*
laticeps, 424, 431, *573*
multivirgatus, 422, *574*
m. gaigeae, *574*
m. multivirgatus, *574*
obsoletus, 432, *575*
septentrionalis, *575*
s. obtusirostris, 428, *576*
s. septentrionalis, 423, *576*
skiltonianus, 421, 441, 442, *576*
s. interparietalis, *577*
s. skiltonianus, *576*
s. utahensis, *577*
tetragrammus, *577*
t. brevilineatus, *578*
t. callicephalus, 439, *578*
t. tetragrammus, *577*

Eurycea
aquatica, *321*
bislineata, *320*
b. bislineata, 88, *321*
b. cirrigera, *321*

b. wilderae, 121, *321*
junaluska, 125, *322*
latitans, *322*
longicauda, 123, *323*
l. guttolineata, 114, *323*
l. longicauda, *323*
l. melanopleura, 113, *323*
lucifuga, 122, *324*
multiplicata, 93, *324*
m. griseogaster, *325*
m. multiplicata, *325*
nana, 5, *325*
neotenes, 4, *326*
quadridigitata, 109, 126, *326*
tridentifera, 2, 11, *327*
troglodytes, 1, *328*
tynerensis, 10, *328*

F

Farancia
abacura, 609
a. abacura, 492, *610*
a. reinwardti, *610*
erytrogramma, 546, *610*
e. erytrogramma, *611*
e. seminola, *611*

Fence Lizard
○ Coast Range, *525*
○ Eastern, *529*
○ Great Basin, *525*
○ Island, *525*
○ Northern, *529*
○ Northwestern, *525*
○ San Joaquin, *525*
○ Sierra, *525*
○ Southern, *529*
○ Western, 379, *525*

Ficimia streckeri, 547, *611*

○ **Fringe-toed Lizard**, *532*
○ Coachella, *532*
○ Desert, 343, *532*
○ Mojave, 344, *533*
○ Sonoran, *532*

Frog
○ African Clawed, 222, *423*
○ Barking, 154, *420*
○ Blanchard's Cricket, *400*
○ Boreal Chorus, *415*
○ Brimley's Chorus, 175, *412*
○ Bronze, 213, *373*
○ California Red-legged, *369*
○ Carolina Crawfish, *367*
○ Carpenter, 205, *381*
○ Cascades, 199, *371*
○ Chorus, *415*
○ Cliff Chirping, 168, *422*
○ Coastal Cricket, *401*
○ Crawfish, *367*
○ Dusky Crawfish, 195, *367*
○ Eastern Barking, *420*
○ Eastern Narrow-mouthed, 219, *383*
○ Florida Crawfish, 194, *367*
○ Florida Cricket, 162, *401*
○ Foothill Yellow-legged, 209, *371*
○ Gopher, *368*
○ Great Plains Narrow-mouthed, 221, *384*
○ Green, 189, *373*
○ Greenhouse, 171, *419*
○ Illinois Chorus, 185, *414*
○ Las Vegas Leopard, *374*
○ Little Grass, 172, *410*
○ Mink, 200, *379*
○ Mountain Chorus, 163, *411*
○ Mountain Yellow-legged, 207, *376*
○ New Jersey Chorus, *415*
○ Northern Chorus, *413*
○ Northern Crawfish, *367*
○ Northern Cricket, 153, *400*
○ Northern Leopard, 192, 203, *377*
○ Northern Red-legged, 208, *369*
○ Ornate Chorus, 176, 177, *413*
○ Pickerel, 201, *377*
○ Pig, 188, *374*
○ Plains Leopard, 197, 202, *370*

- Red-legged, 215, *369*
- Relict Leopard, *376*
- Rio Grande Chirping, 167, *421*
- Rio Grande Leopard, 196, *370*
- River, 210, *375*
- Rough-skinned, *381*
- Sheep, 218, *385*
- Southern Chorus, 183, *413*
- Southern Crawfish, 198, *367*
- Southern Cricket, 161, *401*
- Southern Leopard, 191, 204, *379*
- Spotted, 206, *378*
- Spotted Chirping, *421*
- Spotted Chorus, 180, *412*
- Strecker's Chorus, *414*
- Tailed, 165, *359*
- Tarahumara, 193, *381*
- Upland Chorus, *415*
- Western Barking, *420*
- Western Chorus, 179, *415*
- White-lipped, 166, *420*
- Wood, 211, 212, 214, 216, *380*

Frog Family
Burrowing Toad, *361*
Leptodactylid, *418*
Narrow-mouthed, *383*
Spadefoot Toad, *362*
Tailed, *359*
Toad, *386*
Tongueless, *423*
Tree-, *400*
True, *367*

Frogs and Toads, *357*

G

Gambelia
silus, 347, *507*
wislizenii, 357, *508*
w. copei, 508
w. maculosus, 509
w. punctatus, 509
w. wislizenii, 508

Garter Snake
- Aquatic, *664*
- Black-necked, 536, *666*
- Blue-striped, 538, *674*
- Butler's, 529, *664*
- California Red-sided, 545, *674*
- Checkered, 515, *669*
- Chicago, *674*
- Coast, *667*
- Common, 530, *674*
- Eastern, 539, 576, *674*
- Eastern Black-necked, *666*
- Eastern Plains, *671*
- Giant, *664*
- Klamath, *667*
- Maritime, *674*
- Mexican, 528, *668*
- Mountain, *667*
- Narrow-headed, 548, *672*
- New Mexico, *674*
- Northwestern, 512, *669*
- Oregon, *664*
- Plains, *671*
- Puget Sound, *674*
- Red-sided, 542, *674*
- Red-spotted, 535, *674*
- San Francisco, 541, *674*
- Santa Cruz, *664*
- Short-headed, *663*
- Sierra, *664*
- Texas, 533, *674*
- Two-striped, 504, *664*
- Valley, *674*
- Wandering, 511, *667*
- Western Aquatic, *664*
- Western Black-necked, *666*
- Western Plains, 534, 543, *671*
- Western Terrestrial, *667*

Gastrophryne
carolinensis, 219, *383*
olivacea, 221, *384*

Gecko
- Antillean, *494*
- Ashy, 396, 398, *495*

○ Banded, **392**, *491*
○ Big Bend, *490*
○ California Leaf-toed, *494*
○ Desert Banded, *491*
○ Florida Reef, *496*
○ Indo-Pacific, **401**, *492*
○ Leaf-toed, **391**, *494*
○ Mediterranean, **397**, *493*
○ Ocellated, *494*
○ Reef, **399**, **400**, *496*
○ San Diego Banded, *491*
○ Texas Banded, **393**, **394**, *490*
○ Tucson Banded, **395**, *491*
○ Utah Banded, *491*
○ Yellow-headed, **402**, *492*

Gecko Family, *489*

Gekkonidae, *489*

Gerrhonotus
coeruleus, **448**, *539*
c. coeruleus, *540*
c. palmeri, *540*
c. principis, *540*
c. shastensis, *540*
kingi, **446**, *541*
k. nobilis, *541*
liocephalus, *541*
l. infernalis, *541*
multicarinatus, **445**, **449**, *542*
m. multicarinatus, *542*
m. scincicauda, *542*
m. webbi, *542*
panamintinus, **447**, *543*

Giant Salamander
○ Cope's, *300*
○ Pacific, **23**, **41**, *300*

Giant Salamander Family, *269*

○ **Gila Monster**, **332**, *546*
○ Banded, *546*
○ Reticulate, *546*

Gila Monster Family, *546*

Glass Lizard
○ Eastern, **453**, **456**, *544*

○ Eastern Slender, *543*
○ Island, **454**, *544*
○ Slender, *543*
○ Western Slender, **455**, *543*

Gonatodes albogularis, **402**, *492*

Gopherus
agassizii, **328**, *471*
berlandieri, **329**, *472*
polyphemus, **330**, *473*

Graptemys
barbouri, **283**, *458*
caglei, **279**, *459*
flavimaculata, **276**, *460*
geographica, **280**, *461*
kohni, **278**, *462*
nigrinoda, **281**, *462*
n. delticola, *463*
n. nigrinoda, *463*
oculifera, **275**, *463*
o. ouachitensis (see
p. ouachitensis)
o. sabinensis (see
p. sabinensis)
pseudogeographica, *464*
p. kohni (see *kohni*)
p. ouachitensis, **274**, *464*
p. pseudogeographica, *464*
p. sabinensis, **277**, *464*
pulchra, **284**, *465*
versa, **282**, *466*

Green Snake
○ Eastern Smooth, *640*
○ Rough, **477**, *639*
○ Smooth, **475**, *640*
○ Western Smooth, **476**, *640*

Gyalopion
canum, **588**, *612*
quadrangularis, *612*

Gyrinophilus
palleucus, **6**, **8**, *329*
p. gulolineatus, *329*
p. necturoides, *329*
p. palleucus, *329*

porphyriticus, 329
p. danielsi, 101, 127, *330*
p. dunni, 330
p. duryi, 130, *330*
p. porphyriticus, 330
subterraneus, 330

H

Haideotriton wallacei, 7,
331

○ Hawksbill, 266, *478*
○ Atlantic, *478*
○ Pacific, *478*

○ Hellbender, 24, *269*
○ Allegheny, *269*
○ Ozark, *269*

Heloderma
suspectum, 332, *546*
s. cinctum, 547
s. suspectum, 547

Helodermatidae, *546*

Hemidactylium scutatum,
103, *331*

Hemidactylus
garnoti, 401, *492*
turcicus, 397, *493*

Heterodon
nasicus, 613
n. gloydi, 613
n. kennerlyi, 614
n. nasicus, 572, *613*
platyrhinos, 485, 563, 565,
615
simus, 585, *615*

Hognose Snake
○ Dusty, *613*
○ Eastern, 485, 563, 565,
614
○ Mexican, *613*
○ Plains, 572, *613*
○ Southern, 585, *615*
○ Western, *613*

Holbrookia
lacerata, 509

l. lacerata, 348, *509*
l. subcaudalis, 509
maculata, 510
m. approximans, 346, *510*
m. maculata, 366, *510*
m. perspicua, 510
m. ruthveni, 370, *510*
m. thermophila, 510
propinqua, 365, *511*
p. propinqua, 511

Hook-nosed Snake
○ Desert, *612*
○ Mexican, 547, *611*
○ Western, 588, *612*

Horned Lizard
○ California, 342, *514*
○ Coast, *514*
○ Desert, 339, *517*
○ Desert Short-, 338, *515*
○ Eastern Short-, *515*
○ Flat-tailed, 334, *516*
○ Mountain Short-, 337, *515*
○ Northern Desert, *517*
○ Pygmy Short-, *515*
○ Regal, 335, *518*
○ Round-tailed, 336, *516*
○ Salt Lake Short-, *515*
○ San Diego, *514*
○ Short-, *515*
○ Southern Desert, 341, *517*
○ Texas, 340, *513*

Horned Toad, *514*

Hydromantes
brunus, 97, *332*
platycephalus, 86, *333*
shastae, 87, *333*

Hyla
andersoni, 149, *402*
arenicolor, 159, *402*
avivoca, 156, *403*
a. avivoca, 403
a. ogechiensis, 403
cadaverina, 158, *404*
chrysoscelis, 160, *404*
cinerea, 146, *405*
crucifer, 173, *406*

c. bartramiana, 406
c. crucifer, 406
eximia, 150, 406
femoralis, 164, 407
gratiosa, 145, 151, 408
regilla, 148, 170, 182, 408
squirella, 147, 174, 409
versicolor, 152, 157, 404
wrightorum, 407

Hylactophryne
augusti, 154, 420
a. cactorum, 420
a. latrans, 420

Hylidae, 400

Hypopachus variolosus, 218, 385

Hypsiglena
torquata, 586, 616
t. deserticola, 616
t. jani, 616
t. klauberi, 616
t. loreala, 617
t. nuchalata, 617
t. ochrorhyncha, 617

I

Iguana
◯ Common, 388, 511
◯ Desert, 345, 507
◯ Spiny-tailed, 333, 506

Iguana iguana, 388, 511

Iguanidae, 497

Iguanid Family, 497

J

◯ **Jungle Runner,** 553
◯ Amazon, 553

K

Kingsnake
◯ Arizona Mountain, 598, 621

◯ Black, 483, 618
◯ California, 522, 592, 618
◯ California Mountain, 599, 625
◯ Coast Mountain, 625
◯ Common, 618
◯ Coral, 626
◯ Desert, 561, 618
◯ Eastern, 594, 618
◯ Florida, 590, 618
◯ Gray-banded, 601, 602, 603, 620
◯ Huachuca Mountain, 621
◯ Mexican, 618
◯ Prairie, 557, 569, 617
◯ St. Helena Mountain, 625
◯ San Bernardino Mountain, 625
◯ San Diego Mountain, 625
◯ Scarlet, 622
◯ Sierra Mountain, 625
◯ Sonora Mountain, 621
◯ Speckled, 560, 618
◯ Utah Mountain, 621

Kinosternidae, 438

Kinosternon
bauri, 317, 438
flavescens, 313, 439
f. arizonense, 440
f. flavescens, 440
f. spooneri, 440
hirtipes, 315, 440
h. murrayi, 440
sonoriense, 314, 441
subrubrum, 441
s. hippocrepis, 320, 441
s. steindachneri, 321, 442
s. subrubrum, 318, 442

L

Lacerta
viridis, 386, 566
v. viridis, 567

Lacertidae, 566

Lacertilia, 487

Lampropeltis
calligaster, 557, 569, *617*
c. calligaster, *618*
c. rhrombomaculata, *618*
getulus, *618*
g. californiae, 522, 592, *619*
g. floridana, 590, *619*
g. getulus, 594, *619*
g. holbrooki, 560, *619*
g. niger, 483, *619*
g. nigritus, *620*
g. splendida, 561, *620*
mexicana, 601, 602, 603, *620*
m. alterna, *620*
pyromelana, *621*
p. infralabialis, *621*
p. pyromelana, 598, *621*
p. woodini, *621*
triangulum, *622*
t. amaura, 614, *623*
t. annulata, *623*
t. celaenops, 613, *623*
t. elapsoides, *623*
t. gentilis, 600, *623*
t. multistrata, *623*
t. syspila, 615, *624*
t. taylori, *624*
t. triangulum, 597, *622*
zonata, 599, *625*
z. multicincta, *625*
z. multifasciata, *625*
z. parviruba, *626*
z. pulchra, *626*
z. zonata, *625*

Leaf-nosed Snake
○ Clouded, *643*
○ Maricopa, *642*
○ Pima, **589**, *642*
○ Saddle, *642*
○ Spotted, *643*
○ Western, **571**, **583**, *643*

○ Leatherback, 263, *481*
○ Atlantic, *481*
○ Pacific, *481*

Leatherback Turtle Family, *481*

Legless Lizard
○ Black, *548*
○ California, 452, *548*
○ Silvery, *548*

Leiocephalus
carinatus, 364, *512*
c. armouri, *512*

Leopard Frog
○ Las Vegas, *374*
○ Northern, 192, 203, *377*
○ Plains, 197, 202, *370*
○ Relict, *376*
○ Rio Grande, 196, *370*
○ Southern, 191, 204, *379*

○ Leopard Lizard, 357, *508*
○ Blunt-nosed, 347, *507*
○ Cope's, *508*
○ Lahontan Basin, *508*
○ Long-nosed, *508*
○ Pale, *508*

Lepidochelys
kempi, 262, *479*
olivacea, 264, *480*

Leptodactylidae, *418*

Leptodactylid Frog Family, *418*

Leptodactylus labialis, 166, *420*

Leptodeira
septentrionalis, *626*
s. septentrionalis, 606, *626*

Leptotyphlopidae, *583*

Leptotyphlops
dulcis, *583*
d. dissectus, 464, *584*
d. dulcis, *584*
humilis, 457, *584*
h. cahuilae, *585*
h. humilis, *585*
h. segregus, *585*
h. utahensis, *585*

Leurognathus marmoratus,
135, 334

Lichanura
trivirgata, 525, 587
t. gracia, 508, 588
t. roseofusca, 588
t. trivirgata, 588

Limnaoedus ocularis, 172,
410

Lizard
○ Arizona Alligator, 446,
541
○ Arizona Long-tailed Brush,
534
○ Arizona Night, 403, 551
○ Arizona Zebra-tailed, 502
○ Baja Collared, 504
○ Banded Rock, 358, 513
○ Barred Spiny, 523
○ Big Bend Canyon, 524
○ Big Bend Tree, 535
○ Black Legless, 548
○ Bleached Earless, 370, 510
○ Blue Spiny, 352, 520
○ Blunt-nosed Leopard, 347,
507
○ Bunch Grass, 367, 378,
528
○ California Alligator, 542
○ California Horned, 342,
514
○ California Legless, 452,
548
○ California Side-blotched,
537
○ Canyon, 368, 524
○ Canyon Tree, 535
○ Chihuahuan Collared, 503
○ Clark's Spiny, 349, 372,
519
○ Coachella Valley Fringe-
toed, 532
○ Coast Horned, 514
○ Coast Range Fence, 525
○ Collared, 355, 356, 503
○ Colorado River Tree, 535

○ Colorado Side-blotched,
537
○ Common Tree, 369, 535
○ Cope's Leopard, 508
○ Curly-tailed, 364, 512
○ Crevice Spiny, 354, 527
○ Desert Collared, 359, 504
○ Desert Fringe-toed, 343,
532
○ Desert Horned, 339, 517
○ Desert Night, 406, 551
○ Desert Short-horned, 338,
515
○ Desert Side-blotched, 537
○ Desert Spiny, 350, 523
○ Dunes Sagebrush, 521
○ Eastern Collared, 503
○ Eastern Earless, 510
○ Eastern Fence, 529
○ Eastern Glass, 453, 456,
544
○ Eastern Short-horned, 515
○ Eastern Slender Glass, 543
○ Eastern Tree, 535
○ Flat-tailed Horned, 334,
516
○ Florida Scrub, 376, 531
○ Fringe-toed, 532
○ Granite Night, 404, 405,
407, 550
○ Granite Spiny, 351, 527
○ Great Basin Fence, 525
○ Greater Earless, 361, 503
○ Green, 386, 566
○ Island Fence, 525
○ Island Glass, 454, 544
○ Island Night, 408, 551
○ Keeled Earless, 365, 511
○ Lahontan Basin Leopard,
508
○ Leopard, 357, 508
○ Lesser Earless, 510
○ Lined Tree, 535
○ Long-nosed Leopard, 508
○ Long-tailed Brush, 380,
534
○ Merriam's Canyon, 524
○ Mesquite, 371, 522
○ Mojave Collared, 504

○ Mojave Fringe-toed, **344**, *533*
○ Mojave Zebra-tailed, *502*
○ Mountain Short-horned, **337**, *515*
○ Nevada Side-blotched, *537*
○ Nevada Zebra-tailed, *502*
○ Northern Alligator, **448**, *539*
○ Northern Desert Horned, *517*
○ Northern Curly-tailed, *512*
○ Northern Earless, **366**, *510*
○ Northern Fence, *529*
○ Northern Mesquite, *522*
○ Northern Plateau, *529*
○ Northern Prairie, **375**, *529*
○ Northern Ruin, *567*
○ Northern Sagebrush, **377**, *521*
○ Northern Side-blotched, *537*
○ Northern Spot-tailed Earless, **348**, *509*
○ Northern Tree, *535*
○ Northwestern Fence, *525*
○ Orange-headed Spiny, *523*
○ Oregon Alligator, *542*
○ Pale Leopard, *508*
○ Panamint Alligator, **447**, *543*
○ Plateau Spiny, *519*
○ Presidio Canyon, *524*
○ Pygmy Short-horned, *515*
○ Red-lipped Prairie, *529*
○ Regal Horned, **335**, *518*
○ Reticulate Collared, **360**, *505*
○ Rose-bellied, *530*
○ Round-tailed Horned, **336**, *516*
○ Ruin, **387**, *567*
○ Sagebrush, *521*
○ Salt Lake Short-horned, *515*
○ San Diego Alligator, *542*
○ San Diego Horned, *514*

○ San Francisco Alligator, *539*
○ San Joaquin Fence, *525*
○ Shasta Alligator, *539*
○ Short-horned, *515*
○ Side-blotched, **363**, *537*
○ Sierra Alligator, *539*
○ Sierra Fence, *525*
○ Sierra Night, *551*
○ Silvery Legless, *548*
○ Slender Glass, *543*
○ Small-scaled Tree, **382**, *535*
○ Sonoran Collared, *503*
○ Sonoran Fringe-toed, *532*
○ Sonoran Spiny, *519*
○ Southern Alligator, **445**, **449**, *542*
○ Southern Desert Horned, **341**, *517*
○ Southern Fence, *529*
○ Southern Plateau, *529*
○ Southern Prairie, *529*
○ Southern Ruin, *567*
○ Southern Sagebrush, *521*
○ Southern Spot-tailed Earless, *509*
○ Southwestern Earless, *503*
○ Speckled Earless, **346**, *510*
○ Spot-tailed Earless, *509*
○ Striped Plateau, **374**, *531*
○ Texas Alligator, *541*
○ Texas Earless, *503*
○ Texas Horned, **340**, *513*
○ Texas Keeled Earless, *511*
○ Texas Rose-bellied, **373**, *530*
○ Texas Spiny, **381**, *526*
○ Twin-spotted Spiny, *523*
○ Utah Night, *551*
○ Western Collared, *503*
○ Western Earless, *510*
○ Western Fence, **379**, *525*
○ Western Long-tailed Brush, *534*
○ Western Slender Glass, **455**, *543*
○ White Sands Prairie, *529*
○ Worm, **451**, *580*

○ Yarrow's Spiny, 353, *522*
○ Yellow-backed Spiny, *523*
○ Yellow-headed Collared, *503*
○ Zebra-tailed, 362, *502*

Lizard Family
Anguid, *539*
California Legless, *548*
Gecko, *489*
Iguanid, *497*
Night, *550*
Skink, *568*
Typical Old World, *566*
Whiptail, *553*

Lizards, *487*

○ **Loggerhead**, 265, *475*
○ Atlantic, *475*
○ Pacific, *475*

Lungless Salamander
Family, *302*

M

Macroclemys temmincki,
325, 326, 327, *436*

Malaclemys
terrapin, 466
t. centrata, 467
t. littoralis, 467
t. macrospilota, 298, *467*
t. pileata, 467
t. tequesta, 300, *467*
t. terrapin, 299, *467*
t. rhizophorarum, 467

○ **Map Turtle**, 280, *461*
○ Alabama, 284, *465*
○ Barbour's, 283, *458*
○ Cagle's, 279, *459*
○ False, *464*
○ Mississippi, 278, *462*
○ Ouachita, 274, *464*
○ Sabine, 277, *464*
○ Texas, 282, *466·*

Marsh Turtles, *446*

○ **Massasauga**, *696*
○ Desert, 632, *696*

○ Eastern, 633, *696*
○ Western, 638, *696*

Masticophis
bilineatus, 517, *627*
b. bilineatus, 627
b. lineolatus, 627
flagellum, 628
f. cingulum, 628
f. flagellum, 469, *628*
f. fulginosus, 628
f. lineatulus, 558, *629*
f. piceus, 491, 553, *629*
f. ruddocki, 629
f. testaceus, 554, 556, *629*
lateralis, 630
l. euryxanthus, 630
l. lateralis, 518, *630*
taeniatus, 521, *631*
t. ornatus, 631
t. ruthveni, 631
t. schotti, 631
t. taeniatus, 631

Microhylidae, 383

Micruroides
euryxanthus, 616, *680*
e. euryxanthus, 680

Micrurus
fulvius, 618, *681*
f. fulvius, 681
f. tenere, 617, *681*

Mole Salamander Family,
288

Mudpuppies, *281*

○ **Mudpuppy**, 16, 20, *283*
○ Lake Winnebago, *283*

○ **Mud Turtle**, *441*
○ Eastern, 318, *441*
○ Florida, 321, *441*
○ Illinois, *439*
○ Mexican, 315, *440*
○ Mississippi, 320, *441*
○ Sonora, 314, *441*
○ Southwestern, *439*
○ Striped, 317, *438*
○ Yellow, 313, *439*

Mud Turtles, *438*

Musk Turtle, *441*
○ Flattened, 316, *443*
○ Loggerhead, 311, *444*
○ Razor-backed, 310, *443*
○ Stripe-necked, 312, *444*

Musk Turtles, *438*

N

Narrow-mouthed Frog
○ Eastern, 219, *383*
○ Great Plains, 221, *384*

Narrow-mouthed Frog
Family, *383*

Natrix (see *Nerodia*)

Necturus
alabamensis, 281
beyeri, 282
lewisi, 21, 282
maculosus, 16, 20, 283
m. louisianensis, 283
m. maculosus, 283
m. stictus, 283
punctatus, 19, 284

Neoseps reynoldsi, 450, 578

Nerodia
cyclopion, 632
c. cyclopion, 632
c. floridana, 482, 632
erythrogaster, 633
e. erythrogaster, 490, 633
e. flavigaster, 481, 633
e. neglecta, 634
e. transversa, 634
fasciata, 634
f. clarki, 513, 635
f. compressicauda, 579, 635
f. confluens, 562, 635
f. fasciata, 552, 635
f. pictiventris, 635
f. taeniata, 578, 635
harteri, 636
h. harteri, 549, 636
h. paucimaculata, 636

rhombifera, 574, 636
r. rhombifera, 637
sipedon, 580, 637
s. insularum, 638
s. pleuralis, 638
s. sipedon, 638
s. williamengelsi, 638
taxispilota, 567, 639

Newt
○ Black-spotted, 28, *275*
○ Broken-striped, 27, 30, *276*
○ California, 32, 33, *280*
○ Central, *276*
○ Coast Range, *280*
○ Crater Lake, *278*
○ Eastern, *276*
○ Northern Rough-skinned, *278*
○ Peninsula, *276*
○ Red-bellied, 34, 36, *279*
○ Red-spotted, 26, 29, *276*
○ Rough-skinned, 31, 35, *278*
○ Sierra, *280*
○ Striped, 25, *276*

Newt Family, *275*

Night Lizard
○ Arizona, 403, *551*
○ Desert, 406, *551*
○ Granite, 404, 405, 407, *550*
○ Island, 408, *551*
○ Sierra, *551*
○ Utah, *551*

Night Lizard Family, *550*

Notophthalmus
meridionalis, 28, 275
m. meridionalis, 276
perstriatus, 25, 276
viridescens, 276
v. dorsalis, 27, 30, 277
v. louisianensis, 277
v. piaropicola, 277
v. viridescens, 26, 29, 277

O

Opheodrys
aestivus, 477, 639
vernalis, 475, 640
v. blanchardi, 476, 641
v. vernalis, 641

Ophisaurus
attenuatus, 543
a. attenuatus, 455, 543
a. longicaudus, 544
compressus, 454, 544
ventralis, 453, 456, 545

Osteopilus septentrionalis,
155, 178, 410

Oxybelis aeneus, 641

P

Patch-nosed Snake
○ Big Bend, 514, 650
○ Coast, 651
○ Desert, 527, 651
○ Mojave, 651
○ Mountain, 650
○ Texas, 516, 650
○ Western, 651

Peeper
○ Northern Spring, 406
○ Southern Spring, 406
○ Spring, 173, 406

Pelobatidae, 362

Petrosaurus mearnsi, 358,
513

Phaeognathus hubrichti, 68,
335

Phrynosoma
cornutum, 340, 513
coronatum, 514
c. blainvillei, 514
c. frontale, 342, 514
douglassi, 515
d. brevirostre, 515
d. douglassi, 515
d. hernandesi, 337, 515
d. ornatissimum, 338, 515

d. ornatum, 515
m'calli, 334, 516
modestum, 336, 516
platyrhinos, 339, 517
p. calidiarum, 341, 517
p. platyrhinos, 517
solare, 335, 518

Phyllodactylus
xanti, 391, 494
x. nocticolus, 494

Phyllorhynchus
browni, 642
b. browni, 589, 642
b. lucidus, 642
decurtatus, 643
d. nubilis, 643
d. perkinsi, 571, 583, 643

Pilot, Rattlesnake, 607

Pipidae, 423

Pituophis
melanoleucus, 644
m. affinis, 644
m. annectans, 644
m. catenifer, 537, 645
m. deserticola, 575, 645
m. lodingi, 488, 644
m. melanoleucus, 591, 644
m. mugitus, 645
m. pumilis, 645
m. ruthveni, 645
m. sayi, 573, 645

Pit Viper Family, 682

Plethodon
caddoensis, 144, 335
cinereus, 71, 117, 336
dorsalis, 96, 337
d. angusticlavius, 337
d. dorsalis, 337
dunni, 110, 338
elongatus, 67, 338
fourchensis, 339
glutinosus, 140, 141, 340
g. albagula, 340
g. glutinosus, 340
hoffmani, 70, 341

jordani, 133, 136, 138, 341
larselli, 84, 342
longicrus, 143, 343
neomexicanus, 69, 344
nettingi, 344
n. hubrichti, 345
n. nettingi, 102, 345
ouachitae, 115, 345
punctatus, 139, 346
richmondi, 72, 346
serratus, 118, 347
shenandoah, 348
stormi, 79, 348
vandykei, 349
v. idahoensis, 111, 349
v. vandykei, 349
vehiculum, 119, 349
websteri, 338
wehrlei, 80, 350
welleri, 82, 351
yonahlossee, 116, 352

Plethodontidae, 302

Podarcis
sicula, 387, 567
s. campestris, 567
s. sicula, 567

Pond Turtles, 446

Proteidae, 281

Pseudacris
brachyphona, 163, 411
brimleyi, 175, 412
clarki, 180, 412
nigrita, 183, 413
n. nigrita, 413
n. verrucosa, 413
ornata, 176, 177, 413
streckeri, 414
s. illinoensis, 185, 414
s. streckeri, 414
triseriata, 415
t. feriarum, 415
t. kalmi, 415
t. maculata, 415
t. triseriata, 179, 415

Pseudemys (see *Chrysemys*)

Pseudobranchus
striatus, 17, 271
s. axanthus, 272
s. belli, 272
s. lustricolus, 272
s. spheniscus, 271
s. striatus, 271

Pseudotriton
montanus, 352
m. diastictus, 131, 353
m. flavissimus, 98, 99, 128, 353
m. floridanus, 353
m. montanus, 353
ruber, 353
r. nitidus, 354
r. ruber, 129, 354
r. schencki, 132, 354
r. vioscai, 354

Pternohyla fodiens, 186, 416

Python Family, 586

R

◯ **Racer,** 596
◯ Alameda Striped, 630
◯ Brown-chinned, 596
◯ Black-masked, 596
◯ Blue, 596
◯ Buttermilk, 596
◯ California Striped, 518, 630
◯ Eastern Yellow-bellied, 480, 596
◯ Everglades, 596
◯ Mexican, 596
◯ Northern Black, 486, 596
◯ Southern Black, 468, 596
◯ Speckled, 559, 603
◯ Striped, 630
◯ Tan, 596
◯ Western Yellow-bellied, 478, 596

◯ **Racerunner,** 561
◯ Prairie Lined, 561

○ Six-lined, 411, *561*

Racerunners, *553*

Rana
areolata, 367, *368*
a. aesopus, 194, *368*
a. areolata, 198, *368*
a. capito, *368*
a. circulosa, *368*
a. sevosa, 195, *368*
aurora, 215, *369*
a. aurora, 208, *369*
a. draytoni, *369*
berlandieri, 196, *370*
b. berlandieri, *370*
blairi, 197, 202, *370*
boylei, 209, *371*
capito, *368*
cascadae, 199, *371*
catesbeiana, 187, 190, *372*
clamitans, 189, *373*
c. clamitans, 213, *373*
c. melanota, *373*
fisheri, *374*
grylio, 188, *374*
heckscheri, 210, *375*
muscosa, 207, *376*
onca, *376*
palustris, 201, *377*
pipiens, 192, 203, *377*
pretiosa, 206, *378*
pustulosa, *381*
septentrionalis, 200, *379*
sphenocephala, 191, 204, *379*
sylvatica, 211, 212, 214, 216, *380*
tarahumarae, 193, *381*
virgatipes, 205, *381*

Ranidae, *367*

○ **Rat Snake**, 484, *605*
○ Baird's **509**, *605*
○ Black, *605*
○ Everglades, *605*
○ Gray, 524, 581, *605*
○ Green, 479, *608*
○ Texas, *605*
○ Trans-Pecos, 523, *607*

○ Yellow, **526**, 540, *605*

Rattlesnake
○ Animas Ridge-nosed, *695*
○ Arizona Black, 621, *694*
○ Arizona Ridge-nosed, **643**, *695*
○ Banded Rock, 640, *689*
○ Black-tailed, 626, *690*
○ Canebrake, *689*
○ Carolina Pigmy, 641, *697*
○ Eastern Diamondback, 624, *685*
○ Dusky Pigmy, 642, *697*
○ Grand Canyon, 648, *694*
○ Great Basin, 623, *694*
○ Hopi, 631, *694*
○ Midget Faded, *694*
○ Mojave, 622, *692*
○ Mottled Rock, 636, *689*
○ Northern Black-tailed, *690*
○ Northern Pacific, 630, *694*
○ Panamint Speckled, 646, *690*
○ Pigmy, 645, *697*
○ Prairie, 629, *694*
○ Red Diamond, **644**, *692*
○ Ridge-nosed, *695*
○ Rock, *689*
○ Southern Pacific, 627, *694*
○ Southern Speckled, 635, *690*
○ Speckled, *690*
○ Tiger, 628, *693*
○ Timber, 619, 620, 653, *688*
○ Twin-spotted, 637, *691*
○ Western, *694*
○ Western Diamondback, 639, *686*
○ Western Pigmy, 625, *697*

○ **Red-bellied Turtle**, 295, *451*
○ Alabama, 292, **446**
○ Florida, 296, *449*
○ Plymouth, *451*

Regina
alleni, 510, *646*
grahami, 519, *646*

rigida, 474, 647
  r. deltae, 647
  r. rigida, 647
  r. sinicola, 647
septemvittata, 503, 648

Reptiles, *425*

Reptilia, *425*

Rhadinaea flavilata, 462, 465, 648

Rhineura floridana, 451, 580

Rhinocheilus
lecontei, 609, 649
  l. lecontei, 593, 649
  l. tessellatus, 649

Rhinophrynidae, *361*

Rhinophrynus dorsalis, 217, 220, *361*

Rhyacotriton
olympicus, 100, *301*
  o. olympicus, *301*
  o. variegatus, *301*

Ribbon Snake
○ Arid Land, *670*
○ Blue-striped, 520, *672*
○ Eastern, 532, *672*
○ Gulf Coast, *670*
○ Northern, *672*
○ Peninsula, *672*
○ Red-striped, 544, *670*
○ Western, 531, *670*

Ridley
○ Atlantic, 262, *479*
○ Olive, 264, *480*

Runner
○ Amazon, *553*
○ Jungle, *553*

S

Salamander
○ Alabama, *311*
○ Appalachian Seal, *315*
○ Appalachian Woodland, 133, 136, 138, *341*
○ Arboreal, 106, *305*
○ Arizona Tiger, *298*
○ Barred Tiger, 46, 48, *298*
○ Berry Cave, *329*
○ Big Mouth, *329*
○ Black, *304*
○ Black-bellied, 64, *316*
○ Black-bellied Slender, *307*
○ Black-chinned Red, 132, *353*
○ Black Mountain Dusky, 92, *317*
○ Blotched Tiger, 39, *298*
○ Blue Ridge Red, *353*
○ Blue Ridge Spring, 101, 127, *329*
○ Blue Ridge Two-lined, 121, *320*
○ Blue-spotted, 58, *292*
○ British Columbia, *290*
○ Brown, 57, *290*
○ Buck Knob, *339*
○ Caddo Mountain, 144, *335*
○ California Slender, 76, *307*
○ California Tiger, 43, *289*
○ Carolina Spring, *329*
○ Cascade Cavern, *322*
○ Cave, 122, *324*
○ Channel Islands Slender, 73, *308*
○ Cheat Mountain, 102, *344*
  Cherokee, *311*
○ Clouded, 83, *303*
○ Coeur d' Alene, 111, *349*
○ Comal Blind, 2, 11, *327*
○ Cope's Giant, *300*
○ Crevice, 143, *343*
○ Dark-sided, 113, *323*
○ Del Norte, 67, *338*
○ Desert Slender, *306*
○ Dunn's, 110, *338*
○ Dusky, *313*
○ Dwarf, 109, 126, *326*
○ Eastern Long-toed, 52, *293*
○ Eastern Mud, *352*
○ Eastern Tiger, 37, *298*

○ Eastern Zigzag, *337*
○ Flatwoods, 42, *289*
○ Four-toed, 103, *331*
○ Fourche Mountain, *339*
○ Garden Slender, 75, *307*
○ Georgia Blind, 7, *331*
○ Gray-bellied, *324*
○ Gray Tiger, *298*
○ Green, 85, *302*
○ Grotto, 12, *356*
○ Gulf Coast Mud, 98, 99, 128, *352*
○ Imitator, 134, *314*
○ Jefferson, 56, *291*
○ Jemez Mountains, 69, *344*
○ Junaluska, 125, *322*
○ Kentucky Spring, 130, *329*
○ Kern Canyon Slender, *309*
○ Larch Mountain, 84, *342*
○ Limestone, 97, *332*
○ Long-tailed, 123, *323*
○ Long-toed, *293*
○ Mabee's, 60, *292*
○ Many-lined, 90, *354*
○ Many-ribbed, 93, *324*
○ Marbled, 44, 45, *295*
○ Midland Mud, 131, *352*
○ Mole, 61, *296*
○ Mount Lyell, 86, *333*
○ Mountain Dusky, 95, 112, 120, 124, 137, *316*
○ Mud, *352*
○ Netting's, *344*
○ Northern Dusky, 89, *313*
○ Northern Long-toed, *293*
○ Northern Olympic, *301*
○ Northern Red, 129, *353*
○ Northern Spring, *329*
○ Northern Two-lined, 88, *320*
○ Northwestern, *290*
○ Oklahoma, 10, *328*
○ Olympic, 100, *301*
○ Oregon Slender, 78, *310*
○ Ouachita Dusky, 91, *312*
○ Ozark Zigzag, *337*
○ Pacific Giant, 23, 41, *300*
○ Peaks of Otter, *344*

○ Pygmy, 66, *318*
○ Ravine, 72, *346*
○ Red, *353*
○ Red-backed, 71, 117, *336*
○ Red Hills, 68, *335*
○ Relictual Slender, 77, *309*
○ Rich Mountain, 115, *345*
○ Ringed, 47, *288*
○ Rusty Mud, *352*
○ Sacramento Mountain, 81, *305*
○ San Marcos, 5, *325*
○ Santa Cruz Black, *304*
○ Santa Cruz Long-toed, 50, *293*
○ Seal, 65, *315*
○ Seepage, 94, *311*
○ Shasta, 87, *333*
○ Shenandoah, *348*
○ Shovel-nosed, 135, *334*
○ Silvery, 59, *296*
○ Sinking Cove, *329*
○ Siskiyou Mountain, 79, *348*
○ Slimy, 140, 141, *340*
○ Small-mouthed, 62, *297*
○ Sonora Tiger, *298*
○ Southern Dusky, 63, *312*
○ Southern Long-toed, 53, *293*
○ Southern Olympic, *301*
○ Southern Red, *353*
○ Southern Red-backed, 118, *347*
○ Southern Two-lined, *320*
○ Speckled Black, 142, *304*
○ Spotted, 51, 54, *294*
○ Spotted Dusky, *313*
○ Spring, *329*
○ Tehachapi Slender, 74, *310*
○ Tennessee Cave, 6, 8, *329*
○ Texas, 4, *326*
○ Texas Blind, 3, 9, *355*
○ Three-lined, 114, *323*
○ Tiger, 22, 38, 40, *298*
○ Tremblay's, 55, *299*
○ Two-lined, *320*
○ Valdina Farms, 1, *328*

○ Valley and Ridge, 70, *341*
○ Van Dyke's, *349*
○ Virginia Seal, *315*
○ Washington, *349*
○ Wehrle's, **80**, *350*
○ Weller's, **82**, *351*
○ Western Long-toed, 49, *293*
○ Western Red-backed, 119, *349*
○ West Virginia Spring, *330*
○ White-spotted, 139, *346*
○ White-throated, *340*
○ Yonahlossee, 116, *352*
○ Zigzag, 96, *337*

Salamander Family
Amphiuma, *285*
Giant, *269*
Lungless, *302*
Mole, *288*
Mudpuppy, *281*
Newt, *275*
Siren, *271*

Salamanders, *267*

Salamandridae, *275*

Salientia, *357*

Salvadora
deserticola, **514**, *650*
grahamiae, *650*
g. grahamiae, *651*
g. lineata, **516**, *651*
hexalepis, *651*
h. hexalepis, **527**, *652*
h. mojavensis, *652*
h. virgultea, *652*

Sauromalus
obesus, **331**, *518*
o. multiforaminatus, *519*
o. obesus, *519*
o. tumidus, *519*

Sawback
○ Black-knobbed, **281**, *462*
○ Ringed, **275**, *463*
○ Southern Black-knobbed, *462*
○ Yellow-blotched, **276**, *460*

Scaphiopus
bombifrons, **231**, *362*
couchi, **252**, *363*
hammondi, 229, **253**, *364*
h. hammondi, *364*
h. multiplicata, **230**, *364*
holbrooki, **233**, *365*
h. holbrooki, *366*
h. hurteri, *365*
intermontanus, **232**, *366*

Sceloporus
clarki, *349*, **372**, *519*
c. clarki, *520*
c. villaris, *520*
cyanogenys, *352*, *520*
graciosus, *521*
g. arenicolus, *521*
g. graciosus, **377**, *521*
g. vandenburghianus, *521*
grammicus, **371**, *522*
g. disparilis, *522*
jarrovi, *353*, *522*
j. jarrovi, *523*
magister, *350*, *523*
m. bimaculosus, *523*
m. cephaloflavus, *524*
m. magister, *523*
m. transversus, *523*
m. uniformis, *524*
merriami, **368**, *524*
m. annulatus, *525*
m. longipunctatus, *525*
m. merriami, *524*
occidentalis, **379**, *525*
o. becki, *526*
o. biseriatus, *526*
o. bocourti, *526*
o. longipes, *526*
o. occidentalis, *525*
o. taylori, *526*
olivaceus, **381**, *526*
orcutti, *351*, *527*
o. orcutti, *527*
poinsetti, *354*, *527*
p. poinsetti, *528*
scalaris, **367**, **378**, *528*
s. slevini, *528*

undulatus, 529
u. consobrinus, 529
u. cowlesi, 529
u. elongatus, 529
u. erythrocheilus, 529
u. garmani, **375,** *530*
u. hyacinthinus, 530
u. tristichus, 530
u. undulatus, 529
variabilis, 530
v. marmoratus, **373,** *531*
virgatus, **374,** *531*
woodi, **376,** *531*

Scincella lateralis, **433,** *578*

Scincidae, *568*

Sea Turtle Family, *475*

Seminatrix
pygaea, **487,** *652*
p. cyclas, 652
p. paludis, 653
p. pygaea, **494,** *652*

Serpentes, *581*

Shovel-nosed Snake
○ Colorado Desert, *594*
○ Mojave, *594*
○ Nevada, *594*
○ Organ Pipe, **610,** *595*
○ Sonoran, *595*
○ Tucson, **612,** *594*
○ Western, **604,** *594*

○ Sidewinder, *687*
○ Colorado Desert, **634,** *687*
○ Mojave Desert, *687*
○ Sonora, **647,** *687*

Siren
○ Broad-striped Dwarf, *271*
○ Dwarf, **17,** *271*
○ Eastern Lesser, *272*
○ Everglades Dwarf, *271*
○ Greater, **14,** *273*
○ Gulf Hammock Dwarf, *271*
○ Lesser, **13,** *272*
○ Narrow-striped Dwarf, *271*

○ Rio Grande Lesser, *272*
○ Slender Dwarf, *271*
○ Western Lesser, *272*

Siren
intermedia, **13,** *272*
i. intermedia, 273
i. nettingi, 273
i. texana, 273
lacertina, **14,** *273*

Siren Family, *271*

Sirenidae, *271*

Sistrurus
catenatus, 696
c. catenatus, **633,** *697*
c. edwardsi, **632,** *697*
c. tergeminus, **638,** *697*
miliarius, **645,** *697*
m. babouri, **642,** *698*
m. miliarius, **641,** *698*
m. streckeri, **625,** *698*

Skink
○ Arizona, **434,** *571*
○ Blue-tailed Mole, *438,* **440,** *569*
○ Broad-headed, **424, 431,** *573*
○ Cedar Key Mole, *569*
○ Coal, **425,** *568*
○ Coronado Island, *576*
○ Five-lined, **427, 437, 443,** *570*
○ Florida Keys Mole, **435,** *569*
○ Four-lined, *577*
○ Gilbert's, **430,** *571*
○ Great Basin, *576*
○ Great Plains, **432,** *575*
○ Greater Brown, *571*
○ Ground, **433,** *578*
○ Many-lined, **422,** *574*
○ Mole, **436,** *569*
○ Mountain, **439,** *577*
○ Northern Brown, *571*
○ Northern Coal, *568*
○ Northern Mole, *569*
○ Northern Prairie, **423,** *575*

○ Peninsula Mole, 569
○ Prairie, 575
○ Sand, **450**, 578
○ Short-lined, 577
○ Southeastern Five-lined, **426**, **444**, 572
○ Southern Coal, **429**, 568
○ Southern Prairie, **428**, 575
○ Variable, 574
○ Variegated, 571
○ Western, **421**, **441**, **442**, 576
○ Western Red-tailed, 571

Skink Family, 568

Slender Blind Snake Family, 583

Slender Salamander
○ Black-bellied, 307
○ California, **76**, 307
○ Channel Islands, **73**, 308
○ Desert, 306
○ Garden, **75**, 307
○ Kern Canyon, 309
○ Oregon, **78**, 310
○ Relictual, **77**, 309
○ Tehachapi, **74**, *310*

Slider
○ Missouri, 448
○ Pond, 452

Smilisca baudini, **181**, **184**, 416

Snake
○ Aquatic Garter, 664
○ Arid Land Ribbon, 670
○ Arizona Coral, **616**, 680
○ Arizona Glossy, 590
○ Atlantic Salt Marsh, **578**, 634
○ Baird's Rat, **509**, 605
○ Banded Sand, **605**, 593
○ Banded Water, **552**, 634
○ Big Bend Black-headed, 661
○ Big Bend Patch-nosed, **514**, 650

○ Black Hills Red-bellied, 655
○ Blackhood, 661
○ Black-necked Garter, **536**, 666
○ Black Pine, **488**, 644
○ Black Rat, 605
○ Black Striped, 599
○ Blotched Water, 633
○ Blue-striped Garter, **538**, 674
○ Blue-striped Ribbon, **520**, 672
○ Brazos Water, **549**, 636
○ Broad-banded Water, **562**, 634
○ Brown, **550**, 654
○ Brown Water, **567**, 639
○ Butler's Garter, **529**, 664
○ California Black-headed, **500**, 659
○ California Glossy, 590
○ California Lyre, 676
○ California Night, 616
○ California Red-sided Garter, **545**, 674
○ Carolina Salt Marsh, 637
○ Cat-eyed, 626
○ Central Florida Crowned, **461**, 660
○ Central Lined, 677
○ Central Plains Milk, **600**, 622
○ Checkered Garter, **515**, 669
○ Chicago Garter, 674
○ Chihuahuan Black-headed, 662
○ Clouded Leaf-nosed, 643
○ Coastal Dunes Crowned, 660
○ Coast Garter, 667
○ Coast Patch-nosed, 651
○ Coast Plains Milk, 624
○ Colorado Desert Shovel-nosed, 594
○ Common Garter, **530**, 674
○ Concho Water, 636
○ Copper-bellied Water, 633

○ Coral-bellied Ringneck, 600
○ Corn, 570, 608, 604
○ Delta Crayfish, 647
○ Desert Black-headed, 659
○ Desert Blind, 584
○ Desert Glossy, 590
○ Desert Hook-nosed, 612
○ Desert Night, 616
○ Desert Patch-nosed, 527, 651
○ Devils River Black-headed, 661
○ Diamondback Water, 574, 636
○ Dusty Hognose, 613
○ Eastern Black-necked Garter, 666
○ Eastern Coral, 618, 681
○ Eastern Earth, 679
○ Eastern Fox, 564, 608
○ Eastern Garter, 539, 576, 674
○ Eastern Hognose, 485, 563, 565, 614
○ Eastern Indigo, 489, 602
○ Eastern Milk, 597, 622
○ Eastern Mud, 492, 609
○ Eastern Plains Garter, 671
○ Eastern Ribbon, 532, 672
○ Eastern Smooth Green, 640
○ Eastern Worm, 591
○ Everglades Rat, 605
○ Flat-headed, 463, 658
○ Florida Brown, 654
○ Florida Crowned, 660
○ Florida Green Water, 482, 632
○ Florida Pine, 644
○ Florida Red-bellied, 655
○ Florida Scarlet, 592
○ Florida Water, 634
○ Fox, 608
○ Giant Garter, 664
○ Glossy, 566, 587, 590
○ Glossy Crayfish Water, 474, 647

○ Graham's Crayfish, 519, 646
○ Gray Rat, 524, 581, 605
○ Great Basin Gopher, 575, 644
○ Great Plains Rat, 604
○ Green Rat, 479, 608
○ Green Water, 632
○ Ground, 459, 499, 502, 555, 611, 653
○ Gulf Salt Marsh, 513, 634
○ Gulf Coast Ribbon, 670
○ Gulf Crayfish, 647
○ Harter's Water, 636
○ Huachuca Black-headed, 662
○ Indigo, 602
○ Kansas Glossy, 577, 590
○ Key Ringneck, 600
○ Kirtland's, 551, 596
○ Klamath Garter, 667
○ Lake Erie Water, 637
○ Lined, 507, 677
○ Long-nosed, 609, 649
○ Louisiana Milk, 614, 622
○ Louisiana Pine, 644
○ Lyre, 676
○ Mangrove Water, 579, 634
○ Maricopa Leaf-nosed, 642
○ Maritime Garter, 674
○ Marsh Brown, 654
○ Mesa Verde Night, 616
○ Mexican Black-headed, 656
○ Mexican Garter, 528, 668
○ Mexican Hognose, 613
○ Mexican Hook-nosed, 547, 611
○ Mexican Milk, 622
○ Mexican Vine, 641
○ Midland Brown, 654
○ Midland Water, 637
○ Midwest Worm, 591
○ Milk, 622
○ Mississippi Ringneck, 600
○ Mojave Glossy, 590
○ Mojave Patch-nosed, 651
○ Mojave Shovel-nosed, 594

736 *Index*

- Mole, *617*
- Monterey Ringneck, *600*
- Mountain Earth, *679*
- Mountain Garter, *667*
- Mountain Patch-nosed, *650*
- Mud, *609*
- Narrow-headed Garter, *548, 672*
- Nevada Shovel-nosed, *594*
- New Mexico Blind, *464, 583*
- New Mexico Garter, *674*
- New Mexico Lined, *677*
- New Mexico Milk, *613, 622*
- Night, *586, 616*
- North Carolina Swamp, *652*
- Northern Brown, *654*
- Northern Cat-eyed, *606, 626*
- Northern Lined, *677*
- Northern Pine, *591, 644*
- Northern Red-bellied, *505, 506, 655*
- Northern Ribbon, *672*
- Northern Ringneck, *600*
- Northern Scarlet, *596, 592*
- Northern Water, *580, 637*
- North Florida Swamp, *494, 652*
- Northwestern Garter, *512, 669*
- Northwestern Ringneck, *600*
- Oregon Garter, *664*
- Organ Pipe Shovel-nosed, *610, 595*
- Pacific Gopher, *537, 644*
- Pacific Ringneck, *496, 600*
- Painted Desert Glossy, *590*
- Pale Milk, *622*
- Peninsula Crowned, *458, 660*
- Peninsula Ribbon, *672*
- Pilot Black, *607*
- Pima Leaf-nosed, *589, 642*
- Pine-gopher, *644*
- Pine Woods, *462, 465, 648*
- Plain-bellied Water, *633*
- Plains Black-headed, *460, 658*
- Plains Blind, *583*
- Plains Garter, *671*
- Plains Hognose, *572, 613*
- Prairie Ringneck, *497, 600*
- Puget Sound Garter, *674*
- Queen, *503, 648*
- Rainbow, *546, 610*
- Rat, *484, 605*
- Red-bellied, *501, 655*
- Red-bellied Water, *490, 633*
- Red Milk, *615, 622*
- Red-sided Garter, *542, 674*
- Red-spotted Garter, *535, 674*
- Red-striped Ribbon, *544, 670*
- Regal Ringneck, *600*
- Rim Rock Crowned, *659*
- Ringneck, *498, 600*
- Rough Earth, *470, 473, 678*
- Rough Green, *477, 639*
- Saddle Leaf-nosed, *642*
- San Bernardino Ringneck, *600*
- San Diego Gopher, *644*
- San Diego Night, *616*
- San Diego Ringneck, *600*
- San Francisco Garter, *541, 674*
- Santa Cruz Garter, *664*
- Santa Cruz Gopher, *644*
- Scarlet, *595, 607, 592*
- Sharp-tailed, *471, 599*
- Short-headed Garter, *663*
- Short-tailed, *584, 654*
- Sierra Garter, *664*
- Smooth Earth, *679*
- Smooth Green, *475, 640*
- Sonora Gopher, *644*

○ Sonora Lyre, **582**, *676*
○ Sonoran Shovel-nosed, *595*
○ Southeastern Crowned, *466*, *657*
○ Southern Hognose, **585**, *615*
○ Southern Ringneck, *495*, *600*
○ Southern Water, *634*
○ South Florida Rainbow, *610*
○ South Florida Swamp, *652*
○ Southwestern Blind, *584*
○ Spotted Leaf-nosed, *643*
○ Spotted Night, *616*
○ Striped Crayfish, *510*, *646*
○ Swamp, **487**, *652*
○ Texas Black-headed, *658*
○ Texas Blind, *583*
○ Texas Brown, *654*
○ Texas Coral, **617**, *681*
○ Texas Garter, *533*, *674*
○ Texas Glossy, *590*
○ Texas Indigo, *602*
○ Texas Lined, *677*
○ Texas Long-nosed, *649*
○ Texas Lyre, **568**, *676*
○ Texas Night, *616*
○ Texas Patch-nosed, *516*, *650*
○ Texas Rat, *605*
○ Texas Scarlet, *592*
○ Trans-Pecos Blind, *584*
○ Trans-Pecos Rat, *523*, *607*
○ Tucson Shovel-nosed, *612*, *594*
○ Two-striped Garter, *504*, *664*
○ Utah Black-headed, *659*
○ Utah Blind, *584*
○ Utah Milk, *622*
○ Valley Garter, *674*
○ Wandering Garter, *511*, *667*
○ Western Aquatic Garter, *664*
○ Western Black-headed, *659*

○ Western Black-necked Garter, *666*
○ Western Blind, *457*, *584*
○ Western Earth, *467*, *679*
○ Western Fox, *608*
○ Western Hognose, *613*
○ Western Hook-nosed, **588**, *612*
○ Western Leaf-nosed, *571*, **583**, *643*
○ Western Long-nosed, *593*, *649*
○ Western Mud, *609*
○ Western Patch-nosed, *651*
○ Western Plains Garter, *534*, *543*, *671*
○ Western Ribbon, *531*, *670*
○ Western Shovel-nosed, *604*, *594*
○ Western Smooth Green, *476*, *640*
○ Western Terrestrial Garter, *667*
○ Western Worm, *493*, *591*
○ Worm, *591*
○ Yaqui Black-headed, *663*
○ Yellow-bellied Water, *481*, *633*
○ Yellow Rat, *526*, *540*, *605*

Snake Family
Boa and Python, *586*
Colubrid, *589*
Coral, *680*
Pit Viper, *682*
Slender Blind, *583*

Snakes, *581*

○ **Snapping Turtle, 322, 323**, *435*
○ Alligator, **325, 326, 327,** *436*
○ Common, *435*
○ Florida, *324*, *435*

Snapping Turtle Family, *435*

Softshell
- Eastern Spiny, *485*
- Florida, 272, 273, *483*
- Guadalupe Spiny, *485*
- Gulf Coast Smooth, 269, *484*
- Gulf Coast Spiny, *485*
- Midland Smooth, 268, *484*
- Pallid Spiny, *485*
- Smooth, *484*
- Spiny, 270, 271, *485*
- Texas Spiny, *485*
- Western Spiny, *485*

Softshell Turtle Family, *483*

Sonora
episcopa, *653*
semiannulata, 459, 499, 502, 555, 611, *653*

Spadefoot
- Couch's, 252, *363*
- Eastern, 233, *365*
- Great Basin, 232, *366*
- Hurter's, *365*
- New Mexico, 230, *364*
- Plains, 231, *362*
- Western, 229, 253, *364*

Spadefoot Toad Family, *362*

Sphaerodactylus
argus, *494*
a. argus, *495*
elegans, 396, 398, *495*
notatus, 399, 400, *496*
n. notatus, *496*

Spiny Lizard
- Barred, *523*
- Blue, 352, *520*
- Clark's, 349, 372, *519*
- Crevice, 354, *527*
- Desert, 350, *523*
- Granite, 351, *527*
- Orange-headed, *523*
- Plateau, *519*
- Sonoran, *519*

- Texas, 381, *526*
- Twin-spotted, *523*
- Yarrow's, 353, *522*
- Yellow-backed, *523*

Squamata, *487, 581*

Stereochilus marginatus, 90, *354*

Sternotherus
carinatus, 310, *443*
depressus, 316, *443*
minor, 311, *444*
m. minor, *444*
m. peltifer, 312, *444*
odoratus, 319, *445*

Stilosoma
extenuatum, 584, *654*

- Stinkpot, 319, *445*

Storeria
dekayi, 550, *654*
d. dekayi, *655*
d. limnetes, *655*
d. texana, *655*
d. victa, *655*
d. wrightorum, *655*
occipitomaculata, 501, *655*
o. obscura, *656*
o. occipitomaculata, 505, 506, *656*
o. pahasapae, *656*

Syrrhophus
cystignathoides, 167, *421*
guttilatus, *421*
marnocki, 168, *422*

T

- Tailed Frog, 165, *359*

Tailed Frog Family, *359*

Tantilla
atriceps, *656*
coronata, 466, *657*
gracilis, 463, *658*
nigriceps, 460, *658*
n. fumiceps, *659*

n. nigriceps, 659
oolitica, 659
planiceps, 659
p. eiseni, 500, 660
p. transmontana, 660
p. utahensis, 660
relicta, 660
r. neilli, 461, 661
r. pamlica, 661
r. relicta, 458, 661
rubra, 661
r. cucullata, 662
r. diabola, 662
wilcoxi, 662
w. wilcoxi, 662
yaquia, 663

Taricha
granulosa, 31, 35, 278
g. granulosa, 278
g. mazamae, 278
rivularis, 34, 36, 279
torosa, 32, 33, 280

Teiidae, 553

Terrapene
carolina, 306, 468
c. bauri, 308, 469
c. carolina, 468
c. major, 309, 468
c. triunguis, 304, 468
ornata, 469
o. luteola, 305, 470
o. ornata, 307, 470

Terrapin
○ Diamondback, 466
○ Carolina Diamondback, 466
○ Florida East Coast, 300, 466
○ Mangrove Diamondback, 466
○ Mississippi Diamondback, 466
○ Northern Diamondback, 299, 466
○ Ornate Diamondback, 298, 466
○ Texas Diamondback, 466

Testudines, 433

Testudinidae, 471

Thamnophis
brachystoma, 663
butleri, 529, 664
couchi, 664
c. aquaticus, 665
c. atratus, 665
c. couchi, 665
c. gigas, 665
c. hammondi, 504, 665
c. hydrophilus, 665
cyrtopsis, 536, 666
c. cyrtopsis, 666
c. ocellatus, 666
elegans, 667
e. biscutatus, 667
e. elegans, 667
e. terrestris, 667
e. vagrans, 511, 667
eques, 528, 668
e. megalops, 668
marcianus, 515, 669
m. marcianus, 669
ordinoides, 512, 669
proximus, 531, 670
p. diabolicus, 670
p. proximus, 670
p. orarius, 671
p. rubrilineatus, 544, 671
radix, 671
r. haydeni, 534, 543, 672
r. radix, 671
rufipunctatus, 548, 672
sauritus, 532, 672
s. nitae, 520, 673
s. sauritus, 673
s. sackeni, 673
s. septentrionalis, 673
sirtalis, 530, 674
s. annectans, 533, 674
s. concinnus, 535, 674
s. dorsalis, 675
s. fitchi, 675
s. infernalis, 545, 675
s. pallidula, 675
s. parietalis, 542, 675
s. pickeringi, 675

s. *semifasciatus*, 675
s. *similis*, **538**, 675
s. *sirtalis*, **539**, 576, 674
s. *tetrataenia*, **541**, 676

Toad
○ Amargosa, *388*
○ American, 237, *387*
○ Arizona, *394*
○ Arroyo, *394*
○ Black, 245, *391*
○ Boreal, 226, *388*
○ California, 243, *388*
○ Canadian, 239, *391*
○ Colorado River, 225, *386*
○ Dwarf American, *387*
○ Eastern Green, *390*
○ East Texas, *398*
○ Fowler's, 248, *398*
○ Giant, 242, *393*
○ Great Plains, 247, *389*
○ Green, 254, 255, *390*
○ Gulf Coast, 246, *397*
○ Houston, 228, *392*
○ Hudson Bay, *387*
○ Mexican Burrowing, 217, 220, *361*
○ Oak, 250, *395*
○ Red-spotted, 234, *394*
○ Sonoran Green, 251, *396*
○ Southern, 236, 241, *397*
○ Southwestern, 223, 235, *394*
○ Southwestern Woodhouse's, 249, *398*
○ Texas, **238**, *396*
○ Western, 240, *388*
○ Western Green, *390*
○ Woodhouse's, 224, *398*
○ Wyoming, *391*
○ Yosemite, 227, 244, *389*

Toad Family, *386*
Burrowing, *361*
Spadefoot, *362*

Toads, *357*

Tongueless Frog Family, *423*

Tortoise
○ Berlandier's 329, *472*
○ Desert, 328, *471*
○ Gopher, 330, *473*

Tortoise Family, *471*

Treefrog
○ Barking, 145, 151, *408*
○ Bird-voiced, 156, *403*
○ Burrowing, 186, *416*
○ California, 158, *404*
○ Canyon, 159, *402*
○ Common Gray, 152, 157, *404*
○ Cope's Gray, 160, *404*
○ Cuban, 155, 178, *410*
○ Eastern Bird-voiced, *403*
○ Green, 146, *405*
○ Mexican, 181, 184, *416*
○ Mountain, 150, *406*
○ Pine Barrens, 149, *402*
○ Pine Woods, 164, *407*
○ Pacific, 148, 170, 182, *408*
○ Squirrel, 147, 174, *409*
○ Western Bird-voiced, *403*

Treefrog Family, *400*

Tree Lizard
○ Big Bend, *535*
○ Canyon, *535*
○ Colorado River, *535*
○ Common, 369, *535*
○ Eastern, *535*
○ Lined, *535*
○ Northern, *535*
○ Small-scaled, **382**, *535*

*Trimorphodon
biscutatus*, 676
b. lambda, **582**, *677*
b. vandenburghi, *677*
b. vilkinsoni, **568**, *677*

Trionychidae, *483*

*Trionyx
ferox*, 272, 273, *483*
muticus, *484*

m. calvatus, 269, 485
m. muticus, 268, 485
spiniferus, 270, 271, 485
s. asper, 486
s. emoryi, 486
s. guadalupensis, 486
s. hartwegi, 486
s. pallidus, 486
s. spiniferus, 486

Tropidoclonion
lineatum, 507, 677
l. annectens, 678
l. lineatum, 678
l. mertensi, 678
l. texanum, 678

True Frog Family, 367

Turtle
○ Alabama Map, 284, 465
○ Alabama Red-bellied, 292, 446
○ Alligator Snapping, 325, 326, 327, 436
○ Atlantic Green, 476
○ Barbour's Map, 283, 458
○ Big Bend, 452
○ Blanding's, 291, 458
○ Bog, 301, 456
○ Cagle's Map, 279, 459
○ Chicken, 285, 457
○ Common Snapping, 435
○ Cumberland, 452
○ Desert Box, 305, 469
○ Eastern Box, 306, 468
○ Eastern Chicken, 457
○ Eastern Mud, 318, 441
○ Eastern Painted, 294, 450
○ False Map, 464
○ Flattened Musk, 316, 443
○ Florida Box, 308, 468
○ Florida Chicken, 457
○ Florida Mud, 321, 441
○ Florida Red-bellied, 296, 449
○ Florida Snapping, 324, 435
○ Green, 267, 476
○ Gulf Coast Box, 309, 468
○ Illinois Mud, 439

○ Loggerhead Musk, 311, 444
○ Map, 280, 461
○ Mexican Mud, 315, 440
○ Midland Painted, 450
○ Mississippi Map, 278, 462
○ Mississippi Mud, 320, 441
Muhlenberg's, 456
○ Mud, 441
Musk, 445
○ Northwestern Pond, 455
○ Ornate Box, 307, 469
○ Ouachita Map, 274, 464
○ Pacific Green, 476
○ Painted, 450
○ Plymouth Red-bellied, 451
○ Razor-backed Musk, 310, 443
○ Red-bellied, 295, 451
○ Red-eared, 286, 452
○ Sabine Map, 277, 464
○ Snapping, 322, 323, 435
○ Sonora Mud, 314, 441
○ Southern Painted, 297, 450
○ Southwestern Mud, 439
○ Southwestern Pond, 455
○ Spotted, 290, 453
○ Striped Mud, 317, 438
○ Stripe-necked Musk, 312, 444
○ Texas Map, 282, 466
○ Three-toed Box, 304, 468
○ Western Box, 469
○ Western Chicken, 457
○ Western Painted, 293, 450
○ Western Pond, 303, 455
○ Wood, 302, 454
○ Yellow-bellied, 289, 452
○ Yellow Mud, 313, 439

Turtle Family
Box, 446
Leatherback, 481
Marsh, 446
Mud, 438
Musk, 438
Pond, 446

Sea, 475
Softshell, 483
Snapping, 435
Tortoise, 471

Turtles, 433

Typhlomolge rathbuni, 3, 9, 355

Typhlotriton spelaeus, 12, 356

Typical Old World Lizard Family, 566

U

Uma
inornata, 532
notata, 532
n. notata, 343, 533
n. rufopunctata, 533
scoparia, 344, 533

Urosaurus
graciosus, 380, 534
g. graciosus, 534
g. shannoni, 534
microscutatus, 382, 535
ornatus, 369, 535
o. levis, 536
o. linearis, 536
o. ornatus, 536
o. schmidti, 536
o. symmetricus, 536
o. wrighti, 536

Uta
stansburiana, 363, 537
s. elegans, 537
s. nevadensis, 537
s. stansburiana, 537
s. stejnegeri, 537
s. uniformis, 537

V

Viperidae, 682

Virginia
striatula, 470, 473, 678
valeriae, 679

v. elegans, 467, 679
v. pulchra, 679
v. valeriae, 679

W

Waterdog
○ Alabama, 281
○ Dwarf, 19, 284
○ Gulf Coast, 282
○ Louisiana, 283
○ Neuse River, 21, 282

Waterdogs, 281

Water Snake
○ Banded, 552, 634
○ Blotched, 633
○ Brazos, 549, 636
○ Broad-banded, 562, 634
○ Brown, 567, 639
○ Concho, 636
○ Copper-bellied, 633
○ Diamondback, 574, 636
○ Florida, 634
○ Florida Green, 482, 632
○ Green, 632
○ Harter's, 636
○ Lake Erie, 637
○ Mangrove, 579, 634
○ Midland, 637
○ Northern, 580, 637
○ Plain-bellied, 633
○ Red-bellied, 490, 633
○ Southern, 634
○ Yellow-bellied, 481, 633

Whipsnake
○ Ajo Mountain, 627
○ Central Texas, 630
○ Desert Striped, 631
○ Ruthven's, 631
○ Schott's, 631
○ Sonora, 517, 627
○ Striped, 521, 631

Whiptail
○ Arizona Striped, 558
○ California, 563
○ Checkered, 419, 562
○ Chihuahuan Spotted, 418, 555

○ Coastal, 563
○ Desert-Grassland, 412, 564
○ Giant Spotted, 416, 554
○ Gila, 556
○ Gray Checkered, 555
○ Great Basin, 563
○ Laredo Striped, 559
○ Little Striped, 414, 558
○ Marbled, 420, 563
○ New Mexican, 415, 559
○ Northern, 563
○ Orange-throated, 409, 557
○ Plateau Spotted, 417, 560
○ Plateau Striped, 413, 565
○ Red-backed, 554
○ Sonoran Spotted, 410, 561
○ Southern, 563
○ Texas Spotted, 556
○ Trans-Pecos Striped, 558
○ Western, 563

Whiptail Family, 553

X

Xantusia
henshawi, 404, 405, 407, 550
riversiana, 408, 551
vigilis, 406, 551
v. arizonae, 403, 552
v. sierrae, 552
v. utahensis, 552
v. vigilis, 552

Xantusiidae, 550

Xenopus laevis, 222, 423

NATIONAL AUDUBON SOCIETY
FIELD GUIDE SERIES

Also available in this unique all-color,
all-photographic format:

**African Wildlife • Birds *(Eastern Region)* • Birds
(Western Region) • Butterflies • Fishes • Fossils
• Insects and Spiders • Mammals • Mushrooms
• Night Sky • Rocks and Minerals • Seashells •
Seashore Creatures • Trees *(Eastern Region)* •
Trees *(Western Region)* • Tropical Marine Fishes
• Weather • Wildflowers *(Eastern Region)* •
Wildflowers *(Western Region)***

Prepared and produced by Chanticleer Press, Inc.

Founding Publisher: Paul Steiner
Publisher: Andrew Stewart

Staff for this book:

Editor-in-Chief: Gudrun Buettner
Executive Editor: Susan Costello
Managing Editor: Jane Opper
Guides Editor: Susan Rayfield
Project Editor: Barbara Williams
Assistant Editor: Richard Christopher
Production Manager: Helga Lose
Art Director: Carol Nehring
Picture Library: Edward Douglas
Range Maps and Silhouettes: Paul Singer
Drawings: Sy Barlowe
Visual Key: Carol Nerhing, Susan Rayfield,
Barbara Williams

Original series design by Massimo Vignelli

All editorial inquiries should be addressed to:
Chanticleer Press
500 Fifth Avenue, Suite 3620
New York, NY 10110

To purchase this book or other National Audubon Society
illustrated nature books, please contact:
Alfred A. Knopf
1745 Broadway
New York, NY 10019
(800) 733-3000
www.randomhouse.com